ADVANCED NANOMATERIALS

Synthesis, Properties, and Applications

ADVANCED NANOMATERIALS

Synthesis, Properties, and Applications

Edited by

**Sabu Thomas, PhD, Nandakumar Kalarikkal, PhD,
A. Manuel Stephan, PhD, B. Raneesh, and A. K. Haghi, PhD**

Apple Academic Press

TORONTO   NEW JERSEY

Apple Academic Press Inc.	Apple Academic Press Inc.
3333 Mistwell Crescent	9 Spinnaker Way
Oakville, ON L6L 0A2	Waretown, NJ 08758
Canada	USA

©2015 by Apple Academic Press, Inc.

First issued in paperback 2021

Exclusive worldwide distribution by CRC Press, a member of Taylor & Francis Group
No claim to original U.S. Government works

ISBN 13: 978-1-77463-309-0 (pbk)
ISBN 13: 978-1-926895-79-6 (hbk)

Library of Congress Control Number: 2014937848

Library and Archives Canada Cataloguing in Publication

Advanced nanomaterials : synthesis, properties and applications/edited by Sabu Thomas, PhD, Nandakumar Kalarikkal, PhD, A. Manuel Stephan, PhD, B. Raneesh and A. K. Haghi, PhD.

"This book is the collection of peer reviewed papers presented at the Second International Conference on Nanomaterials: Synthesis, Characterization and Application in January 2012 (ICN 2012) in Kottayam, India."--Preface.

Includes bibliographical references and index.
ISBN 978-1-926895-79-6 (bound)

1. Nanostructured materials--Congresses. 2. Nanotechnology--Congresses. I. Thomas, Sabu, editor II. Kalarikkal, Nandakumar, author, editor III. Stephan, A. Manuel, editor IV. Raneesh, B., author, editor V. Haghi, A. K., editor VI. International Conference on Nanomaterials: Synthesis, Characterization and Application (2nd : 2012 : Kottayam, India)

| TA418.9.N35A38 2014 | 620.1'15 | C2014-902470-3 |

Apple Academic Press also publishes its books in a variety of electronic formats. Some content that appears in print may not be available in electronic format. For information about Apple Academic Press products, visit our website at **www.appleacademicpress.com** and the CRC Press website at **www.crcpress.com**

ABOUT THE EDITORS

Sabu Thomas, PhD

Dr. Sabu Thomas is the Director of the School of Chemical Sciences, Mahatma Gandhi University, Kottayam, India. He is also a full professor of polymer science and engineering and the Director of the International and Inter University Centre for Nanoscience and Nanotechnology of the same university. He is a fellow of many professional bodies. Professor Thomas has authored or co-authored many papers in international peer-reviewed journals in the area of polymer processing. He has organized several international conferences and has more than 420 publications, 11 books and two patents to his credit. He has been involved in a number of books both as author and editor. He is a reviewer to many international journals and has received many awards for his excellent work in polymer processing. His h Index is 42. Professor Thomas is listed as the 5th position in the list of Most Productive Researchers in India, in 2008.

Nandakumar Kalarikkal, PhD

Dr. Nandakumar Kalarikkal obtained his master's degree in physics with a specialization in industrial physics and his PhD in semiconductor physics from Cochin University of Science and Technology, Kerala, India. He was a postdoctoral fellow at NIIST, Trivandrum, and later joined with Mahatma Gandhi University. Currently he is the Joint Director of the International and Inter University Centre for Nanoscience and Nanotechnology as well as Associate professor in the School of Pure and Applied Physics at Mahatma Gandhi University. His current research interests include synthesis, characterization, and applications of nanophosphors, nanoferrites, nanoferroelctrics, nanomultiferroics, nanocomposites, and phase transitions.

A. Manuel Stephan, PhD

A. Manuel Stephan is a Scientist at the Electrochemical Power Systems Division of the Central Electrochemical Research Institute in Tamilnadu, India. He is a life member of the Society for the Advancement of Electrochemical Science and Technology (SAEST) and the Materials Society of India.

B. Raneesh

B. Raneesh is currently working as an Analytical Engineer (HRTEM Division) at International and Inter University Centre for Nanoscience and Nanotechnology, Mahatma Gandhi University, Kottayam, India.

A. K. Haghi, PhD

Dr. A. K. Haghi holds a BSc in urban and environmental engineering from The University of North Carolina (USA); a MSc in mechanical engineering from North Carolina A&T State University (USA); a DEA in applied mechanics, acoustics and materials from The Université de Technologie de Compiègne (France); and a PhD in engineering sciences from Université de Franche-Comté (France). He is the author and editor of 65 books as well as 1000 published papers in various journals and conference proceedings. Dr. Haghi has received several grants, consulted for a number of major corporations, and is a frequent speaker to national and international audiences. Since 1983, he served as a professor at several universities. He is currently Editor-in-Chief of the *International Journal of Chemoinformatics and Chemical Engineering* and *Polymers Research Journal* and on the editorial boards of many international journals. He is a member of the Canadian Research and Development Center of Sciences and Cultures (CRDCSC), Montreal, Quebec, Canada.

CONTENTS

LIST OF CONTRIBUTORS

P. M. Aneesh
Optoelectronic Devices Laboratory, Department of Physics, Cochin University of Science and Technology, Kochi, India-682 022.
Present address:School of Physics, Indian Institute of Science Education and Research (IISER) Trivandrum Campus, Trivandrum, India-695016.

A. Balakrishnan
Amrita centre for Nanosciences and Molecular medicine, Kochi, Kerala, India-682041.
E-mail: avinash.balakrishnan@gmail.com
Telephone and Fax No.: +91-484-2801234

I. Banerjee
Department of Applied Physics, Birla Institute of Technology, Mesra, Ranchi-835215, India.
E-mail: indranibanerjee@bitmesra.ac.in and indrani@seas.ucla.edu

P. K. Barhai
Department of Applied Physics, Birla Institute of Technology, Mesra, Ranchi-835215, India.
E-mail: pkbarhai@bitmesra.ac.in

Yu. O. Barmenkov
Centro de Investigaciones en Optica, Loma del Bosque 115, Leon 37150, Guanajuato, Mexico,
E-mail: yuri@cio.mx

S. K. Bhadra
Fiber Optics and Photonics Division, CSIR-Central Glass & Ceramic Research Institute, 196, Raja S. C. Mullick Road, Kolkata-700 032, India.
E-mail: skbhadra@cgcri.res.in

B. A. Bhanvase
Chemical Engineering Department, Vishwakarma Institute of Technology, 666, Upper Indiranagar, Bibwewadi, Pune-411037, Maharashtra, India.
E-mail: bharatbhanvase@gmail.com

MayankBhushan
Research Scholar, Centre for Nanoscience and Technology, Madanjeeth School of Green Energy Technologies, Pondicherry University, Puducherry, India.
Centre for Nanoscience and Technology, Pondicherry University, Kalapet, Puducherry-605014, India.
E-mail: mayank.bhshn@gmail.com

R. Biswas
Amrita centre for Nanosciences and Molecular medicine, Kochi, Kerala, India-682041.
E-mail: rajabiswas@aims.amrita.edu,
Telephone and Fax No.: +91-484-2801234

A. K. Das

Department of Applied Physics, Birla Institute of Technology, Mesra, Ranchi-835215, India.
E-mail: akdas@barc.gov.in
Laser & Plasma Technology Division, Bhabha Atomic Research Center, Mumbai-400085, India.

S. Das

Fiber Optics and Photonics Division, CSIR-Central Glass & Ceramic Research Institute,
196, Raja S. C. Mullick Road, Kolkata-700 032, India.
E-mail: dshyamal@cgcri.res.in

Nirmal Ghosh O. S.

Research Scholar, Centre for Nanoscience and Technology, Madanjeeth School of Green Energy Technologies, Pondicherry University, Puducherry, India.
Centre for Nanoscience and Technology, Pondicherry University, Kalapet, Puducherry-605014, India.

Sandhya Gopalakrishnan

Govt. Dental college, Kottayam, Kerala 686 008, India.
International and Inter University Centre for Nanoscience and Nanotechnology, Mahatma Gandhi University, Kottayam, Kerala 686 560, India.

Arvind Gupta

N.S.N. Research Centre for Nanotechnology and Bionanotechnology, SICES College, Jambhul Phata,
Kalyan - Badlapur Road, Ambernath (W), 421505, Maharashtra, India.
E-mail: arvindgupta106@gmail.com

D. V. S. Jain

Panjab University, Chandigarh, India

M. K. Jayaraj

Optoelectronic Devices Laboratory, Department of Physics, Cochin University of Science and Technology,
Kochi, India-682 022.
E-mail: mkj@cusat.ac.in

Nanda-kumarKalarikkal

School of Pure and Applied Physics
International and Inter University Centre for Nanoscience and Nanotechnology, Mahatma Gandhi University, Kottayam, Kerala 686 560, India.

B. Kalska-Szostko

Institute of Chemistry, University of Bialystok, Hurtowa 1, 15-399 Bialystok, Poland.
Faculty of Physics, University of Bialystok, Lipowa 41, 15-424 Bialystok, Poland.
E-mail: kalska@uwb.edu.pl

S. E. Karekar

Chemical Engineering Department, Vishwakarma Institute of Technology, 666, Upper Indiranagar,
Bibwewadi, Pune-411037, Maharashtra, India.
E-mail: ekarekar@gmail.com

T. N. Kim

Department of Materials Engineering, Paichai University, Daejeon, S. Korea-302-735.
E-mail: tnkim@mail.pcu.ac.kr
Telephone and Fax No.: +91-484-2801234

A. V. Kir'yanov

Centro de Investigaciones en Optica, Loma del Bosque 115, Leon 37150, Guanajuato, Mexico,
E-mail: alejandrokir@gmail.com
A.M. Prokhorov General Physics Institute (Russian Academy of Sciences) Vavilov Str. 38, Moscow 119991,
Russian Federation.
E-mail: mcpal@cgcri.res.in

Sujith. K
Amrita centre for Nanosciences and Molecular medicine, Kochi, Kerala, India-682041.

Vinod Kumar
Assir specialist dental centre, Abha,Kingdom of Saudi Arabia.

P. Laha
Department of Applied Physics, Birla Institute of Technology, Mesra, Ranchi-835215, India.
E-mail: pinaki.laha@yahoo.com

Asha Anish Madhavan
Amrita centre for Nanosciences and Molecular medicine, Kochi, Kerala, India-682041.

S. K.Mahapatra
Department of Applied Physics, Birla Institute of Technology, Mesra, Ranchi-835215, India.
E-mail: skmahapatra@bitmesra.ac.in

A. Martinez-Gamez
Centro de Investigaciones en Optica, Loma del Bosque 115, Leon 37150, Guanajuato, Mexico,
E-mail: mamg@cio.mx
J. L. LucioMartíne, Instituto de Fisica de la Universidad de Guanajuato, Loma del Bosque 113, Leon 37150,
Guanajuato, Mexico
E-mail: lucio@fisica.ugto.mx

V. K. Meena
Central Scientific Instruments Organisation (CSIR-CSIO), Chandigarh, India.

Apurba Krishna Mitra
Department of Physics, NIT Durgapur-713209, West Bengal, India.
E-mail: akmrecdgp@yahoo.com

BalachandranUnni Nair
Chemical Laboratory, Central Leather Research Institute, Council of Scientific and Industrial Research,
Adyar, Chennai, 600 020 India.
E-mail: bunair@clri.res.in
Telephone: +91 44 2441 1630
Fax. +91 44 2491 1589

Manitha Nair
Amrita centre for Nanosciences and Molecular medicine, Kochi, Kerala, India-682041.

S. V. Nair
Amrita centre for Nanosciences and Molecular medicine, Kochi, Kerala, India-682041.

Marimuthu Nidhin
Chemical Laboratory, Central Leather Research Institute, Council of Scientific and Industrial Research,
Adyar, Chennai, 600 020 India.
E-mail: nidhinscs@gmail.com
Telephone No.: +91 44 2441 1630
Fax No.: +91 44 2491 1589

Goldie Oza
N.S.N. Research Centre for Nanotechnology and Bionanotechnology, SICES College, Jambhul Phata,
Kalyan - Badlapur Road, Ambernath (W), 421505, Maharashtra, India.
E-mail: goldieoza@gmail.com

M. Pal
Fiber Optics and Photonics Division, CSIR-Central Glass & Ceramic Research Institute,
196, Raja S. C. Mullick Road, Kolkata-700 032, India.
E-mail:mpal@cgcri.res.in

A. B. Panda
Department of Applied Physics, Birla Institute of Technology, Mesra, Ranchi-835215, India.
E-mail: atalabit@gmail.com

Priyanka. P
Amrita centre for Nanosciences and Molecular medicine, Kochi, Kerala, India-682041.

Sunil Pandey
N.S.N. Research Centre for Nanotechnology and Bionanotechnology, SICES College, Jambhul Phata, Kalyan - Badlapur Road, Ambernath (W), 421505, Maharashtra, India.
E-mail: gurus.spandey@gmail.com

M. C. Paul
Fiber Optics and Photonics Division, CSIR-Central Glass & Ceramic Research Institute, 196, Raja S. C. Mullick Road, Kolkata-700 032, India.
E-mail: mcpal@cgcri.res.in

A. Pigiel
Institute of Chemistry, University of Bialystok, Hurtowa 1, 15-399 Bialystok, Poland.
Faculty of Physics, University of Bialystok, Lipowa 41, 15-424 Bialystok, Poland.
E-mail: kalska@uwb.edu.pl

Indu Raj
Govt. Dental college, Kottayam, Kerala 686 008, India.
International and Inter University Centre for Nanoscience and Nanotechnology, Mahatma Gandhi University, Kottayam, Kerala 686 560, India.

B. Raneesh
School of Pure and Applied Physics
International and Inter University Centre for Nanoscience and Nanotechnology, Mahatma Gandhi University, Kottayam, Kerala 686 560, India.

M. Rogowska
Institute of Chemistry, University of Bialystok, Hurtowa 1, 15-399 Bialystok, Poland.
Faculty of Physics, University of Bialystok, Lipowa 41, 15-424 Bialystok, Poland.
E-mail: kalska@uwb.edu.pl

J. K. Sahu
Optoelectronic Research Centre, University of Southampton, Southampton So17 1BJ, United Kingdom.
E-mail: jks@orc.soton.ac.uk

D. Satula
Institute of Chemistry, University of Bialystok, Hurtowa 1, 15-399 Bialystok, Poland.
Faculty of Physics, University of Bialystok, Lipowa 41, 15-424 Bialystok, Poland.
E-mail: kalska@uwb.edu.pl

D. S. Schmool
IFIMUP - IN and Departamento de Física e Astronomia, Universidade do Porto, Rua Campo Alegre 687, 4169 007 Porto, Portugal.
E-mail: dschmool@fc.up.pt

Madhuri Sharon
N.S.N. Research Centre for Nanotechnology and Bionanotechnology, SICES College, Jambhul Phata, Kalyan - Badlapur Road, Ambernath (W), 421505, Maharashtra, India.
E-mail: sharonmadhuri@gmail.com

M. L. Singla
Central Scientific Instruments Organisation (CSIR-CSIO), Chandigarh, India.

Suman Singh
Central Scientific Instruments Organisation (CSIR-CSIO), Chandigarh, India.
Panjab University, Chandigarh, India.
E-mail: sumansingh01@gmail.com
Telephone: 91-0172-265-17-87
Fax: 91-0172-265.70.82

S. H. Sonawane
Department of Chemical Engineering National Institute of Technology, Warangal, 506004 AP India.
E-mail: shirishsonawane09@gmail.com

Kalarical Janardhanan Sreeram
Chemical Laboratory, Central Leather Research Institute, Council of Scientific and Industrial Research, Adyar, Chennai, 600 020 India.
E-mail: kjsreeram@clri.res.in
Telephone No.: +91 44 2441 1630
Fax No.: +91 44 2491 1589

Keka Talukdar
Department of Physics, NIT Durgapur-713209, West Bengal, India.
E-mail: keka.talukdar@yahoo.co.in

Kannan Vaidyanathan
Pushpagiri Institute of Medical Sciences and Research Centre, Tiruvalla, Kerala 689 101, India.

Dr. A. Kasi Viswanath
Associate Professor, Centre for Nanoscience and Technology, Madanjeeth School of Green Energy Technologies, Pondicherry University, Puducherry, India.
Centre for Nanoscience and Technology, Pondicherry University, Kalapet, Puducherry-605014, India.

S. Yoo
School of Electrical and Electronic Engineering, Nanyang Technological University.
E-mail: SEON.YOO@ntu.edu.sg
Optoelectronic Research Centre, University of Southampton, Southampton So17 1BJ, United Kingdom.

LIST OF ABBREVIATIONS

AFM	Atomic force microscopy
AuNPs	Gold nanoparticles
BBB	Blood brain barrier
CET	Cell encapsulation therapy
CG	Colloidal gold
CHPP	Continuous hyperthermic peritoneal perfusion
CLEM	Correlative light and electron microscopy
CNT	Carbon nanotubes
CPMV	Cowpea mosaic virus
CPPs	Cell-penetrating peptides
DBT	Department of Biotechnology
DCs	Dendritic cells
DDI	Dipole-dipole interactions
DEG	Zero-dimensional electron gas
DMIM	Discontinuous metal- insulator magnetic multilayers
DNP	Dielectric Nanoparticles
DOS	Density of states
DSSC	Dye sensitized solar cell
DST	Department of Science and Technology
ECAP	Equal channel angular pressing
ECM	Native extracellular matrix
EDX	Energy dispersive X-ray analyses
EEDF	Electron Energy Distribution Function
EIS	Electrochemical Impedance Spectroscopy
ENM	Electroactive nanostructured membranes
EPMA	Electron probe micro analyzer
EPR	Enhanced permeability and retention
FBGs	Fibers Bragg gratings
FFF	Field flow fractionation
FHV	Flock house virus
FMR	Ferromagnetic resonance
FRET	Forster Resonant Energy Transfer
FWHM	Full width at the half- maximum
GD	Gene delivery
GMR	Magnetoresistance systems
HCP	Hexagonal close packing
HRTEM	High resolution transmission electron microscopic
HTEM	High-resolution transmission electron microscope

ICMR	Indian Council of Medical Research
IDEG	One-dimensional electron gas
INMs	Inorganic nanomaterials
IR	Infrared spectroscopy
LDC	Lipid drug conjugate
LIBAD	Low incident beam angle diffraction
LMA	Large mode area
LO	Longitudinal optical
LOD	Limit of detection
LUMO	Lowest unoccupied molecular orbital
MCVD	Modified chemical vapor deposition
MDT	Magnetic drug targeting
MERAM	Electric-field controlled magnetic random access memory
MNPs	Magnetic nanoparticles
MQW	Multiple quantum well
MR	Magnetoresistive
MRI	Magnetic resonance imaging
NCI	National Cancer Institute
NIR	Near-infrared
NLCs	Nanostructured lipid carriers
NMOF	Nanoscale metal organic frameworks
NMR	Nuclear magnetic resonance
NPs	Nanoparticles
OES	Optical emission spectroscopy
OSA	Optical spectrum analyzer
PALM	Photo-activated localizing microscopy
PCR	Polymerase chain reaction
PDLSCs	Periodontal ligament stem cells
PDT	Photodynamic therapy
PET	Positron emission tomography
PLA	Pulsed laser ablation
PM	Powder metallurgy
PV	Photovoltaic
QCM	Quartz crystal microbalance
QDs	Quantum dots
QW	Quantum well
RES	Reticulo-endothelial system
RI	Refractive index
RSV	Respiratory syncytial virus
SAED	Selected-area electron diffraction
SAMs	Self-assembled monolayers
SBF	Simulated body fluid
SD	Solution doping
SEM	Scanning electron microscopy
SERS	Surface-enhanced Raman scattering

SLNs	Solid lipid nanoparticles
SPB	Surface Plasmon band
SPD	Severe plastic deformation
SPECT	Single photon emission computed tomography
SPIONs	Super paramagnetic iron oxide nanoparticles
SPR	Surface Plasmon resonance
SS	Stainless steels
SWCNTs	Single wall carbon nanotubes
SWNT	Single Waal Nano Tube
TCOs	Transparent conducting oxides
TEM	Transmission electron microscopy
UCNs	Upconversion nanoparticles
VLP	Virus-like particles
VNPs	Viral nanoparticles
WCR	Water confined regime
WDM	Wavelength division multiplexer
XRD	X-ray diffraction
YFs	Ytterbium fibers

PREFACE

Recently, nanotechnology has attracted much attention because of its application in chemistry, physics, materials science, and biotechnology to create novel materials that have unique properties. The combination of a high surface area, flexibility, and superior directionality makes nanostructures suitable for many applications. Difference in properties at bulk and nano-level has provided new insight and direction to researchers to meet the ever-increasing demand of smaller and more efficient devices. Nanotechnology offers potential solutions to many problems using emerging nanotechniques. Depending on the strong interdisciplinary character of nanotechnology, there are many research fields and several potential applications that involve nanotechnology.

This book is the collection of peer reviewed-papers presented at the second international conference on nanomaterials: synthesis, characterization, and application in January 2012 (ICN 2012) in Kottayam, India, that showcased the research of an international roster of scientists and students. It was organized by the Centre for Nanoscience and Nanotechnology, Mahatma Gandhi University, Kottyam, Kerala. The conference had over 350 delegates, from all across the world, with good representation from France, Australia, USA, Netherlands, Germany, Iran, Sweden, Spain, and Libya as well as a substantial number of eminent Indian scientists. The conference was designed as the perfect opportunity for international researchers interested in nanoscience and functional materials to meet, present, and discuss issues related to current developments in the field of nanotechnology. The goal of the conference was to promote interdisciplinary research on synthesis, characterization, and application of nanoparticles, their composites and their applications in medicine, automotive, optoelectronic, sensors, chemical, memory device, and engineering. The three-day conference included discussion on recent advances, difficulties, and breakthroughs in the field of nanotechnology.

This book collects together a selection of 15 papers presented during the conference. The papers include a wide variety of interdisciplinary topics in nanoscience with emphasis on synthesis, characterization and applications. In this important work, an excellent team of international experts provides an exploration of the emerging nanotechnologies that are poised to make the nano-revolution become a reality in the manufacturing sector. We trust that the work will stimulate new ideas, methods, and applications in ongoing advances in this growing area of strong international interest.

We would like to thank all who kindly contributed their papers for this issue and the editors of Apple Academic Press for their kind help and cooperation. We are also indebted to the Apple Academic Press editorial office and the publishing and production teams for their assistance in preparation and publication of this issue.

— **Sabu Thomas, PhD, Nandakumar Kalarikkal, PhD,**
A. Manuel Stephan, PhD, B. Raneesh, and A. K. Haghi, PhD

CHAPTER 1

RECENT STUDIES OF SPIN DYNAMICS IN FERROMAGNETIC NANOPARTICLES

D. S. SCHMOOL

CONTENTS

1.1 INTRODUCTION

In recent years, the study of magnetic nanoparticle systems has received growing attention. With view to many applications, this area of scientific study shows enormous potential for growth. One central theme of this line of investigation is how the magnetic properties of nanoparticle assemblies are modified with respect to the bulk. Such differences in magnetic behavior are expected to arise from interparticle interactions, which will depend on the intervening non-magnetic matrix in which the particles are suspended, and from the size confinement effects, essentially due to the surface anisotropy of the particles. In addition to these fundamental effects, we also need to consider the superparamagnetic (SPM) behavior that is of great importance to the overall magnetic behavior of these systems. The SPM effects are strongly related to the overall magnetic anisotropy of the individual particles and have strong temperature dependences (Dormann et al., 1997).

Ferromagnetic resonance (FMR) can provide useful information not only on "bulk" magnetic properties, but is also useful in evaluating surface magnetic properties and interactions (Vonsovskii, 1966; Puszkarski, 1979; Maksymowicz, 1986). Much of the previous FMR studies on nanoparticle systems have concentrated on the temperature dependence of the resonance field. Raikher and Stepanov (1994) considered the case of a suspension of single domain ferromagnetic particle ferrofluids. In this work the authors consider the SPM thermal fluctuation and anisotropy effects and particle mobility in an applied magnetic field. The double peaked spectra are explained in terms of the alignment of the particles' anisotropy axes in the field and from the random orientation of the grains at low temperatures. Antoniak et al. (2005) considered nanoparticle assemblies of Fe_xPt_{1-x}. In this study the authors show that the size distribution of the particles is an important parameter in the evaluation of the blocking temperature and the overall effective anisotropy. The particles were assumed to be noninteracting, where a surface anisotropy was enhanced by the formation of an oxide ($\gamma - Fe_2O_3$) surface layer. The same system was studied by Ulmeanu et al., (2004), where frequency dependent FMR measurements were used to assess the g factor of the nanoparticles. Berger et al. (2001) considered the temperature dependence of superparamagnetic resonance in the maghemite ($\gamma - Fe_2O_3$) system, where again the importance of the size distribution is shown to be of fundamental importance in the evaluation of blocking temperatures and the existence of double peaked spectra. Dipole-dipole interactions (DDI) are discussed, but without calculating their effects. In a recent study we have also studied this system where we use DDI to evaluate the angular dependence of the resonance lines (Schmool and Schmalzl, 2007).

Nanogranular layered magnetic films are a special class of magnetoresistive (MR) system which consists of separate layers of ferromagnetic nanoparticles embedded in an insulating matrix and are usually called discontinuous metal- insulator magnetic multilayers (DMIM) (Sankar et al., 1997). Due to their transport properties (tunnel magnetoresistance properties; TMR ~ 10%) and multilayered structure, they possess a general similarity with the well known magnetic multilayered giant magnetoresistance (GMR) systems, (Baibich et al., 1988) see also GMR in magnetic metal/ nonmagnetic metal discontinuous GMR multilayers (Hylton et al., 1993).

Our main objective will be to consider some of the main aspects of the behavior of magnetic nanoparticles and in particular how ferromagnetic resonance (FMR) can be applied to such assemblies. The FMR is a highly sensitive technique, which essentially measures the internal effective magnetic field of ferromagnetic systems:

$$H_{eff} = -\frac{\partial E}{\partial M} \qquad (1)$$

where E represents the energy density of the magnetic system and M the saturation magnetization. (In all cases where energy is considered we have parameterized such that it is in the form of an energy density.) This is a convenient way of considering the magnetic behavior since we can include only those contributions to the magnetic properties, which are appropriate to a particular situation. For example, the Zeeman energy, magnetostatic energies from sample shape, magnetocrystalline, and surface anisotropies can all be easily included into the total magnetic energy.

In this chapter, we outline the basic study of the effects of the magnetic dipolar interaction in an assembly of ferromagnetic nanoparticles, its effect on magnetic properties of nanoparticle assemblies and in particular with respect to ferromagnetic resonance studies. In the theory of dipole-dipole interactions, we show the basic physics of the dipolar interaction and illustrate the overall effects using computer simulations of regular and random arrays of particles. The theory is presented for random general assemblies including an explanation of the nature of the simulations for arrays of particles. We will introduce the basic theory of FMR and then show how this is adapted to the study of nanoparticle systems; the relation between interparticle interactions in NP assemblies and angular studies by ferromagnetic resonance (FMR) will be discussed. An introduction to surface anisotropy is made as well as the multispin approach to the problem. A brief overview of linewidth of resonance is also given. We show some experimental results for random particle assemblies, and illustrate how information can be extracted from the FMR spectra. Some studies with respect to sample temperature are also given.

1.2 THEORY OF DIPOLE–DIPOLE INTERACTIONS IN FERROMAGNETIC NANOPARTICLE ASSEMBLIES

1.2.1 THE DIPOLAR INTERACTION

The fundamental interparticle interactions in ferromagnetic nanoparticle assemblies (whatever form they may take) in which the particles are embedded in a non-conducting non-magnetic matrix will occur via the dipole – dipole interaction.

The basic magnetic dipolar interaction can be expressed by the equation:

$$E_{ij}^{DDI}(\mathbf{r}_{ij}) = \frac{1}{4\pi r_{ij}^3}\left[\mathbf{m}_i \cdot \mathbf{m}_j - \frac{3(\mathbf{m}_i \cdot \mathbf{r}_{ij})(\mathbf{m}_j \cdot \mathbf{r}_{ij})}{r_{ij}^2}\right] \qquad (2)$$

where $m_{i,j}$ represents the magnetic moment of particle i,j and $\mathbf{r}_{ij}$ is the displacement vector between them. We are principally interested in the effects of interactions in assemblies under conditions of ferromagnetic resonance. We consider the case in which the applied magnetic field is sufficient to saturate the sample and as such all magnetic moments will be aligned (as in the case of an FMR experiment).

Under this state, we can simplify (2) using the spherical coordinate system as:

$$E_{ij}^{DDI}(\theta,\vartheta,\phi,\varphi) = \frac{m_i m_j}{4\pi r_{ij}^3}\left[1 - 3(\sin\theta\sin\vartheta\cos(\phi-\varphi) + \cos\theta\cos\vartheta)^2\right] \quad (3)$$

All angles are defined in Figure 1 (a) and (b). This equation can be further simplified for the case, where the vector between magnetic moments is aligned along one of the principal axes. Importantly, this equation demonstrates the introduction of a magnetic anisotropy into the system. The overall anisotropy will depend explicitly on the spatial distribution of the magnetic nanoparticle assembly.

We can sum all interactions in an assembly using the following equation:

$$E_{TOT}^{DDI} = \sum_{i,j}E_{ij}^{DDI} = \frac{1}{2}\sum_i\sum_{j\neq i}E_{ij}^{DDI}(\theta,\vartheta,\phi,\varphi) \qquad (4)$$

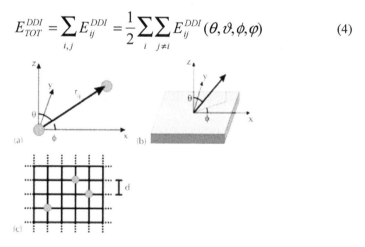

FIGURE 1 (a) Basic coordinate geometry for the dipolar interaction between two magnetic particles. (b) Spherical coordinate system for the magnetic sample. (c) Schematic illustration of a section of the nanoparticle assembly.

The factor of 1/2 is required so as to not count the interactions twice. Equation (4) can be used for a random array of particles, though positions of all particles must be known for the calculation to be performed. We illustrate simulations on systems of this type to show how such calculations can be made. For systems in which the ferromagnetic nanoparticles are arranged in layers, such as discontinuous magnetic multilayer systems (Sankar et al., 1997), then we can separate the interactions to those in the same plane and those in adjacent layers, where we neglect non-adjacent layer interactions. Figure 2 illustrates the geometry of such a system. We consider contributions arising from dipolar interactions within each (discontinuous) layer in - plane (or intraplanar interactions, IP) and those between the various planes (out of plane, OP), or interplanar interactions.

Within the layer, the total (in - plane, IP) dipolar energy, E_{IP}^{DDI}, is the sum over all the magnetic moments, where $\Theta_{ij} = 90$, which gives:

$$E_{IP}^{DDI} = \frac{1}{2}\sum_{\substack{i=1}}^{N}\sum_{\substack{j=1 \\ j\neq i}}^{N} E_{ij}^{DDI} = \Gamma_1^{IP} + \Gamma_2^{IP} \sin^2 \vartheta \tag{5}$$

Here $\Gamma_{1,2}^{IP}$ are constants depending on the distribution of the magnetic moments in the layer:

$$\Gamma_1^{IP} = \frac{1}{2V_m}\sum_{\substack{i=1}}^{N}\sum_{\substack{j=1 \\ j\neq i}}^{N} \frac{m_i m_j}{4\pi r_{ij}^3}$$

$$\Gamma_2^{IP} = \frac{1}{2V_m}\sum_{\substack{i=1}}^{N}\sum_{\substack{j=1 \\ j\neq i}}^{N} \frac{m_i m_j}{4\pi r_{ij}^3} \sin^2 \Phi_{ij} \tag{6}$$

The summation over j considers one particles' interaction with all the other (N - 1) particles within the layer, while the summation over i considers all of these interactions for all N particles. The value of the constants (m_i, m_j, and $<r_{ij}>$) will determine the energy scale, or in the FMR spectra, the field scale, of the total angular variation. V_m represents the total magnetic volume of the particles, which can be expressed as:

$$V_m = \sum_{i=1}^{N} V_i \tag{7}$$

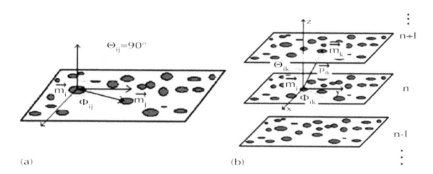

FIGURE 2   Schematic diagram of the planar geometry for layered particle systems illustrating (a) intraplanar interactions and (b) interplanar particle interactions.

We note that the magnetic volume and volume fraction are related via the expression, $f = V_m/V$, where V is the total volume of the sample. With regards to the dipolar interaction between layers; only interactions between adjacent layers are considered, while those between non-adjacent layers are considered as negligible. In this case, we consider m_i and m_k as two magnetic moments from two different but adjacent layers. Summing over all the magnetic moments of the two layers, we obtain:

$$E_{OP}^{DDI} = \Gamma_1^{OP} + \Gamma_2^{OP}\sin^2\vartheta + \Gamma_3^{OP}\cos^2\vartheta \tag{8}$$

where $\Gamma_{1,2,3}^{OP}$ are constants depending on the distribution of the magnetic moments inside the two layers.

These are given by:

$$\Gamma_1^{OP} = \frac{1}{2V_m}\sum_{i=1}^{N}\sum_{k=1}^{N'}\frac{m_i m_k}{4\pi r_{ik}^3}$$

$$\Gamma_2^{OP} = \frac{1}{2V_m}\sum_{i=1}^{N}\sum_{k=1}^{N'}\frac{3m_i m_k}{4\pi r_{ik}^3}\sin^2\Theta_{ik}\sin^2\Phi_{ik}$$

$$\Gamma_3^{OP} = \frac{1}{2V_m}\sum_{i=1}^{N}\sum_{k=1}^{N'}\frac{3m_i m_k}{4\pi r_{ik}^3}\cos^2\Theta_{ik} \tag{9}$$

The sum over k corresponds to the combined interactions of a magnetic particle in one layer with all N' particles in an adjacent layer. The sum over i is required to consider all the (N) particles of the layer (See Figure. 2). The total dipolar energy for n layers will be:

$$E_{TOT}^{DDI} = \sum_{L=1}^{n} (E_{IP}^{DDI})_L + \sum_{L'=1}^{n} (E_{OP}^{DDI})_{L'} = \Gamma_1^{eff} + \Gamma_2^{eff} \sin^2 \vartheta + (n-1)\Gamma_3^{OP} \cos^2 \vartheta \quad (10)$$

with $\Gamma_1^{eff} = n\Gamma_1^{IP} + (n-1)\Gamma_1^{OP}$ and $\Gamma_2^{eff} = n\Gamma_2^{IP} + (n-1)\Gamma_2^{OP}$.

We note that the summation for the first term is over the n layers (intraplanar term) while the second summation (interplanar term) is only between adjacent layers, of which there are (n–1) pairs, care has been taken not to count interactions twice. In this consideration, it is assumed that the interplanar and intraplanar interactions are effectively identical for each (discontinuous) layer.

1.2.2 SIMULATIONS FOR ASSEMBLIES OF NANOPARTICLES

To perform the simulations, we have constructed a regular simple cubic lattice, with lattice constant d, as illustrated in Figure 1(c). For all simulations, we use a slab shaped spatial distribution (Figure 1(b)), this being the shape of the magnetic NP samples measured by FMR.

We have performed two sets of simulations, which will be reported here:

(i) Variation of the aspect ratio, R, (thickness/length; $R = z/x$; where we maintain $x = y$) of the particle distribution, that is varying the number of particles in the z and x - directions.

(ii) Variation of the particle density, which can be controlled by simply varying the lattice parameter d, where the particle density can be expressed as: $\rho_N = N/V = 1/d^3$. For the former, we have used random arrays of particles, while in the latter we have performed simulations for both regular and random arrays of NPs.

To assess these energy variations, we consider the angular dependence of Equation (4), maintaining the azimuthal orientation constant. This illustrates the magnetic anisotropy introduced by the DDI, which is also proportional to the average interaction. In Figure 3 (a), we show this variation for a regular array of $(50 \times 50 \times 5)$ particles.

In this case the DDI variation can be written in the form:

$$E_{RA}^{DDI}(\vartheta) = E_0 \cos^2 \vartheta + E_1 \quad (11)$$

The size of the variation will be given by E_0 and an offset E_1 is also included. Both constants will depend on the magnetic moment (magnetization) of the particles as well as their average effective separation. The energy E_0 is a measure of the maximum to minimum energy of the angular dependence.

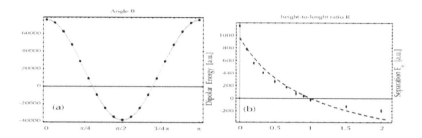

FIGURE 3 (a) Angular variation of the DDI energy for a regular cubic array of NPs (50 × 50 × 5) as evaluated from (3) (points) and a fit to (4) (solid line). (b) Aspect ratio (sample shape) dependence of E_0 on $R = z/x$, the dashed line corresponds to the demagnetizing energy of an uniformly magnetised sample.

In a series of simulations we have varied the aspect ratio R, from a thin layer (R ~ 0; R< 1) through a cube (R = 1) to a columnar shape (R >1). For such a shape variation, we see the dependence as illustrated in Figure 3 (b). We note that for slab type geometries, $E(0) > (\pi/2)$, that is positive E_0, while for the cube shaped distribution, we have $E_0 = 0$ and E_0 becomes negative for columnar samples distributions, that is, DDI energy curve becomes inverted. The dashed line is based on the demagnetizing energy with the dimensions of the NP distribution; $E(R) = K(1 - R)/(1 + R)$, where K is a constant (Schmool and Barandiarán, 1999). In Figure 4, we show the results of simulations for both regular and random arrays of NPs, where we have varied the density (or equivalently the filling factor, $f = V_m/V_{TOT}$) of NPs in the assembly. For the case of the random array calculations, a computer generated random number program partially "occupies" a set number of lattice sites (depending on the density to be evaluated). In these calculations, we repeat the simulation 20 times and obtain an average to eliminate any distorted assembly distributions that may occur; the error in the random distributions reflects the variance in these results. In both cases, we observe a density-squared dependence.

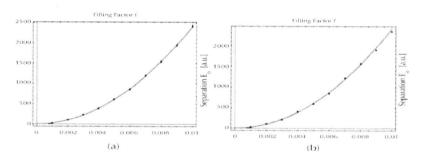

FIGURE 4 Variation of DDI energy E_0 as a function of particle density, ρ_N, for (a) random particle arrays and (b) regular NP assemblies.

1.3 FERROMAGNETIC RESONANCE IN NANOPARTICLE ARRAYS

1.3.1 INTRODUCTION

While the basic theory of FMR in nanoparticle systems will use the usual formalism common to ferromagnetic resonance experiments, we need to consider how the energy contributions must be selected to account for the particular behavior of nanoparticle assemblies. To obtain the resonance condition used in FMR we generally start by considering the Landau–Lifshitz equation with some form of damping. From this, the so-called Smit–Beljers equation can be derived. This can be suitably modified to include the possible existence of spin wave resonance modes, important in confined magnetic systems, such as thin films and nanoparticles. This equation will require the energy equation with full angular dependence from which we evaluate the second derivatives with respect to the polar and azimuthal angles. We also require the equilibrium conditions, which are used in conjunction with the resonance equation and are obtained by setting the first derivatives of the energy, with respect to the angles, to zero.

Using the uniform FMR mode our resonance equation takes the form:

$$\left(\frac{\omega}{\gamma}\right)^2 = \frac{(1+\alpha)}{M_0 \sin^2 \vartheta}\left[\left(\frac{\partial^2 E}{\partial \vartheta^2}\right)\left(\frac{\partial^2 E}{\partial \varphi^2}\right) - \left(\frac{\partial^2 E}{\partial \vartheta \partial \varphi}\right)^2\right] \tag{12}$$

where ω is the angular frequency of the exciting field (microwave radiation), g is the magnetogyric ratio and α is the Gilbert damping parameter. For spin wave resonance additional terms must be added, see for example (Schmool and Barandiarán, 1998).

1.3.2 DIPOLAR INTERACTIONS

The essential ingredient in the evaluation of the resonance equation is the form of the free energy density used in Equation (12). In previous studies, granular systems have been considered by separating the magnetostatic energy into two components; one due to the particle shape and another arising from the overall samples shape. Such a consideration leads to the modified Netzelmann equation (Netzelmann, 1990; Dubowik, 1996; Kakazei et al., 1999):

$$E = \frac{1}{2}f(1-f)\mathbf{M}\cdot \tilde{\mathbf{N}}_p \cdot \mathbf{M} + \frac{1}{2}f^2 \mathbf{M}\cdot \tilde{\mathbf{N}}_s \cdot \mathbf{M} - f\mathbf{M}\cdot \mathbf{H} \tag{13}$$

where f represents the volume fraction of particles; $f = V_m/V$, V being the total volume of the sample, $\tilde{\mathbf{N}}_p$ is the particle shape tensor while $\tilde{\mathbf{N}}_s$ is that of the sample. In a recent article we have shown that a more appropriate treatment would be to substitute the sample shape energy with a summation of the dipolar interactions between all the magnetic particles (Schmool et al., 2007).

In this case Equation (13) is replaced by:

$$E = \frac{1}{2}f(1-f)\mathbf{M}\cdot\tilde{\mathbf{N}}_p\cdot\mathbf{M} + E_{TOT}^{DDI} - f\mathbf{M}\cdot\mathbf{H} \tag{14}$$

where the dipolar energy term can be expressed as given in the dipole energy equations shown above; Equation. (4) or (10), depending on the type of nanoparticle assembly under consideration. We note that the form of the dipolar energy used is only valid for the case where all magnetic moments are aligned, for example by an applied external field. In FMR this is a valid assumption for weak magnetic anisotropies. Using this analysis we can obtain the resonance equation in the following form (Schmool et al., 2007; Schmool and Schmalzl, 2009):

$$\frac{\sin\theta\cos(\theta-\vartheta_0)}{\sin\vartheta_0}H_{appl}^2 - 2C\frac{\sin\theta\cos2\vartheta_0}{\sin\vartheta_0}H_{appl} - \left(\frac{\omega}{\gamma}\right)^2 = 0 \tag{15}$$

The angle ϑ_0 is defined by the equilibrium condition:

$$C\sin2\vartheta_0 = H_{appl}\sin(\theta-\vartheta_0) \tag{16}$$

where C is a constant which depends on details of the particle shape and assembly type. For example, for spherical particles takes the form:

$$C = \frac{\pi\langle r\rangle^3}{6}MV \tag{17}$$

<r> is the average particle radius. For non-spherical particles this constant will have an additional term related to the particle shape anisotropy (Schmool et al., 2007).

In the case of the layered particle system we considered above we can write:

$$C = \frac{3MV_m}{8\pi}\left| \frac{n}{N(N-1)}\sum_{i=1}^{N}\sum_{\substack{j=1\\j\neq i}}^{N}\frac{1}{r_{ij}^3}\sin^2\Phi_{ij} - \frac{(n-1)}{NN'}\sum_{i=1}^{N}\sum_{k=1}^{N'}\frac{1}{r_{ik}^3}\cos^2\Theta_{ik} \right| \tag{18}$$

where it has been assumed that all particles have the same volume (monodispersion) and therefore magnetic moment, $V_i = V_j = V_p$ such that $\sum_i V_i = NV_p = V_m$,

$\sum_j V_j = (N-1)V_p \approx V_m$ and $M = m_i/V_i$. We see more clearly from this expression the geometric factors, which must be evaluated in order to estimate the expected angular variation of the resonance field.

1.3.3 SURFACE ANISOTROPY

The perturbation of the crystal symmetry at the surface of nanoparticles should lead to a magnetocrystalline anisotropy, which differs in magnitude and symmetry from that of the bulk. The overall influence of surface anisotropy in a nanoparticle will depend on the strength of the anisotropy constant itself and on the size of the particles. Such effects will be more pronounced in smaller particles. With regards to the energy related to surface anisotropy we can distinguish between cases where the surface spins have a preferred orientation with respect to the crystalline axes, as described by (Kachkachi and Bonet, 2006):

$$E_{Surf} = -\sum_i K_s (\mathbf{m}_i \cdot \hat{\mathbf{u}}_i)^2 \qquad (19)$$

where m_i is the unit vector in the direction of the magnetization and $\hat{\mathbf{u}}_i$ a unit vector in the direction of the easy axis. Here K_s will be the anisotropy strength. This model is known as the transverse surface anisotropy. The summation in Equation. (19) will account for the combined energy of all the spins (i) in the nanoparticle. In this case we can use a one-spin approach whereby all individual spins in the nanoparticle are approximated by a macro-spin. A more realistic description is given by the Néel model for surface anisotropy and can be expressed as (Kachkachi and Bonet, 2006):

$$E_{Surf}^{N\acute{e}el} = -K_s \sum_i \sum_{j \neq i} (\mathbf{m}_i \cdot \hat{\mathbf{e}}_{ij})^2 \qquad (20)$$

where $\hat{\mathbf{e}}_{ij} = \mathbf{r}_{ij}/r_{ij}$ is an interatomic vector between surface atoms. This will naturally account for the change in symmetry when a surface atom is encountered and is a microscopic model, which can be distinguished from the one-spin approximation. Such expressions can be readily introduced into the FMR equation by summing the surface anisotropy energy to the other components of the energy and applying the usual approach via Equation (12). Extension to a full multi-spin calculation is complex since the N spins in the nanoparticle will have to be assessed for their individual resonance condition giving rise to an N coupled resonance equation.

1.3.4 THEORY OF MULTISPIN SYSTEMS

Recent simulations based on a new model by (Sousa et al. 2009; Schmool, 2009) aim to clarify the role of surface anisotropy in FM nanoparticle systems, where the theory has been developed to study the resonance conditions in a multispin system. In this model, a single nanoparticle is constructed of $N-$ spins, which are exchange coupled, the resonance condition for each spin (i) is then considered in its own specific effective field, H_i^{eff}.

As such, the reduced free energy takes the form:

$$-\varepsilon_i = -\frac{\hat{H}_i}{J} = h(\hat{\mathbf{e}}_h \cdot \mathbf{s}_i) + k_i A(s_i) + \mathbf{s}_i \cdot \sum_{j=1}^{z_i} \lambda_{ij} \mathbf{s}_j \tag{21}$$

In this equation, the first term represents the Zeeman energy, the second term is the reduced anisotropy energy, which can be modified to include any relevant symmetries including spins at the particle surface and the final term is the reduced exchange energy which accounts for the local coordination, z_i. Using a similar approach to that when considering the classical FMR theory, we obtain the N - spin resonance equation, which is analogous to the Smit - Beljers equation, of the form:

$$\sum_{\beta=\vartheta,\varphi} \sum_{k=1}^{N} \left\{ [\mathbf{H}_{ik} - \Omega_{ik}(\eta)]\mathbf{I}_{-\eta} \right\}_{\alpha,\beta} \delta\mathbf{s}_{k,\beta}(0) = 0 \tag{22}$$

where the following substitutions are used:

$$\mathbf{H}_{ik}(\varepsilon_i) = \begin{pmatrix} \partial^2_{\vartheta_k \vartheta_i} \varepsilon_i & \dfrac{1}{\sin \vartheta_i} \partial^2_{\vartheta_k \varphi_i} \varepsilon_i \\ \dfrac{1}{\sin \vartheta_k} \partial^2_{\varphi_k \vartheta_i} \varepsilon_i & \dfrac{1}{\sin \vartheta_k \sin \vartheta_i} \partial^2_{\varphi_k \varphi_i} \varepsilon_i \end{pmatrix} \tag{23}$$

$$\mathbf{I}_\eta = \begin{pmatrix} -\eta & -1 \\ 1 & -\eta \end{pmatrix} \tag{24}$$

$$\Omega_{ik}(\eta) = \frac{i\omega}{1+\eta^2} \delta_{ik}\mathbf{I}_\eta \tag{25}$$

Solutions for the resonance frequency are then obtained from the pure imaginary

values of the matrix $[\mathbf{H}_{ik}\mathbf{I}_{-\eta}]_{\alpha\beta}$. It is note worthy that the form of the kernel has a particular symmetry; the leading diagonal block terms (of 2 x 2) are of the form of the Smit–Beljers equation for each spin, then off - diagonal (2 x 2) blocks will be interaction terms between the various spins, where the further off-diagonal the further spins are physically. Preliminary simulations illustrate the variation of the resonance (frequency) spectra with possible indications of the influence of surface spins, which are illustrated in Figure 5 for cubic- and spherical-shaped particles. The spectral analysis is made as a function of the particle size (number of spins), this should show the influence of the surface spins which will be most important for the smaller particles. The direct computed spectrum is comprised of the resonance frequencies of the individual spins, which will be different for each spin environment (corner, edge, face, and so on.). The corresponding correlation function is then evaluated to obtain the statistically most probable frequencies. The size dependence of the resonance frequency is illustrated in Figure 5 (c) and 5 (d) for the cubic and spherical particles, respectively. As the particles increase in size the resonance frequency tends to saturate. It will be noted that the main resonance frequency saturates rapidly in the case of the cubic particles, which have a much larger surface area to volume ratio. Also, we note a small contribution (shoulder) which appears to correspond to the resonance of the smallest particle, which could be related to the surface spins resonance (Sousa et al. 2009, Schmool, 2009).

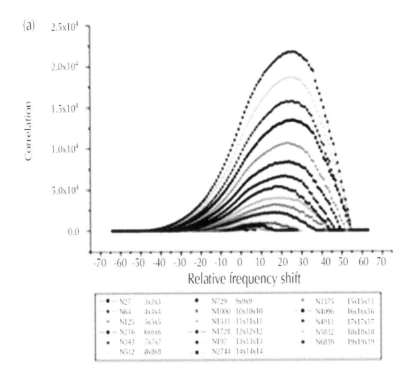

FIGURE 5 *(Continued)*

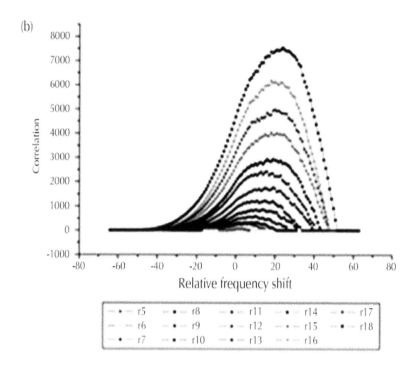

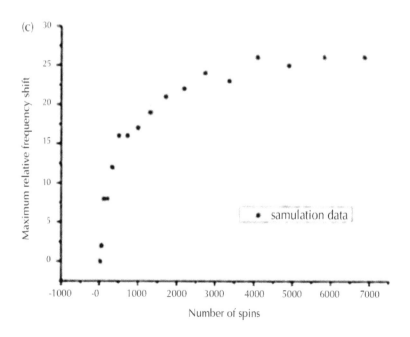

FIGURE 5   *(Continued)*

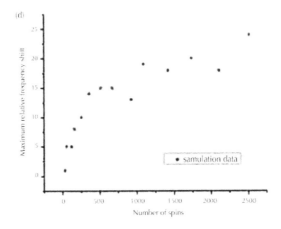

FIGURE 5 Simulated frequency spectra for (a) cubic and (b) spherical shaped Fe nanoparticles. The evolution of the dominant resonance peak for these are given in (c) and (d), respectively.

1.3.5 RESONANCE LINEWIDTH

The resonance linewidth can provide valuable qualitative information. The inclusion of a damping term in the Landau–Lifshitz equation of motion gives, for the case of the so- called Gilbert damping:

$$\frac{1}{\gamma}\frac{\partial \mathbf{M}}{\partial t} = -(\mathbf{M}\times \mathbf{H}_{eff}) + \frac{\alpha}{\gamma M}\left(\mathbf{M}\times \frac{\partial \mathbf{M}}{\partial t}\right) \tag{26}$$

The Gilbert parameter, α, represents the damping strength and is related to the intrinsic linewidth by the relation (Vonsovskii, 1966; Farle, 1998):

$$\Delta H_0 = \frac{\alpha}{M}\left(\frac{\partial^2 E}{\partial \vartheta^2} + \frac{1}{\sin \vartheta}\frac{\partial^2 E}{\partial \varphi^2}\right) \tag{27}$$

There are other contributions to the linewidth that are related to material properties providing several broadening mechanisms. In general, we consider that anything which gives rise to varying microscopic environments for individual spins will cause some form of extrinsic broadening due to effective modification caused in the effective field felt by the spins. This will produce a range of resonance conditions which, when taken together in an experimental spectrum, will give rise to a superposition of the various resonances resulting in an overall linewidth broadening. The various contributions, following Vittoria et al. (1967), can be expressed as:

$$\Delta H = \Delta H_0 + \left(\frac{\partial H}{\partial \phi}\right)\Delta\phi + \left(\frac{\partial H}{\partial H_i}\right)\Delta H_i + \left(\frac{\partial H}{\partial V}\right)\Delta V + \left(\frac{\partial H}{\partial S}\right)\Delta S \tag{28}$$

Here the first term on the right-hand side represents the intrinsic linewidth given by Equation (27), the second term is a contribution due to a spread in the crystalline axes, the third term arises from magnetic in homogeneities, the fourth term arises from variations in grain size for particulate systems, and the final term is the contribution due to surface spins of the magnetic particles. We have extended this approach to include variations in particle volume and surface contributions in nanoparticle systems. Here ΔV will be a contribution from the distribution in particle volume e.g., log-normal behavior, which is frequently observed in nanoparticle systems (Tronc et al., 2000).

1.4 EXPERIMENTAL STUDIES

1.4.1 ANGULAR DEPENDENCE OF THE RESONANCE FIELD

Samples of $\gamma - Fe_2O_3$ NP assemblies, with mean particle diameters of 2.7, 4.6, and 7.3 nm, were measured by FMR as a function of the direction of the applied magnetic field, from the perpendicular (0°) to the parallel (90°) orientations. In figure 6, we show the angular dependence of the FMR resonance field of NPs with mean diameter 4.6 nm. Also shown is the theoretical variation of the resonance field as predicted by Equation (15) and has only an angular dependence due to the DDI, where we assume the particles are spherical. The points refer to experimental data and the line is the theoretical fit. This curve is the representative of all samples measured, with other sizes. We see an excellent agreement between the theory and experimental measurements. This implies that the main influence on the angular dependent magnetic properties is due to DDI. The model assumes that all particles are of the same size with weak magnetic anisotropy.

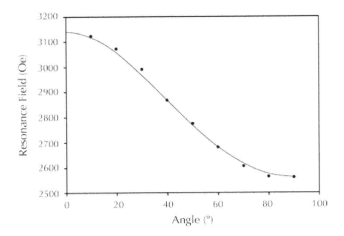

FIGURE 6 Angular dependence of the resonance field for samples of $\gamma - Fe_2O_3$ nanoparticles with an average diameter of 4.6 nm. The line is a fit to Equation (15).

The layered type structures indicated can be realized with the deposition of discontinuous magnetic multilayers, in which the magnetic films are sub-coalescence and effectively form a layer of nanoparticles. A multilayer system composed of ten insulating/ferromagnetic bilayers given by $[Al_2O_3(40 \text{ Å})/ Co_{80}Fe_{20} (t)]_{10}/Al_2O_3(30 \text{ Å})$, where t is the effective thickness of the CoFe layer. This corresponds to an effective thickness of a continuous layer with the same quantity of CoFe. We need to use this effective thickness instead of an absolute thickness because below 18 Å, CoFe films become discontinuous (Popplewell and Sakhnini, 1995; Kakazei et al. 2001). It will be noted that the Al_2O_3 layers are insulating and only dipolar interactions between the nanoparticles will be expected.

Figure 7 displays a sequence of representative FMR spectra as a function of the angle of the applied field, θ, for the sample with an effective thickness of 13 Å. As the angle increase from 0° to 90°, the resonance lines shift to lower fields. Figure 8 shows the resonance fields, as well as theoretical simulations for these results, as a function of θ, for the four samples studied. We can see that increasing the effective thickness, the angular dependence of the resonance field becomes stronger. The difference $(H_\perp - H_\parallel)$ for the 13 Å sample is about 8750 Oe, while for the 7 Å sample it is only 2850 Oe. Simulations are based on a numerical solution of Equation (15), where the fitting parameters are given in Table 1.

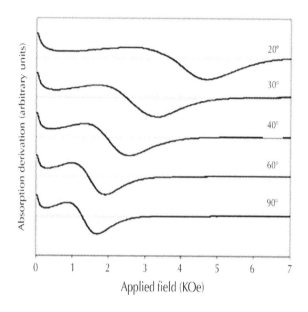

FIGURE 7   Representative spectra obtained for the $[Al_2O_3(40 \text{ Å})/ Co_{80}Fe_{20} (t)]_{10}/Al_2O_3(30 \text{ Å})$, sample with $t = 13$ Å for various angles.

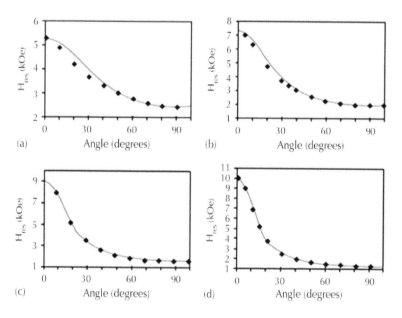

FIGURE 8   Angular variation of the resonance field from out-of-plane 0° to in-plane 90° for (a) $t = 7$ Å, (b) $t = 9$ Å, (c) $t = 11$ Å, and (d) $t = 13$ Å.

TABLE 1   Fitting parameters for the $[Al_2O_3(40\ \text{Å})/\ Co_{80}Fe_{20}\ (t)]_{10}/Al_2O_3(30\ \text{Å})$, sample with $t = 13\ \text{Å}$ used for angular measurements

t (Å)	C (Oe)	(ω/γ) (kOe)
7	990	3.30
9	1950	3.40
11	2800	3.45
13	3450	3.25

1.4.2 TEMPERATURE DEPENDENT STUDIES

In addition to room temperature measurements, measurements at low temperature can also yield useful information. Angular studies were performed on the $\gamma - Fe_2O_3$ nanoparticles at fixed temperatures (5, 100 and, 295 K) where similar fitting procedures have been performed. We have not explicitly included the presence of 'bulk' magnetocrystalline anisotropy into the energy. Given that the system we are considering has randomly oriented particles, that is, there is no preferential direction; we would expect that this contribution would have no explicit angular dependence. Despite this we do expect a constant contribution for all configurations, as prescribed by the random anisotropy model (Sankar et al. 1997). This can be simply incorporated

into the resonance Equation (15) by taking $(\omega/\gamma)^2 \rightarrow [(\omega/\gamma) - H_K]^2$, where H_K is the anisotropy field given by $2<K>/M$, where $<K>$ represents the average anisotropy constant. The angular variation of H_{res} will in no way depend on this term and reflects the variation of magnetization M in the superparamagnetic (SPM) regime, which can be expressed using a weighted Langevin function (Goya et al., 2003):

$$M = M_s \int L\left(\frac{HMV_m}{k_B T}\right) f(V) \mathrm{d}V \qquad (29)$$

$f(V)$ represents the log-normal distribution (Schmool and Schmalzl, 2007). This can be expressed as:

$$f(V) = \frac{1}{\sqrt{2\pi}\sigma V} \exp\left\{-\frac{[\ln(V/V_m)]^2}{2\sigma^2}\right\} \qquad (30)$$

In this expression, V_m represents the mean volume of the particles and σ is the distribution width, obtained experimentally. In Figure 9 we illustrate the comparison of the Langevin function with the experimental data. While this does not constitute a proof, it shows the trend expected for this behavior. In Table 2 we give the fitting parameters used for the principal resonance of these data. Note that we are using a fixed $(\omega/\gamma) = 3353$ Oe, which corresponds to a frequency of 9.35 GHz and $g = 2$. We note that the room temperature value for K compares reasonably well with the bulk value of 4.7×10^5 erg cm^{-3} (Krupicka and Zaveta, 1975).

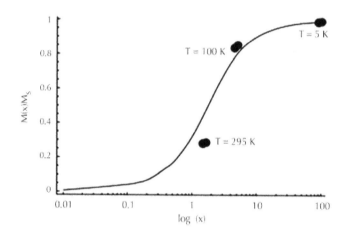

FIGURE 9 Comparison of experimental data points at the measured temperatures with the Langevin function (line) for $\gamma - Fe_2O_3$ nanoparticles of mean diameter $<D> = 4.6$ nm.

TABLE 2 Fitting parameters for the $\gamma - Fe_2O_3$ nanoparticles of mean diameter $<D> = 4.6$ nm

T (K)	C (Oe)	H_K (Oe)	$<K>$ (x 10^6 erg cm^{-3})
5	210	2223	3.034
100	180	713	1.13
295	61	463	0.239

Measurements were also performed in the discontinuous magnetic multilayer system as a function of the sample temperature. The resonance field as a function of the temperature for these samples is shown in Figure 10. Here we see that the resonance field, H_{res} , reduces with T, from 300 to 100 K, in a manner expected from classical FMR behavior of ferromagnetic materials. At approximately 90 K there is a significant enhancement of H_{res}, which is very marked for lowest effective thicknesses.

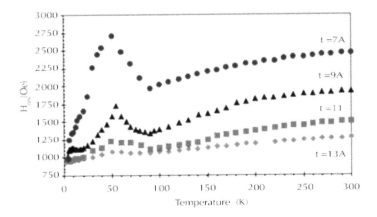

FIGURE 10   Temperature dependence of the resonance field for the $[Al_2O_3(40 \text{ Å})/ Co_{80}Fe_{20}$ $(t)]_{10}/Al_2O_3(30 \text{ Å})$ discontinuous multilayers.

In Figure 11, we plot H_{enh} against $1/t$, where $H_{enh} = H_{res} - H_{res}^{expt}$ and H_{res}^{expt} is the resonance field expected from a normal classical dependence, extrapolating the high temperature trend to lower temperatures. $H_{enh} (1/t)$ shows a linear dependence indicating that this enhancement may have its origin in surface anisotropy, which scales as $1/t$. It is clear that the surface contribution should grow with decreasing particle size, since the surface area to volume ratio increases. Therefore, it appears that, although the resonance equation used at room temperature is valid, for lower temperatures, < 90 K, we need to include an extra term corresponding to surface anisotropy to account for the effective field. From Figure 10, it appears that the enhancement vanishes at 15K.

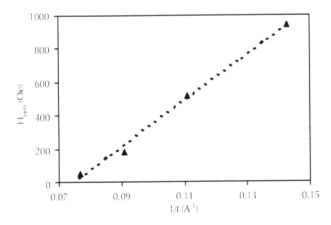

FIGURE 11   Maximum resonance field enhancement observed as a function of inverse thickness for the discontinuous multilayers.

1.5   CONCLUSION

We have presented a theory for ferromagnetic resonance in magnetic nanoparticle systems, in which we use a direct approach to account for the dipolar interactions between the magnetic particles. We have shown that the dipole–dipole interaction in ferromagnetic nanoparticle assemblies is of great importance when considering the magnetic properties of the assembly as a whole. By summing all interactions in the assembly, it is possible to illustrate how this introduces a magnetic anisotropy into the system. Results show that the spatial distribution will have a great influence on this overall anisotropy and that random arrays of particles can be easily approximated to regular arrays of NPs as they exhibit virtually identical results for the same particle density. This is much simpler to treat from a theoretical point of view. The important factor is the particle density and through this, the magnetic filling factor.

Measurements by FMR on NP assemblies of $\gamma-Fe_2O_3$ illustrate that the theory presented here shows an excellent agreement with experiment. We are currently investigating the effects of the volume dispersion on the DDI and its influence in FMR. We have also applied the theory to the discontinuous multilayer system $[Al_2O_3(40\ \text{Å})/Co_{80}Fe_{20}\ (t)]_{10}/Al_2O_3(30\ \text{Å})$, where the effective magnetic layer thickness, t, was varied between 7 and 13 Å. We observe that the DDI theory adequately explains the angular dependence of the resonance field and is directly related to the average dipole interaction between the magnetic nanoparticles. From angular dependent FMR measurements there appears to be a change in magnetic regime from thinner to thicker samples. For samples with effective thicknesses of 7 and 9 Å there is an overall superparamagnetic behavior, where dipolar coupling between particles almost vanishes, with each particle acting in an isolated manner. On the other hand, thicker samples with effective thicknesses of 11 and 13 Å act as super ferromagnetic, through the mechanism of magnetic dipolar coupling between nanoparticles. In this way, the magnetic dipolar interaction reveals itself to play an important role in nanoparticulate systems.

SUMMARY

Magnetic nanoparticle assemblies present novel magnetic properties with respect to their bulk constituent components. In addition to the surface effects produced by the modified atomic symmetry in such low dimensional systems, the magnetic coupling between the particles also plays a significant role in determining the overall magnetic behavior of a magnetic nanoparticle assembly. In this Chapter, we describe a theoretical model that accounts for the dipolar magnetic interaction between particles. There are two fundamental aspects of interest in our studies: the spatial distribution of the particles and density of the particle. These aspects have been addressed in our simulations, where we have performed simulations for regular and random arrays of particles. We have discussed the general theory of ferromagnetic resonance (FMR) applied to such systems and how the specific dipolar interactions can be incorporated for nanoparticle systems. The spatial distribution of particles can give valuable information of the strength of the dipolar interaction between them and we will demonstrate how this can be used in real systems. We have performed FMR experiments on nanoparticle assemblies of $\gamma - Fe_2O_3$ nanoparticles with different average particle sizes (2.7–7.3 nm) and particle densities, where samples are in rectangular slab shape. We have also performed detailed ferromagnetic resonance studies of $[Al_2O_3(40\ \text{Å})/ Co_{80}Fe_{20}(t)]_{10}/Al_2O_3(30\ \text{Å})$, $t = 7$–13 Å, discontinuous multilayers, with measurements taken as a function of the angle of the applied static magnetic field with respect to the sample at room and low temperatures. Measurements were performed as a function of the angle between the sample plane and applied magnetic field. Recent studies have shown that the incorporation of the dipolar interactions into FMR theory can explain the experimental results for angular studies in magnetic nanoparticle assemblies. A brief overview is also given of the general theory of FMR in a multispin particle.

KEYWORDS

- **Ferromagnetic Nanoparticles**
- **Ferromagnetic Resonance**
- **Dipole-dipole Interactions**
- **Magnetic Anisotropy**
- **Spherical Coordinate System**

REFERENCES

1. Antoniak, C., Lindner J., and Farle, M. Magnetic anisotropy and its temperature dependence in iron-rich x1-x nanoparticles. *Europhys. Lett*, **70**, 250 (2005).
2. Baibich, M. N. *et al.* Giant Magnetoresistance of (001) Fe/(001)Cr Magnetic Superlattice. *Phys. Rev. Lett*, **61**, 2472 (1988).
3. Berger, R., Bissey J. C., Kliava, J., Daubric, H. and Estournès, C. Temperature dependence of superparamagnetic resonance of iron oxide nanoparticles. *J. Magn. Magn. Mater*, **234**, 535 (2001).
4. Dormann, J. L., Fiorani, D., and Tronc, E. Magnetic relaxation in fine-particle system. Adv. *Chem. Phys.*, **98**, 283 (1997).
5. Dubowik, J. Shape anisotropy of magnetic heterostructures. *Phys. Rev. B*, **54**, 1088 1996.
6. Farle, M. Ferromagnetic resonance of ultrathin metallic layers. *Rep. Prog. Phys.*, **61**, 75 (1998).
7. Goya, G. F., Berquó, T. S., Fonseca, F. C., and Morales, M. P. Static and dynamic magnetic properties of spherical magnetite nanoparticles. *J. Appl. Phy*, **94**, 3520 (2003)
8. Hylton, T. L., Coffey, K. R., Parker, M. A., and Howard, J. K. Giant Magnetoresistance at Low Fields in Discontinuous NiFe-Ag Multilayer Thin Films. *Science*, **261**, 1021. (1993).
9. Kachkachi, H. and Bonet, E. Surface-induced cubic anisotropy in nanomagnets. *Phys. Rev, B*, **73**, 224402 (2006).
10. Kakazei, G.N., Kravets, A. F., Lesnik, N. A., Pereira de Azevedo, M. M., Pogorelov, Yu G., and Sousa, J. B.. Ferromagnetic resonance in granular thin films. *J. Appl. Phy.*, **85**, 5654 (1999).
11. Kakazei, G. N., Pogorelov, Yu. G., Lopes, A. M. L., Sousa, J. B., Cardoso, S., Freitas, P. P., Pereira de Azevedo, M. M., and Snoeck, E. Tunnel magnetoresistance and magnetic or-

dering in ion-beam sputtered Co80Fe20/Al2O3 *discontinuous multilayers. J. Appl. Phys.,* **90**, 4044 (2001).

12. Krupicka, S. and Zaveta, K. *Magnetic Oxides, part I.* D.J. Craik (Ed.), *Wiley Interscience* (1975).

13. Maksymowicz, A. Z. Spin-wave spectra of insulating films: Comparison of exact calculations and a single-wave-vector model. *Phys. Rev.* B, **33**, 6045 (1986).

14. Netzelmann, U. Ferromagnetic resonance of particulate magnetic recording tapes. *J. Appl. Phys.* **68,** 1800 (1990)

15. Popplewell, J., and Sakhnini, L. The dependence of the physical and magnetic properties of magnetic fluids on particle size. *J. Magn. Magn. Mater,* **149**, 72 (1995).

16. Puszkarski, H. Theory of Surface States in Spin Wave Resonance.Prog. Surf. Sci. **9**, 191 (1979).

17. Raikher, Yu. L. ,and Stepenov, V. I. Ferromagnetic resonance in a suspension of single-domain particles. *Phys. Rev. B,* **50**, 6250 (1994).

18. Sankar, S., Dieny, B., and Berkowitz, A. Spin-polarized tunneling in discontinuous CoFe/HfO2 multilayers. *J. Appl. Phys.,* **81**, 5512 (1997).

19. Schmool, D. S. and Barandiarán, J. M. Ferromagnetic resonance and spin wave resonance in multiphase materials: theoretical considerations. *J. Phys.: Condens. Matter* **10**, 10679 (1998).

20. Schmool, D. S. and Barandiarán, J. M. New phenomena in the study of shape effects in ferromagnetic resonance. *J. Magn. Magn. Mater.*, **191**, 211 (1999).

21. Schmool, D. S., Rocha, R., Sousa, J. B., Santos, J. A. M., Kakazei, G. N., Garitaonandia, J.S., and Lezama, L. The role of dipolar interactions in magnetic nanoparticles: Ferromagnetic resonance in discontinuous magnetic multilayers. *J. Appl. Phys.,* **101**, 103907. (2007).

22. Schmool, D. S., and Schmalzl, M. Ferromagnetic resonance in magnetic nanoparticle assemblies. *J. Non-Cryst. Solids* **353**, 738 (2007).

23. Schmool, D. S. and Schmalzl, M. Magnetic dipolar interactions in nanoparticle systems: theory, simulations and ferromagnetic resonance. *Advances in Nanomagnetism* B. Aktaş and F. Mikailov (Eds.), Springer, 321–326. (2009)

24. Schmool, D. S. Spin dynamics in nanometric magnetic systems. *The Handbook of Magnetic Materials,* Volume 18, K. H. J. Buschow (Ed.), *Elsevier Science,* pp 111 – 346 (2009).

25. Sousa, N., Schmool, D. S., and Kachkachi, H. Many spin approach to ferromagnetic resonance in magnetic nanostructures. NanoSpain , Zaragoza, Spain (March 9–12, 2009).

26. Tronc, E., Ezzir, A., and Cherkaoui, A. Surface-related properties of γ-Fe_2O_3 nanoparticles. *J. Magn. Magn,* Mater. **221**, 63 (2000).

27. Ulmeanu, M., Antoniak, C., Wiedwald, U., Farle, M., Frait, Z., and Sun, S. Composition-dependent ratio of orbital-to-spin magnetic moment in structurally disordered FexPt1-x nanoparticles. *Phys. Rev. B,* **69**, 054417 (2004).

28. Vittoria, C., Barker, R. C., and Yelon, A. Anisotropic Ferromagnetic Resonance Linewidth in Ni Platelets. *Phys. Rev. Lett.* **19**, 792 (1967).

29. Vonsovskii, S. V. *Ferromagnetic Resonance*, Pergamon, New York (1966).

CHAPTER 2

NOVEL MAGNETIC CARBON BIOCOMPOSITES

B. KALSKA-SZOSTKO, M. ROGOWSKA, A. PIGIEL, and D. SATULA

CONTENTS

2.1   INTRODUCTION

Nanoscience plays with structures as small 1 billionth of a meter (Sun et. al., 2009). These elements could be used as building blocks to create bigger heterostructures such as composites and hybrid materials which contain two or more different nanoscale materials connected together. They show unique properties often not observable for separate nano materials. Nanoparticles have many applications in a number of fields, including nanoelectrics, sensor materials, optical devices, molecular recognition media, or biomedical applications (Sun et. al., 2009; Haratifar et. al., 2009). Magnetic nanoparticles made of iron oxide (Fe_3O_4) and its derivatives (γ- Fe_2O_3) have become extensively studied in nanotechnology and attracted more and more attention due to their interesting properties such as special magnetic targeting and good biocompatibility. They are often used in magnetic resonance imaging (MRI), drug delivery system, treatment of cancer/HIV, or biosensors (Timko et. al., 2004; Laurent et. al., 2008; Sun et. al., 2009). The effective use of magnetic nanoparticles in these areas is possible due to their strong magnetic property, low toxicity, and possible controllable size from a few nanometers up to tens of nanometers. Their sizes are comparable to a virus (20–450 nm), proteins (5–50 nm), or genes (2 nm wide and 10–100 nm long) and situate it on length scale below living cells (10–100 μm). Before any magnetic nanoparticles can be used in the above-mentioned application, their surface should be modified to proper groups which caused them compatible with the matrix material. Functionalization of nanoparticles is based on the attachment of compounds with mainly amine (-NH_2), carboxyl (-COOH), and hydroxide (-OH) groups, or coating by polymers (for example: chitosan, PEG), or SiO_2 shell. The presence of these groups allows immobilization on their surface biological molecules such as proteins, antibodies, vitamins, and so forth (Sun et. al., 2009; Haratifar et. al., 2009). Cooperation of both particles is possible by bond activation between immobilized proteins and surface of magnetic nanoparticles (Woo et. al., 2005). Not only nanoparticles are used in nanotechnology but also carbon structures have shown widespread interest due to their unique properties. The discovery of fullerenes, nanotubes, and nanodiamonds in the last decade of the 20th Century opened new opportunities in nanotechnology (Allen et. al., 2009). Each of these structures shows unique electrical, mechanical and, optical properties, for example: nanotubes are very resistant to stretching and bending, they can also be electrically conductive in the same way as fullerene (C_{60}) (Klumpp, 2006). Similar to magnetic nanoparticles, for good usage of carbon structures, their surface should be chemical modified by the attachment of various functional groups. This opens a possibility for the creation of hierarchical structures, such as composites or hybrid nanostructures. Fabrication of heterostructures causes that carbon nanomaterials can also be used in medicine or electronic science (Bakry et. al., 2007; Markovic et. al., 2008).

In this study, nanocomposites from different nanoparticles were fabricated. In the first step, magnetic nanoparticles (synthesized from three various methods) and carbon nanostructures were functionalized to amine or carboxyl groups and then used as building blocks for composites creation and with vitamin-biotin immobilization.

In the present chapter, magnetite nanoparticles were prepared by three different synthesis methods: condensation from iron chlorides, decomposition of Fe[acac]$_3$

complexes, and structures as core-shell covered by SiO_2. Magnetic nanoparticles were functionalized to amine and carboxyl groups. Modified particles were collaborated into nanocomposite with various carbon structures such as fullerenes C_{60}, nanotubes, and nanodiamonds decorated by the appropriate functional bioparticle. The resulting nanocomposites were tested by infra-red spectroscopy, Transmission Electron Microscope, X-ray diffraction, and Mössbauer spectroscopy.

2.2 EXPERIMENTAL

2.2.1 SYNTHESIS OF MAGNETITE NANOPARTICLES FROM IRON CHLORIDES AND Fe(acac)₃ COMPLEX

Fe_3O_4 magnetite nanoparticles (NPs) obtained from iron chlorides nanoparticles in aqueous solution were prepared by modified Massart's method (Massart et al. 1987, Kalska-Szostko et. al., 2006). The magnetite obtained from Fe[acac]₃ complexes in organic solution was prepared according to modified Sun recipe (Sun et. al., 2002; Kalska Szostko et. al., 2011). 20 ml of diphenyl ether, 10 mmol 1,2-hexadecanediol, 6 mmol oley amine, 6 mmol oleic acid, and 2 mmol of iron III acetylacetonate were placed in a three-necked round-bottom flask. All reagents were stirred by a permanent magnet, deoxygenated by continuously bubbling argon gas through the reaction solution and heated to 250°C. After that, stable temperature was kept for around 35 minutes. In the last step, the solution was cooled down and washed three times by deoxygenate acetone and finally dried at room temperature.

From Sun synthesis, the obtained magnetic nanoparticles already have proper functional groups at the surface so we do not have to modify them further.

2.2.2 SYNTHESIS OF MAGNETITE WITH SiO₂ SHELL NANOPARTICLES

Magnetite nanoparticles obtained from iron chlorides by condensation synthesis were coated by SiO_2 shell according to method (Kalska-Szostko B and M. Rogowska). In the first step, 0.01M citric acid was added to reactive solution- iron (II, III) chlorides until flocculent sediment was visible, then TMAOH (tetramethylammonium hydroxide) solution was added to glass to obtain pH = 7. In the next step, the reactive mixture consisted of 37.5 ml of ethanol and 10 ml of distilled water, and 2.5 ml of ammonia solution (25%) was added to the resulting precipitate. All reagents were mixed under a magnetic stirrer for ten minutes and then left for 24 hr at room temperature. After that time, the magnetic nanoparticles were separated from reaction medium by a permanent magnet and placed in distilled water for one week.

2.2.3 MODIFICATION OF MAGNETITE NANOPARTICLES

Magnetite nanoparticles from Massart's method were functionalized to amine and carboxyl groups to be suitable for further experiments. The changes are based on replacing the solvent and surfactant from polar to non-polar one. Magnetite nanoparticles powder was placed into a flask containing distilled water and combined with organic mixture consisting of toluene, oleyamine, and oleic acid in molar ratio 1:2:2 respectively. Then, the obtained suspension was mixed in a sonication bath, shaken and left for four days. After that, magnetic nanopowder was washed, filtered, and dried at room temperature.

2.2.4   MODIFICATION OF MAGNETITE WITH SiO₂ SHELL NANOPARTICLES

Magnetic nanoparticles obtained at part 2.2.2 were functionalized to amine groups. In this case, reaction mixture consisted of toluene and APTMS ((3-aminopropyl)-tri-methoxysilane) was added to 15 mg nanoparticles with SiO_2 shell. All reagents were stirred vigorously, deoxygenated by bubbling solution with argon gas and heated to 150°C. After that time, modified nanoparticles were washed three times by ethanol and dried at room temperature (Woo et. al., 2005).

2.2.5   MODIFICATION OF CARBON NANOPARTICLES: FULLERENE C_{60}, NANOTUBE, AND NANODIAMOND

Carbon structures like fullerene C_{60}, nanodiamonds, and nanotubes were used for the preparation of nanocomposites with magnetic and biological molecules. Fullerene C_{60} nanoparticles have been functionalized to amine group already described in our previous studies (Kalska-Szostko et al. 2012). Carbon nanotubes were modified by nitrating mixture (of HNO_3:H_2SO_4 in the ratio 1:3) to form on their surface carboxyl groups (-COOH) (Jiang et al. 2003). The sample in the reaction mixture was stirred in a sonication bath for 2 hr. After that time, the modified powder was washed in 2 M NaOH and distilled water to obtain pH = 7, and finally dried at room temperature.

Nanodiamond powder has been modified in two ways:
1) By the same method as nanotubes (nitrating mixture) and
2) By hexamethylenediamine as fullerenes.

In the case of nitrating mixture nanodiamonds were further modified with 5 ml of octylamine. All reagents were deoxygenated by bubbling argon gas through the reaction mixture and heated in 50°C for 4 hr. After that time, nanodiamonds were washed a few times by ethanol and dried at room temperature. In the second case, the reaction mixture consisted of 54 µL of hexamethylenediamine and 8 ml of toluene, and all reagents were deoxygenated and heated at room temperature for 3 hr. In the final step, nanodiamond particles were washed by petroleum ether and dried at room temperature.

Prepared particles were subjected to different characterization methods such as: TEM (Transmission Electron Microscopy), SEM (Scanning Electron Microscopy) XRD (X-ray diffraction), Mössbauer spectroscopy, and IR (Infrared spectroscopy).

2.3   DISCUSSION AND RESULTS

2.3.1   TRANSMISSION ELECTRON MICROSCOPY (TEM)

Transmission electron microscopy illustrated in Figure 1, shows magnetite nanoparticles obtained in three different syntheses.

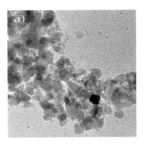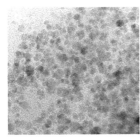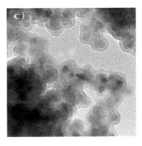

FIGURE 1   The TEM images of magnetite nanoparticles from synthesis: a) Massart's method b) Fe(acac)$_3$ complex, and c) SiO$_2$ shell.

From TEM images, we can observe that the assembly of magnetic nanoparticles obtained from different syntheses is not the same. Fe$_3$O$_4$ nanoparticles prepared from Massart's method (Figure 1(a)) are bigger and form agglomerates, which is not visible in the second case, where magnetic nanoparticles are fabricated from Fe[acac]$_3$ complex (Figure.1(b)). This method gives better (nanometer) size distribution of nanoparticles which governs particles self-assembly process. In the case of nanoparticles formed with silica, shell differences are also clearly visible (Figure.1(c)). Fe$_3$O$_4$ nanoparticles appear as small, darker spots which are surrounded by brighter silica shell. This proves that these nanoparticles formed a core-shell structure.

In all three cases, particles interact with each other differently due to different surface chemistry. The ratio of particle diameter to particle distance governs magnetic interaction between cores but also, electrostatic or Van der Waals forces depend on the nature of nanoparticle assay.

2.3.2   X-RAY DIFFRACTION (XRD)

The crystalline structure of magnetic nanoparticles obtained from synthesis was determined by the use of XRD and the results are shown in Figure 2.

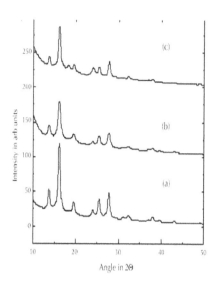

FIGURE 2   XRD pattern of magnetic NPs from synthesis: a) Massart's method b) Fe[acac]$_3$ complex, and c) SiO$_2$ shell.

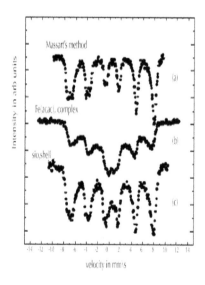

FIGURE 3   Mössbauer spectrum of magnetite NPs from synthesis: a) Massart's method b) Fe[acac]$_3$ complex, and c) SiO$_2$ shell.

From Figure 2, it can be seen that the crystal structure of all three series is exactly the same. The position and ratio of peak intensity is contained in all three batches. Slight modification of one of the peaks is observable in case of particles with SiO$_2$ shell, and the presence of one additional peak is due to the shell structure. The peaks' width of the sample (b) (obtained from Fe[acac]$_3$ complex) is largest as compared to another one, and suggests smaller particles (mean diameter), which are in good agreement with TEM observation.

2.3.3 MÖSSBAUER SPECTROSCOPY

Magnetic properties of the particles core were measured by Mössbauer spectroscopy and depicted in Figure 3.

In Figure 3, a set of Mössbauer spectra collected at room temperature for the particles prepared from different syntheses can be seen. At these figures, a difference in magnetic properties can be marked. The most important observation is that all particles consist mostly of magnetite because spectra are built up by the superposition of minimum two sextets typical for bulk magnetite. This is most clearly indicated in the case of spectrum (a). The possible presence of maghemite cannot be excluded but it can be located only on the particle's surface. A reader can find more discussion on this subject (Satuła et. al., in press).

2.3.4 NANOCOMPOSITE FABRICATION

Successful connection of both ingredients (magnetite NPs and CNT) can be followed by digital camera pictures presented in Figure 4, where in the first bottle, magnetic particles are deposited, in the second one, only carbon nanotubes and in the third one, the composite is placed. Only powder in the first and third glass reacts to external magnetic field, which proves the connection between tested particles.

FIGURE 4   Digital camera images of: a) magnetite nanoparticles (Fe$_3$O$_4$), b) carbon nanotubes (CNT), and c) composite magnetite with carbon nanotubes (Fe$_3$O$_4$+ CNT) in the presence of external magnetic field.

2.3.5   COMPOSITES NANOTUBES AND NANOPARTICLES IN EXTERNAL MAGNETIC FIELD

The fabrication of hybrid materials with magnetic particles has advantage over the other one by the fact that this one can be manipulated by external magnetic field. For such a reason, we have tested the behavior of nanotubes and nanoparticles deposited on the gold surface with and without external magnetic field. Oxidized nanotubes were placed in distilled water and then mixed with modified magnetite nanoparticles. The tested samples were applied onto two gold plates: the first one was placed in external magnetic field and the second one without it and both were dried at room temperature. A schematic drawing of the described scenarios is shown in Figure 5.

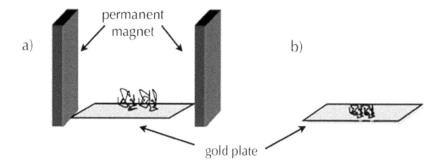

FIGURE 5   Schematic illustration of layer formation in a) and without b) external magnetic field.

The resulting films were characterized by scanning electron microscopy and obtained images are presented in Figure 6.

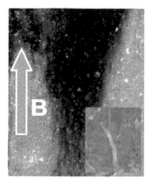

FIGURE 6   SEM images: a) dried film in external magnetic field, b) dried film without external magnetic field. The arrow shows a direction of external magnetic field. In insets are zooms of composite morphology.

Comparing SEM images of both cases, we can observe that they differ with one another. In Figure 6(a) nanotubes have a tendency to be elongated in the direction of external magnetic field, which is particularly evident on the scan at lower magnification as the streaks. At higher magnification, the preferred direction along the external magnetic field is also present. Some structures are not perfectly aligned, which can be caused by several reasons: too few particles attached to their surface; the solution dried too fast in comparison with their movement; visible structures are agglomerates and we do not see in this magnification separated nanostructures which are aligned, which suggests small magnification. In the second case (Figure 6(b)), the situation looks quite different tested structures formed chaotically distributed "islands" and rather uniform film. This proves that we can control the position of the nanotubes combined with magnetic nanoparticles (like magnetite) when both are placed together with external magnetic field. The hybrid material obtained in such a way possesses either magnetic or conducting properties at the same time, and two different fields can be used for manipulation.

I) NANOCOMPOSITES PREPARED FROM CARBON NANOTUBES AND MAGNETITE NANOPARTICLES WITH BIOTIN

In the first step, the carbon nanotubes and magnetic nanoparticles (obtained from different syntheses as described in chapter 2.3–2.5) were attached to each other. In the further step, biomolecule, such as biotin was immobilized to previously obtained structures. Details and schematic reaction are shown in Figure 7.

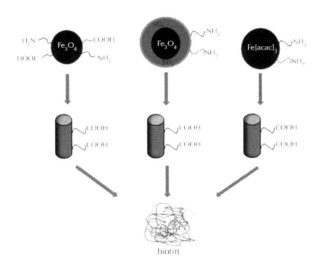

FIGURE 7   Schematic illustration of step-by-step preparation of nanocomposite with different kind of particles.

In the first attempt, oxidized nanotubes were mixed with modified magnetite nanoparticles and dispersed in buffer PBS. The sample with the reaction mixture was shaken in a sonication bath for 2 hr. After that time, modified nanoparticles were divided into two parts for the immobilization biotin (vitamin): 25 µl of these biomolecules was dispersed in 0.5 ml of working buffer PBS at pH=5.5 or pH=7.4 and 100 µl of EDC (N-(3-Dimethylaminopropyl)-N'-ethylcarbodiimide hydrochloride) solution was also added to both reaction samples. The obtained suspension was gently shaken and sonicated from time to time and left for 24 hr. Finally, (magnetite, nanotube, and vitamin) bionanocomposites were washed three times with PBS and dried at room temperature. The resultant nanocomposites at each step of the composite fabrication were analyzed by IR spectroscopy.

II) NANOCOMPOSITES PREPARED FROM FULLERENE C_{60}, NANODIAMONDS WITH MAGNETITE NANOPARTICLES AND BIOTIN

In our studies, we also tested other carbon nanomaterials such as fullerene C_{60} and nanodiamonds to check whether the formation of composite from other carbon ingredients is possible and takes place in the same way as for CNT. A schematic procedure is shown in Figure 8.

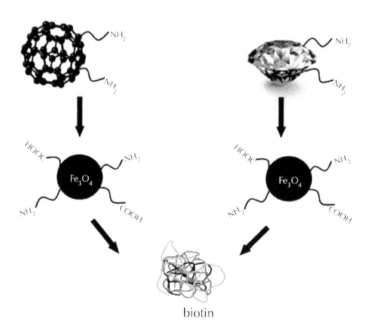

biotin

FIGURE 8 Schematic illustration of step-by-step preparation of magnetite-fullerene/nanodiamond and biotin nanocomposites.

A) CNT (CARBON NANOTUBES)

In the first stage of this study, magnetite particles were synthesized by standard methods, and in the next step, their surface was modified to appropriate functional groups
(Figure 9). Similarly, the surfaces of commercially available CNT were modified to
get chemically active places (Figure 10).

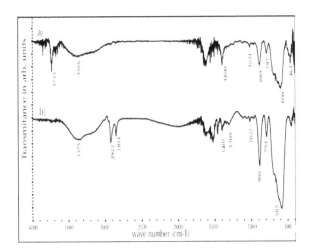

FIGURE 9   The IR spectra of: a) Fe_3O_4 from synthesis before surface modification and b) after
modification.

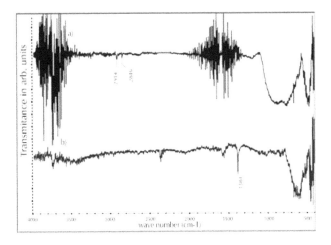

FIGURE 10   IR spectra of nanotube: a) before and b) after modification.

A comparison of IR spectra depicted in Figure 9 and Figure 10 can give an idea about the type of modification at the surface. In Figure 9, two spectra are shown, as made magnetite particles taken from Massart's synthesis (a) and after surface modification (b). The spectrum (Figure (9a)) shows characteristic absorption bands at 600 cm^{-1}, 889 cm^{-1}, 1400 cm^{-1} which come from vibration of Fe-O bonds in iron-oxide (Barilaro et. al., 2005). There are also less important signals (around 3400cm^{-1} and 1300cm^{-1}) which originate in vibration –OH group from water and some residual TBAOH (tetrabutylammonium hydroxide). The main and most pronounced differences in Figure 9(b) appear at around 2900 cm^{-1} – where strong peaks appear, which can be referred to –CH$_2$ group (Dallas et. al., 2007) in alkyl chains coming from oleylamine and oleic acid used in magnetite modification. Here we also observed signals around 1430 cm^{-1} and 1309 cm^{-1}, which could belong to amine or carboxylate anions (Klokkenburg et. al., 2007).

The collected reference to the IR spectrum of carbon nanotubes is shown in Figure 10a. After the treatment and modification of carbon nanotubes with acid mixture, the modified IR spectrum was obtained (Figure 10(b)). Here, a new peak was observed at 1384 cm^{-1} which comes from the vibrations of the carboxyl group. Magnetic nanoparticles (with amine and carboxyl groups) prepared in such a way and functionalized carbon nanotubes were used for binding biotin vitamin. For the immobilization of biomolecule, three kinds of magnetic nanoparticles were tested: from iron chlorides condensation, from Fe[acac]$_3$ complex, and with SiO$_2$ shell. The results confirmed by IR are shown in Figure 11–13, respectively.

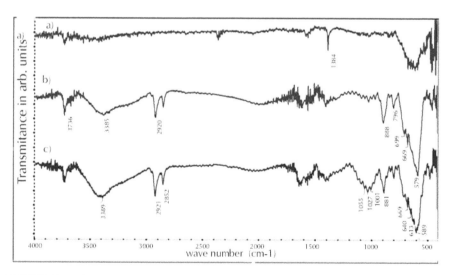

FIGURE 11   The IR spectra of: a) modified carbon nanotubes, b) modified Fe$_3$O$_4$ with carbon nanotubes, and c) biotin immobilization to magnetite-nanotube composite.

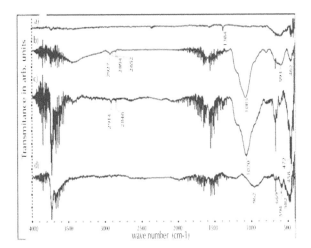

FIGURE 12 The IR spectra of: a) modified carbon nanotubes b) Fe_3O_4-SiO_2, c) Fe_3O_4-SiO_2-nanotubes, and d) biotin immobilization.

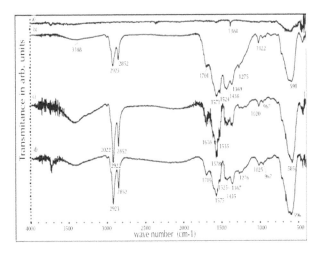

FIGURE 13 The IR spectra of: a) modified nanotubes, b) magnetite from Fe[acac]$_3$ c) Fe[acac]$_3$ and nanotubes, and d) biotin immobilization.

Comparing the spectra in Figure 11–13, we can observe that after each step of the composite creation, IR spectra differing between each other were obtained. The analysis of all three cases (spectra c) shows that characteristic of IR spectra obtained for different particles is not exactly the same. This is due to the fact that the surface chemistry of the presented cases is different. The only alike signals are observed around 3500 cm^{-1}, where characteristic vibration band for -OH in water is present.

There are also detected signals around 2900–2800 cm⁻¹ originated from -CH$_2$ groups. In either sample, Fe-O band can be found in the range of 450–600 cm⁻¹. Other obtained peaks are covered by characteristic signals from different particles. The attachment of biomolecule (spectra d) provided further changes in the IR spectra and proves that the biotin was successfully attached. That fact confirms new signals appearing in the range of 900 cm⁻¹–1400 cm⁻¹, respectively.

B) FULLERENE C$_{60}$

In the same way as CNT, carbon nanodiamonds and fullerenes C$_{60}$ have been tested on the effectiveness of nanocomposites preparation with biological molecule-biotin. The effects of the process for C$_{60}$ are depicted in IR spectra presented in Figure 14.

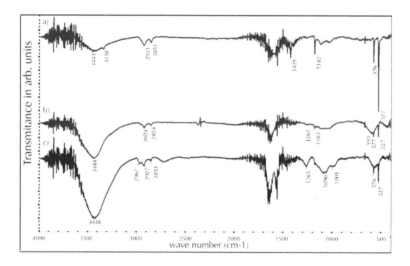

FIGURE 14 The IR spectra of: a) modified C$_{60}$, b) modified C$_{60}$+Fe$_3$O$_4$, and c) modified C$_{60}$-Fe$_3$O$_4$+biotin.

A typical fullerene spectrum after chemical modification (C$_{60}$ was treated with amine compound) is shown in Figure 14a Here we observe the peaks at 3339 cm⁻¹, which can be referred to -NH$_2$ groups vibrations and at 1560 cm⁻¹ characteristic for C=N vibration bond (Coates, 2000). In the next step, magnetite and carbon particles - C$_{60}$ were merged together. The result of this process is shown in Figure 14b, where IR spectrum consists of absorption bands at wave numbers similar to the individual particles spectra, such as signals at 2900 cm⁻¹ (typical for methylene group) and around 588 cm⁻¹ from Fe-O is depicted. At the same time, typical sharp signals for fullerene C$_{60}$ at 1183cm⁻¹ and 527 cm⁻¹ are present The spectra change after the attachment of biotin vitamin (spectrum Figure 14(c)) and a new signal around 1090 cm⁻¹ is observed. This fact confirms our assumptions that biological molecule like biotin can react with different hierarchical ordered structures such as magnetite and carbon nanoparticles in all tested (up to now) cases. The signals from biological molecule (observed at IR spectra) were strong and well visible when it was immobilized at the last step. The fabrication of composite in other order is also successful but much weaker signal from biomolecule is observed.

C) NDS-NANODIAMONDS

In the case of nanodiamonds we have modified their surface by three different methods:
   a) By nitrating mixture
   b) Octylamine, and
   c) Hexamethylenediamine to test which of them is most efficient.

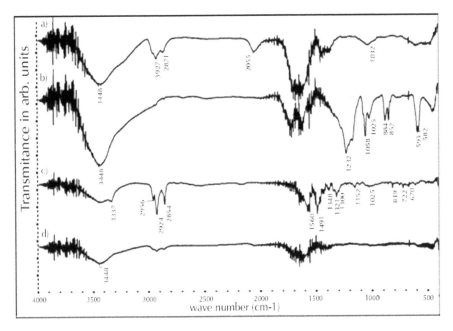

FIGURE 15   IR spectra of: a) NDs basic, b) NDs basic modified by nitrating mixture, c) NDs from (b) treatment by octylamine, and d) NDs with hexamethylenediamine modification.

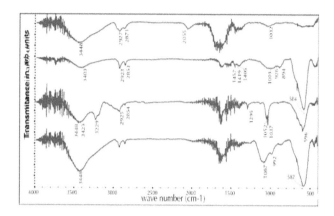

FIGURE 16   IR spectra of: a) NDs basic modified by nitrating mixture, b) modified Fe_3O_4, and c) NDs with Fe_3O_4, d) composite prepared in (c) with biotin immobilization.

After modification of NPs by nitrating mixture, the obtained spectrum has changed significantly (Figure 15 (b)) and few new peaks below 1500 cm^{-1} appear. One wave number 1232 cm^{-1} may be identified as a C=O bond which comes from COOH groups (19). The strong peaks at a wave number 1058–1025 cm^{-1}, 854–852 cm^{-1}, and 593–582 cm^{-1} may be assigned to bisulfate ion adsorbed on the diamond surface (Biermann et. al., 2000). Here, we also observed a typical band at 3448 cm^{-1}, which is identified as -OH group from H_2O. In the next step, modified NDs were functionalized by octyl-amine, which confirms IR spectrum (Figure 15(c)). In that case, we observed number of additional peaks which can be identified as amine vibrations bonds at 3337 cm^{-1}, and a new peak at 1560 cm^{-1} which comes from C=N vibration. Another not strong band at a wave number of 1025 cm^{-1}, which may be identified as a C-N (19). The peaks at around 2900 cm^{-1} can be assigned to the vibration of -CH_2 groups. Nanodiamonds were also tested to be modified with hexamethylenediamine. The resulting IR spectrum is shown at Figure 15d, where we also observe bands from methyl and hydroxyl groups. The new broad peak at 1036 cm^{-1} could be assigned to C-N vibration bond which is shifted towards higher wave numbers.

In Figure 16 further modifications of NPs are presented. In Figure 16(c), the spectrum of composite prepared with modified NDs by nitrating mixture and Fe_3O_4 is depicted. Here, we observe characteristic peaks for magnetite around 600 cm^{-1} and for nanodiamonds modified by nitrating mixture – peaks around 1050–1030 cm^{-1}. In the next step, biotin molecule was immobilized to magnetite-nanodiamod composite and changes in the obtained structure confirm Figure 16(d). The signal from the attached biological particles is observed as a strong and wide band around 1060–1000 cm^{-1}, which is in good agreement with previous results (other carbon particles).

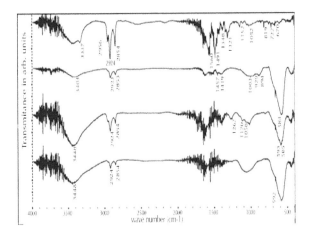

Figure 17 The IR spectra of: a) NDs with octylamine, b) modified Fe_3O_4, c) modified NDs with Fe_3O_4, and d) NDs-Fe_3O_4 with biotin immobilization.

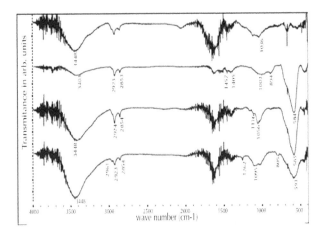

FIGURE 18 The IR spectra of: a) NDs with hexamethylenediamine, b) modified Fe_3O_4, c) modified NDs with Fe_3O_4, and d) NDs-Fe_3O_4 with biotin immobilization.

Nanodiamonds modified by octylamine and hexamethylenediamine were used in the next step to fabricate nanocomposite also with magnetite, which is shown in Figure.17 and Figure 18. In the case of NDs-octylamine coupled with Fe_3O_4 nanoparticles (Figure.17(c)), the characteristic bands from Fe-O at 593 cm^{-1} are observed. The same is obtained for NDs-hexamethylenediamine where these bonds appear at 585 cm^{-1}. In both IR respective spectra c) we observe signals at wave numbers 1058cm^{-1}–1023 cm^{-1}, which may be identified as C-N bonds (19). In the final step of the composite fabrication, vitamin – biotin were immobilized onto magnetite-nanodiamonds composite. The obtained IR spectra differ from the previous one because broad bands from immobilized biotin appear in the range of 1100–1000 cm^{-1}.

Modification of constituent particles and composite fabrication can be easily followed by IR characterization. There can be seen chemical changes of the surface of the obtained nanomaterials and modification of the particle core.

2.4   CONCLUSION

This study has shown preparation and modification of magnetic nanoparticles Fe_3O_4 and carbon structures such as fullerene $C_{60,}$ nanotubes, and nanodiamonds. In the further steps, these nanoparticles were used to fabricate nanocomposites which were hierarchically ordered to contain magnetic particle, carbon structures, and immobilized biological molecules-biotin. In all cases, the particular elements were step-by-step attached with each other until desired composites were not obtained. Those facts confirm IR spectra collected after each step of individual elements attached during the composite fabrication. Morphology of the obtained magnetic nanoparticles -Fe_3O_4 from three different syntheses were verified by TEM and Mössbauer spectroscopy, and the behavior of magnetic-carbon structures (preliminary nanocomposite) formed in/without magnetic field was imaged by SEM.

The study shows that three different carbon structures possessing different physo-chemical properties in almost the same manner can be attached to magnetic particle and further to bioparticle. Different ordering in the composite creation was also tested, which shows that biomolecule is not very tightly bonded with the composite. This observation can be either the advantage or disadvantage of the fabricated composite depending on the application. On the other hand, the performed measurements prove that modification and magnetic particle treatment in this test do not change magnetic core too much. There is no further oxidation process taking place and manipulation by external magnetic field can be performed any time during the modification process.

KEYWORDS

- **Biotin immobilization**
- **Carbon nanoparticles**
- **Functionalization**
- **Magnetite**
- **Nanocomposite**

ACKNOWLEDGMENT

We are grateful to Dr A. T. Dubis for equipment support in IR tests. The Mössbauer experiment was done in close collaboration with Department of Physics University of Białystok. The work was partly supported by the EU under contract number POPW.01.03.00-20-034/09-00 and national funding under contract number N N204246435.

REFERENCES

1. Allen, N. S., Zeynalov E. B., Taylor K., and Birkett, P. Antioxidant capacity of novel amine derivatives of buckminsterfullerene: Determination of inhibition rate constants in a model oxidation system. *Polymer Degradation and Stability*, **94**, 1932–1940 (2009).
2. Bakry, R., Vallant R. M., Najam-ul-Haq, M., Rainer, M., Szabo, Z., Huck, C. W., and Bonnet, G. K. Medical applications of fullerenes. *International Journal of Nanomedicine*, **2**(4) 639–649 (2007)
3. Barilaro, D., Barone, G., Crupi, V., Donato, M. G., Majolino, D, Messina, G., and Poterino, R. Spectroscopic techniques applied to the characterization of decorated potteries from Caltagirone. *Journal of Molecular Structure*, 744–747, 827–831 (2005)
4. Biermann, U. M., Luo, B. P., and Peter, T. Absorption Spectra and Optical Contrasts of Binary and Ternary Solutions of H_2SO_4, HNO_3 in the Mid Infrared at Atmospheric Temperatures. *Journal of Physical Chemistry A*, **104**, 783–793 (2000).
5. Coates, J. Interpretation of Infrared Spectra, a Practical Approach. *Encyclopedia of Analytical Chemistry* (2000).
6. Dallas, P., Bourlinos A. P., Niarchos D., and Petridis D. Synthesis of tunable sized capped magnetic iron oxide nanoparticles highly soluble in organic solvents. *Journal of Material Science*, **42**, 4996–5002 (2007)
7. Van Ewijk, G. A., Vroege, G. J., and Philipse, A. P. Convenient preparation methods for magnetic colloids. *Journal of Magnetism and Magnetic Materials*, **201**, 31–33 (1999).
8. Al din Haratifar, E., Shahverdi, H. R., and Shakibaie, M. Semi-Biosynthesis of Magnetite-Gold Composite Nanoparticles Using an Ethanol Extract of Eucalyptus camaldulensis and Study of the Surface Chemistry. *Journal of Nanomaterials* **2009**, 1–5 (2009).
9. Jiang, K., Eitan, A., Schadler, L. S., Ajayan, P. M., Siegiel, R. W., Grobert, N., Mayne, M., Reyes-Reyes, M., Terrones, H., and Terrones, M. Selective Attachment of gold nanoparticles to nitro gen-doped carbon nanotubes. *Nano Letters*, **3**(3), 275–277 (2003)
10. Kalska-Szostko, B, Cydzik, M., Satuła, D., and Giersig, M. Mössbauer Studies of Core-Shell Nanoparticles. *Acta Physica Polonica A*, **119**, 15–17 (2011).
11. Kalska-Szostko, B, Zubowska, M., and Satuła, D. Studies of Magnetite Nanoparticles by means of Mössbauer Spectroscopy. *Acta Physica Polonica A*, **109**, 365–369 (2006).
12. Kalska-Szostko, B and Rogowska, M. Preparation of magnetite - fullerene nanocomposite with enzyme immobilization. *Journal of Nanoscience and Nanotechnology*, **12**, 6907 (2012).
13. Klokkenburg, B., Hilhorst, J., and Erne, B. H. Surface analysis of magnetite nanoparticles in cyclohexane solutions of oleic acid and oleylamine. *Vibrational Spectroscopy*, **43**, 243–248 (2007).
14. Klumpp, C., Kostarelos, K., Prato, M., and Bianco A. Functionalized carbon nanotubes as emerging nanovectors for the delivery of therapeutics. *Biochimica et Biophysica Acta* **1758**, 404–412 (2006).
15. Laurent, S., Forge, D., Port, M., Roch, A., Robic, C., Vander, Elst L., and Muller, R. N. Magnetic iron oxide nanoparticles: synthesis, stabilization, vectorization, physicochemical characterizations and biological applications. *Chemical Reviews*, **108**, 2064–2110 (2008).
16. Markovic, Z., Trajkovic, V. Biomedical potential of the reactive oxygen species generation and quenching by fullerenes (C_{60}). *Biomaterials* **29**, 3561–3573 (2008)
17. Massart R., Cabuil V. *Journal de Chimie Physique* 84, 967 (1987).

18. Sun, W. H., Suslick, K. S., Stucky, G. D., Suh, Y. H. Nanotechnology, nanotoxicology and neuroscience. *Progress in Neurobiology*, **87**, 133–170 (2009).

19. Sun, S., Zeng, H. Size-controlled synthesis of magnetic nanoparticles. *Journal of the American Chemical Society* **124**, 8204–8205 (2002).

20. Timko, M., Koneracka, M., and Kopcansky, P. Application of magnetizable complex system in biomedicine. *Czechoslovak Journal of Physics* **54**, 599–606 (2004).

21. United States Patent Application Publication (Nov 25, 2010).

22. Woo, K., Hong, J., and Ahn, J. P. Synthesis and surface modification of hydrophobic magnetite to processible magnetite@silica propylamine. *Journal of Magnetism and Magnetic Materials*, **293**, 177–181 (2005).

CHAPTER 3

AuNPs AND BIOSENSORS: A HAND TO HAND JOURNEY

SUMAN SINGH, V. K. MEENA, D. V. S. JAIN, and M. L. SINGLA

CONTENTS

ABSTRACT

The chapters of this book deal mainly with application of Gold Nanoparticles (AuNPs) in biosensors. Synthetic approaches, based on physical and on chemical procedures, as well as structural and optical features, are also described. Owing to the large surface-to-volume ratio, high surface reaction activity, high catalytic efficiency, strong adsorption ability, and their biocompatibility, numerous biosensors have been constructed using AuNPs with higher selectivity, better stability, and a lower detection limits. Not only optical properties, electrochemical properties of AuNPs have also been harnessed. Still, development of new protocols for preparing functionalized AuNPs and using them for biosensing is an active research area.

3.1   INTRODUCTION

The synthesis and application of nanoparticles (NPs) have fueled the growth of nanotechnology, foundation of which is based on their size and shape. Owing to their nanometer size, high surface-to-volume ratio, and ability to couple with surface plasmons of neighboring metal particles or with electromagnetic wave, they have a variety of interesting spectroscopic, electronic, and chemical properties different than their bulk counterparts (Templeton, Wuelfing et al. 1999; Whetten, Shafigullin et al. 1999; Daniel and Astruc 2004). Particular interest has been focused on the noble metal nanoparticles as technologically they play vital role in important fields such as catalysis (Narayanan and El-Sayed 2004), optoelectronic devices (Kamat 2002), and surface-enhanced Raman scattering (SERS) (Campion and Kambhampati 1998; Kwon, Lee et al. 2006). As their size scale is similar to that of biological molecules (e.g., proteins, DNA) and structures (e.g., viruses and bacteria), they are enormously exploited for various biomedical applications also like biosensing (Taton, Mirkin et al. 2000; Karhanek, Kemp et al. 2005), imaging (Jiang, Papa et al. 2004), gene, and drug delivery (Kohler, Sun et al. 2005; Yang, Sun et al. 2005). Among various noble metal NPs, colloidal gold (CG) NPs have gained much more attention and are being extensively studied because of different synthetic approaches, variable size, good biocompatibility, relatively large surface, and their variable optical behavior and catalytic properties (Lahav, Shipway et al. 2000; Thanh and Rosenzweig 2002). Their optical and electrical properties are known to be dramatically affected by their size, shape, and surrounding surface environments (Link and El-Sayed 2000; Norman, Grant et al. 2002; Zhong, Patskovskyy et al. 2004), which in turn depend on the methods used for the synthesis, molar ratio of reductant to gold ion, type of stabilizer (Boopathi, Senthilkumar et al. 2012), reaction time, and temperature (Schmid and Corain 2003), pH (Shou, Guo et al. 2011) and refractive index (Templeton, Pietron et al. 2000; Sau and Murphy 2004; Richardson, Johnston et al. 2006).

3.2   HISTORY OF GOLD NANOPARTICLES

Gold is the first metal discovered by humans, and is the subject of one of the most ancient themes of investigation in science. The extraction of gold started in the 5th millennium B.C. near Varna (Bulgaria) and reached 10 tons per year in Egypt around

1200–1300 B.C. First data on CG can be found in treatises by Chinese, Arabic, and Indian scientists, who prepared CG and used it, in particular, for medical purposes as early as 5 ± 4th centuries B. C. Colloidal gold was used to make ruby glass and for coloring ceramics, and these applications are still continuing now. The most famous example is the Lycurgus Cup (Figure 1) that was manufactured in the 5th to 4th century B.C, using silver and gold NPs in approximate ratio of 7:3. The presence of these metal NPs gives special color display for the glass. When viewed in reflected light, for example in daylight, it appears green. However, when a light is shone into the cup and transmitted through the glass, it appears red. This glass can still be seen in British museum.

FIGURE 1 Lycurgus Cup (a) green color, if light source comes from outside of the cup. (b) red color, if the light source comes from inside of the cup (Freestone, Meeks et al. 2007).

The beginning of scientific research on CG dates back to the mid-19th century, when Michael Faraday published an article on synthesis and properties of CG (Faraday 1857). In this chapter, Faraday described, for the first time, aggregation of CG in the presence of electrolytes, the protective effect of gelatin and other high-molecular-mass compounds, and the properties of thin films of CG. Colloidal gold solutions prepared by Faraday are still stored in the Royal Institution of Great Britain in London. Richard Zsigmondy (Zsigmondy 1898), was the first to describe the methods of synthesis of CG with different particle sizes using different reducing agents. Zsigmondy used colloidal gold as the main experimental object when inventing (in collaboration with Siedentopf) an ultra-microscope. In 1925, Zsigmondy was awarded the Nobel Prize in 'Chemistry for his demonstration of the heterogeneous nature of colloid solutions and for the methods he used, which have since became fundamental in modern colloid chemistry' (R Zsigmondy 1925). Studies by the Nobel Prize laureate Theodor

Svedberg on the preparation, analysis of mechanisms of colloidal gold formation, and their sedimentation properties (with the use of the ultra-centrifuge he had invented) are among important studies (Svedberg 1924). Now days, CG is used by scientists as a perfect model for studies of optical properties of metal particles, the mechanisms of aggregation, and stabilization of colloids. As a result, it has led to the exponential increase in number of research efforts for applications in almost every sphere such as biological detection, controlled drug delivery, low-threshold laser, optical filters, and also sensors among others (Bohren 1983; Andrievskii 2003; Kostoff, Koytcheff et al. 2008). Various researchers have reviewed the synthesis, properties, and role of NPs in various fields (Martin 1996; Bönnemann and Richards 2001; Raveendran, Fu et al. 2003; Daniel and Astruc 2004; Saha, Agasti et al. 2012).

3.3 SYNTHESIS METHODS FOR GOLD NANOPARTICLES

The methods used for the synthesis of metal NPss in colloidal solution are very important as they control the size and shape of NPs, which in turn affects their properties. Moreover, successful utilization of NPs in biological assays relies on the availability of nanomaterials in desired size, their morphology, water solubility, and surface functionality. Several reviews on the synthesis of nanoparticles are available (Masala and Seshadri 2004; Chai, Wang et al. 2010). Some reviews dedicatedly covered the synthesis of AuNPs (Shipway, Lahav et al. 2000; Thomas and Kamat 2003).

Synthesis methods of CG (and other metal colloids) can arbitrarily be divided into following two major categories:

- Dispersion methods (metal dispersion) It has been checked and is correct.
- Condensation methods (reduction of the corresponding metal salts).

Dispersion methods are based on destruction of the crystal lattice of metallic gold in high-voltage electric field (Cigang, Harm van et al. 2006; Choi, Liew et al. 2008). The yield and shape of AuNPs formed under electric current depend not only on the voltage between electrodes and the current strength, but also on the presence of electrolytes in the solution. The addition of even very small amount of alkalis or chlorides and the use of high-frequency alternating current for dispersion can substantially improve the quality of gold hydrosols.

Condensation methods are more commonly employed than dispersion methods. The CG is most often prepared by reduction of gold halides (for example, of $HAuCl_4$) with the use of chemical reducing agents and/or irradiation. The systematic adjustment of the reaction parameters, such as reaction time, temperature, concentration, and the selection of reagents and surfactants can be used to control size, shape, and quality of NPs.

The condensation methods can further be classified into:
- Citrate Reduction Method
- The Brust-Schiffrin Method: Two-Phase Synthesis and Stabilization by Thiols
- Micro–emulsion, Reversed Micelles, Surfactants, Membranes, and Poly-electrolytes
- Seeding-growth Method
- Physical Methods (ultrasonic, UV, IR or ionizing radiation or laser photolysis)

3.3.1 CITRATE REDUCTION METHOD

The simplest and most commonly used method for the preparation of AuNPs is the aqueous reduction of $HAuCl_4$ by sodium citrate at boiling point (Turkevich, Stevenson et al. 1953; Xiulan, Xiaolian et al. 2005). Particles synthesized by this method are nearly mono-dispersed having controlled size which in turn depends on the initial reagent concentrations (Hostetler, Wingate et al. 1998; Pillai and Kamat 2003) and can be easily characterized by their plasmon absorbance band at about 520 nm for 10–15 nm sized particles. The NPs from other noble metals may also be prepared by citrate reduction, such as silver particles from $AgNO_3$, palladium from $H_2[PdCl_4]$, and platinum from $H_2[PtCl_6]$ (Cassagneau and Fendler 1999; Richter, Seidel et al. 2000; Lin, Khan et al. 2005). The similarities in the preparation of these different metal colloids allow the synthesis of mixed-metal particles, which may have functionality different from each individual metal. Although sodium citrate is the most common reducing agent, metal NPs can also be synthesized by the use of both strong and weak reducing agents including borohydride, alcohols, hydrogen gas, gallic acid, glutamic acid, polyvinyl pyrollidone, and so on. Reduction can also be achieved via photo-chemical reduction, thermal reduction, sonochemical reduction and electrochemical reduction, and other reducing agents (Mayer and Mark 1998; Wilson, Hu et al. 2002). Table 1 shows different kinds of reducing and stabilizing agents used for the synthesis of AuNPs.

TABLE 1   References to common reducing agents and stabilizers used in the synthesis of NPs in colloidal solution

Important Factors for NP synthesis	Example
Reducing Agents/Reduction Methods	Alcohols (Li, Boone et al. 2002; Narayanan and El-Sayed 2003; Adlim, Abu Bakar et al. 2004)
	Hydrogen Gas (Fu, Wang et al. 2002; Narayanan and El-Sayed 2004; Bagotsky 2006)
	Sodium Borohydride (Pittelkow, Moth-Poulsen et al. 2003; Tabuani, Monticelli et al. 2003; Bagotsky 2006)
	Sodium Citrate (Link and El-Sayed 1999; Zhu, Vasilev et al. 2003)
	Thermal Reduction (Fleming and Williams 2004)
	Photochemical Reduction (Pal, Esumi et al. 2005; Park, Atobe et al. 2006)
	Sono-chemical Reduction (Zhang, Du et al. 2006)
	Metal Vapor Condensation (Klabunde 1996; Prasad, Stoeva et al. 2003; Swihart 2003)
	Electrochemical Reduction (Klabunde 1996)
Stabilizers	Surfactants (Zhang and Zeng 2008) (Shon, Gross et al. 2000)
	Polymers (Corbierre, Cameron et al. 2004; Ding, Hu et al. 2010; Gibson, Danial et al. 2011)
	Dendrimers (Nijhuis, Oncel et al. 2006; Boisselier, Diallo et al. 2010)
	Block copolymers (Wang, Kawanami et al. 2008; Rahme, Vicendo et al. 2009; Zhai, Hong et al. 2010)
	Other Ligands (Warner, Reed et al. 2000; Meli and Green 2008; Hermes, Sander et al. 2011)

3.3.2   THE BRUST-SCHIFFRIN METHOD: TWO-PHASE SYNTHESIS AND STABILIZATION BY THIOLS

Another procedure that has become extremely popular for smaller AuNPs synthesis in organic solvent is two phase reduction method developed by Brust *et. al.*, (Brust, Walker et al. 1994). This synthesis technique is inspired by Faraday's two-phase system and uses thiol ligands that strongly bind gold due to the soft character of both Au and S (Brust, Fink et al. 1995). The stabilization of AuNPs with alkanethiols was first reported in 1993 by Mulvaney and Giersig, who showed the possibility of using thiols of different chain lengths for stabilization (Giersig and Mulvaney 1993). The method has had a considerable impact on the overall field in less than a decade. The NPs prepared by this method can be repeatedly isolated and re-dissolved in common organic solvents without irreversible aggregation or decomposition. Subsequently, the Brust route was explored for the synthesis of wide range of monolayer protected clusters (Carotenuto and Nicolais 2003; Shimizu, Teranishi et al. 2003; Sharma, Mahima et al. 2004; Bhat and Maitra 2008).

3.3.3 MICRO-EMULSION, REVERSED MICELLES, SURFACTANTS, MEMBRANES, AND POLYELECTROLYTES

The AuNPs can also be prepared by the two-phase micro-emulsion method, first the metal-containing reagent is transferred from an aqueous to an organic phase. After the addition of a surfactant solution to this system, a micro-emulsion, that is, a dispersion of two immiscible liquids, is formed. In micro-emulsion methods, alkane thiols are often added to the reaction solution, and these additives form dense self-assembled monolayers on the gold surface. This method was employed for the preparation of self-assembled two and three-dimensional ensembles of AuNPs (Giersig and Mulvaney 1993; Brust, Walker et al. 1994; Brust and Kiely 2002).

3.3.4   SEEDING-GROWTH METHOD

The seeding-growth procedure is another popular technique that has been used for a century. Recent studies have successfully led to the controlled size distribution (typically 10–15%) in the range of 5–40 nm, where the sizes can be manipulated by varying the ratio of seed to metal salt (Jana, Gearheart et al. 2001; Meltzer, Resch et al. 2001; Sau, Pal et al. 2001). The step-by-step particle enlargement is more effective than a one-step seeding method to avoid secondary nucleation (Carrot, Valmalette et al. 1998) . Gold nanorods have been conveniently fabricated using the seeding-growth method (Busbee, Obare et al. 2003).

3.3.5 PHYSICAL METHODS

Physical methods based on ultrasonic, UV, IR, or ionizing radiation or laser photolysis (V.A Fabrikanos 1963; Mallick, Witcomb et al. 2005) and electrochemical methods (Ma, Yin et al. 2004) of reduction are much less commonly employed than chemical methods. The advantage of the former methods is that impurities of chemical compounds are absent in the resulting sols (on the metal particle surface) (Ershov 2001).

3.4 PROPERTIES OF GOLD NANOPARTICLES

The AuNPs possess many physical and chemical properties owing to their easy synthesis, unique optical properties, high surface to volume ratio, biocompatibility, and tunable shape and size. All these properties make them excellent scaffolds for the fabrication of chemical and biological sensors (Boisselier, Diallo et al. 2010; Bunz and Rotello 2010; Zeng, Yong et al. 2011; Saha, Agasti et al. 2012). Some reviews are available which have discussed the properties of nanomaterials in detail (Daniel and Astruc 2004; Wang 2004). Figure 2 shows different types of properties of AuNPs, which are given as follows,

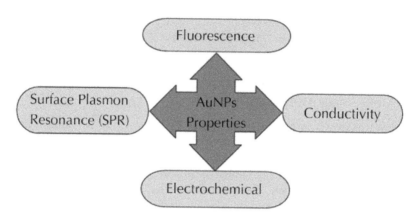

FIGURE 2   Properties of AuNPs (AuNPs).

3.4.1   OPTICAL PROPERTIES

SURFACE PLASMON BAND (SPB)

The SPB is due to the collective oscillations of the electron gas at the surface of NPs (6s electrons of the conduction band for AuNPs) that is correlated with the electromagnetic field of incoming light, that is, the excitation of the coherent oscillation of the conduction band. The nature of SPB was rationalized in a master publication authored by Mie in 1908 (Mie 1908). According to Mie theory, the total cross section composed of the SP absorption and scattering is given as a summation over all electric and magnetic oscillations. Mie theory attributes the plasmon band of spherical particles to the dipole oscillations of free electrons in conduction band occupying the energy states immediately above the Fermi energy level.

The main characteristics of the SPB are:
(i)   Its position around 520 nm
(ii) Its sharp decrease with decreasing core size for AuNPs due to the onset of quantum size effects. The decrease of SPB intensity with decrease in particle size is accompanied by broadening of the Plasmon bandwidth
(iii) Step like spectral structures indicating transitions to the discrete unoccupied levels of the conduction band.

The SPB maximum and bandwidth are also influenced by the particle shape, medium dielectric constant, and temperature. The refractive index of the solvent has been shown to induce a shift of the SPB, as predicted by Mie theory. But since all AuNPs need some kind of stabilizing ligands or polymer, the band energy is rarely exactly as predicted by Mie theory. Applications of the sensitivity of the position of the SPB are known, especially in the fields of sensors and biology. A shift of the SPB of AuNPs has been measured upon adsorption of gelatin and quantitative yield measurements of the adsorbed amount were obtained (Chandrasekharan, Kamat et al. 2000). Phase transfer of dodecylamine-capped AuNPs dispersed in an organic solvent into water containing the surfactant cetyltrimethylammonium bromide (CTAB) was monitored by color changes initiated upon shaking (Thomas, Zajicek et al. 2002). Surface interaction of AuNPs with functional organic molecules was probed by monitoring the shifts of SPB position.

FLUORESCENT PROPERTIES

AuNPs have attracted increasingly attention for the unique nano-optics properties. These fascinating optical properties, including those of SPB, surface-enhanced SERS and Raleighresonance scattering (RRS), have been well documented. In contrast, the studies on photoluminescence from AuNPs are very limited (J. P. Wilcoxon 1998; Huang and Murray 2001; Hwang, Jeong et al. 2002; Link, Beeby et al. 2002; Zheng, Petty et al. 2003; Li, Chen et al. 2011). Shen *et. al.,* (Shen, Jiang et al. 2006) used fluorescent AuNPs for the detection of 6-mercaptopurine. The fluorescent AuNPs with mean diameter of ~15 nm were synthesized in aqueous solution and significant enhancement in fluorescence emission was observed upon AuNPs self-assembly with 6MP. A new fluorescent method for sensitive detection of biological thiols in human plasma was developed using a near-infrared (NIR) fluorescent dye, FR 730 (Shang, Yin et al. 2009). The sensing approach was based on the strong affinity of thiols to gold and highly efficient fluorescent quenching ability of AuNPs. In the presence of thiols, NIR fluorescence would enhance dramatically due to desorption of FR 730 from the surfaces of AuNPs, which allowed the analysis of thiol-containing amino acids in a very simple approach. The size of Au NPs was found to affect the fluorescent assay and the best response for cysteine detection was achieved when using AuNPs of diameter 24 nm. Visible luminescence has been reported for water-soluble AuNPs, for which a hypothetical mechanism involving $5\ d^{10}\ f^6\ (sp)^1$ inter-band transition has been suggested (Huang and Murray 2001).

ELECTROCHEMICAL PROPERTIES

Just as ionic space charges, the electrical double layer exist at all electrified metal/electrolyte solution interfaces, nanoparticles in solutions (colloids, metal sols, regardless of the metal, and semiconductor nanoparticles) have double layers with ionic surface excesses on the solution side that reflect any net electronic charge residing on metal NP surface (or its capping ligand shell). In this light, one can say that all metal-like nanoparticles are intrinsically electroactive and act as electron donor/ acceptors to the quantitative extent of their double-layer capacitances. The electron charge storage capacity per nanoparticle however depends on the nanoparticle size (surface area), nanoparticle double-layer capacitance (DLC), and potential (relative to nanoparticle zero charge). This capacity can be quite substantial; for example, a 10-nm-diameter nanoparticle with CDL=120 aF (equivalent to 40 $\mu F/cm^2$) can store ~750 e/V. This capacity is capable, as a "colloidal microelectrode", of driving electrochemical reactions such as proton reduction to H_2. The quantitative demonstrations of this property by Henglein (Henglein 1979) represented the beginning of modern understanding of the electrochemistry of metal NPs. In the earliest experiments, NPs were charged by chemical reactions. The transition to electrochemical control by Ung et. al.,(Ung, Giersig et al. 1997) was made by showing that solutions of Ag NPs capped with poly (acrylic acid) could also be charged at (macroscopic) working electrodes, diffusing to undergo electron transfer at electrode/electrolyte interfaces. Murray thoroughly reviewed the electrochemical properties of metal nanoclusters, such as Ag nanoclusters and gold nanoclusters (Murray 2008). However, no report so far is available about the quantitative characterization of AuNP population by their redox current. The reason is probably that the faradic current is intrinsically small at the sub-micro ampere or nano-amper level, which yields a very poor sensitivity and limit of detection. When gold clusters are prepared and used for different applications, their populations are usually very low. When the nano-clusters are small enough, they have molecular-like electrochemical voltammograms and it has been observed that the electrochemical current is determined by the diffusion of the molecular-like clusters (Menard, Gao et al. 2006). The diffusion coefficients of these nanoclusters are around or smaller than ~10^{-6} cm^2/s, which is close to or smaller than "usual" inorganic ions in their water solutions (Bagotsky 2006). Therefore, both the Cottrell equation and Levich equation apply to relate the Faradic current and population of gold clusters. An amperometric analysis can be applied because the faradic current is intrinsically proportional to the gold cluster population. When relatively larger gold clusters (14–28 kDa) were examined, quantized double layer charging/discharging voltamograms showed distinguishable peaks of 1e⁻ transfers. In these dispersions, the gold nanoclusters behave as quantum capacitors, instead of electroactive species. When the AuNPs are larger than 5 nm, they are not considered like molecules. Each AuNP consists of a number of gold atoms. For example, assuming the AuNPs as perfect spheres and their densities are identical to bulk gold, AuNP of 5 nm diameter, 10 nmdiameter, or 20 nm diameter contains about 3.8×10^3, 3.1×10^4, and 2.5×10^5 gold atoms, respectively. Therefore, when the average diameter of the AuNPs is a known value, the number of AuNP can be estimated by analyzing the quantity of element Au.

3.5 BIOSENSORS AND NANOMATERIALS

3.5.1 BIOSENSORS

Analytical chemistry plays an important role in our everyday life because almost every sector of industry and public service relies on quality control. Majority of chemical analysis methods are time-consuming and heavily employ expensive reagents and equipment in order to achieve high selectivity and low detection limits. Biosensors emerge as upbeat technology to face this challenge.

A biosensor is generally defined as an analytical device which converts a biological response into a quantifiable and processable signal.

Figure 3 shows schematically the parts comprising a typical biosensor:

a) Bio-receptors that specifically bind to the analyte

b) An interface architecture where a specific biological event takes place and gives rise to a signal picked up by

c) The transducer element; the transducer signal (which could be anything from the in-coupling angle of a laser beam to the current produced at an electrode) is converted to an electronic signal and amplified by a detector circuit using the appropriate reference and sent for processing.

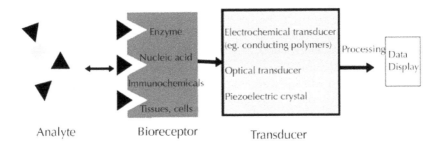

FIGURE 3. Components of a Biosensor.

Successful biosensor should meet the following conditions:

• The biocatalyst must be highly specific for the purpose of the analysis.

• Be stable under normal storage conditions and show a low variation between assays.

• The reaction should be independent of physical parameters such as stirring, pH, and temperature.

• The response should be accurate, precise, reproducible, and linear over the concentration range of interest without dilution or concentration.

• It should also be free from electrical or other transducer induced noise.

• If the biosensor is to be used for invasive monitoring in clinical situations, the probe must be tiny and biocompatible, having no toxic or antigenic effects.

• For rapid measurements of analytes from human samples, biosensor should provide real-time analysis.

• The complete biosensor should be cheap, small, portable, and capable of being used by semi-skilled operators.

Typical recognition elements used in biosensors are: enzymes, nucleic acids, antibodies, whole cells, and receptors. Of these, enzymes are among the most common. (Eggins 2002)

Taking into account the biomolecule that recognizes the target analyte, biosensors can be named as:

(i) Affinity sensors, when the bioreceptor uses non-covalent interactions like antibody-antigen reactions or DNA strand hybridization, and

(ii) Catalytic or enzyme sensors, when the analyte is the enzyme substrate, or it can be detected by measuring the signal produced by one substrate or product of the enzymatic reaction involving the analyte.

Affinity biosensors make use of the specific capabilities of an analyte to bind to a biorecognition element. This group can be further divided into immunosensors (which rely on specific interactions between an antibody and an antigen), nucleic acid biosensors (which make use of the affinity between complementary oligonucleotides), and biosensors based on interactions between an analyte (ligand) and a biological receptor. Some whole-cell biosensors act as recognition elements responding to (trigger) substances by expressing a specific gene. Catalytic biosensors make use of biocomponents capable of recognizing (bio) chemical species and transforming them into a product through a chemical reaction. This type of biosensor is represented mostly by enzymatic biosensors, which make use of specific enzymes or their combinations. Many whole cell biosensors also rely on biocatalytic reactions.

Biosensors are also classified according to the parameter that is measured by the physicochemical transducer of the biological event. Thus, classically biosensors are grouped into optical, electrochemical, acoustic, and thermal ones. Optical transducers of most common enzyme biosensors are based on optical techniques such as absorption, reflectance, luminescence, chemi-luminescence, evanescent wave, surface plasmon resonance, and interferometry.

The early era of biosensing research and development was first sparked with the defining paper by Clark (Clark and Clark 1987) and his invention of the oxygen electrode in 1955/56 (Clark 1956). The subsequent modification of the oxygen electrode led up to another publication in 1962 (Clark and Lyons 1962), which reported the development of the first glucose sensor and the enhancement of electrochemical sensors (e.g. polarographic, potentiometric, and conductometric) with enzyme-based transducers. Soon after Clark's proposal, Updike *et. al.,* (Updike and Hicks 1967) introduced modifications to this first approach to avoid oxygen concentration dependence. Clark's work and the subsequent transfer of his technology to Yellow Spring Instrument Company (Ohio, USA) led to the successful commercial launch of the first dedicated glucose biosensor in 1975. Since then, many serious players in the field of medical diagnostics, such as Bayer, Boehringer Mannheim, Eli Lilly, Lifescan, DKK Corporation, and so on, invested in the development and the mass scale production of biosensors, which are utilized in health care (Alcock and Turner 1994), environmental monitoring (Dennison and Turner 1995), food and drink (Kress-Rogers 1996), the process industries (Turner 1996), and defense and security.

3.5.2 GOLD NANOPARTICLES BASED BIOSENSOR

Biosensors are quickly becoming prevalent in modern society and have many applications, including detecting biological hazards and diagnosing certain diseases. With the recent advances in nanotechnology, nanomaterials have received great interests in the field of biosensors due to their exquisite sensitivity in chemical and biological sensing (Jain 2003). Nanoparticles, with their unique size-dependent properties, are an extremely promising technology for biosensor creation. They offer key advantages through increased biocompatibility and a method of simple visual recognition of sensing, although this takes place at the cost of some of the resolution found in other types of biosensors. Owing to the unique properties of nanomaterials, direct electrochemistry and catalytic activity of many proteins have been observed at electrodes modified with various nanomaterials, semiconductor nanoparticles, and metal nanoparticles (Wang, Liu et al. 2002; Xiao, Patolsky et al. 2003; Luo, Xu et al. 2004; Schierhorn, Lee et al. 2006; Chai, Wang et al. 2010). Various nanostructures have been examined as hosts for protein immobilization via approaches including protein adsorption, covalent attachment, protein encapsulation, and sophisticated combinations of methods. Studies have shown that nanomaterials can not only provide a friendly platform for the assembly of protein molecules but also enhance the electron-transfer process between protein molecules and the electrode.

Metal nanoparticles have many unique properties like large surface-to-volume ratio, high surface reaction activity, high catalytic efficiency, and strong adsorption ability. The conductivity properties of nanoparticles at nanoscale dimensions allow the electrical contact of redox-centers in proteins with electrode surfaces (Hernández-Santos, González-García et al. 2002; Katz, Willner et al. 2004). These nanoparticles have been used to facilitate the electron transfer in nanoelectronic devices owing to the roughening of conductive sensing interface, catalytic properties of the nanoparticles. Among the nanomaterials used as component in biosensors, AuNPs have received greatest interests because they have several kinds of intriguing properties (Wang, Polsky et al. 2001; Wang, Xu et al. 2002). Gold colloid has many advantages for biosensor applications like AuNPs can provide a stable surface for the immobilization of biomolecules, such that the molecules retain their biological activities. Modification of an electrode surface with AuNPs provides a microenvironment similar to that of the redox proteins in native systems and offers the protein molecules more freedom in orientation, which can weaken the insulating property of the protein shell for the direct electron transfer and facilitate the electron transfer through the conducting tunnels of colloidal gold (Liu, Leech et al. 2003). The AuNPs can form conducting electrodes and are the site of electron transfer when anchored to the substrate surface, allowing direct electron transfer between redox proteins and electrode surfaces with no mediators required (Yáñez-Sedeño and Pingarrón 2005). They can act as an electron-conducting pathway between prosthetic groups and electrode surface. Moreover, AuNPs have an ability to permit fast and direct electron transfer between a wide range of electroactive species and electrode materials. In addition, the light-scattering properties and extremely large enhancement ability of the local electromagnetic field enables AuNPs to be used as signal amplification tags in diverse biosensors. The roles that AuNPs have played in the biosensing process and the mechanism of AuNPs for improving the analytical performances are discussed.

AuNPs BASED OPTICAL BIOSENSORS

Optical biosensors generally measure changes in light or photon output. For optical bi-osensing utilizing AuNPs, the optical properties provide a wide range of opportunities, all of which ultimately arise from the collective oscillations of conduction band electrons ("plasmons") in response to external electromagnetic radiation (Murphy 2008). There are several optical sensing modalities for AuNPs, and the Surface Plasmon Resonance (SPR) is the one that attracted most intensive research. The SPR, which is an optical phenomenon arising from the interaction between an electromagnetic wave and the conduction electrons in a metal, is used for probing and characterizing physicochemical changes of thin films on metal surface (Figure 4). This resonance is a coherent oscillation of the surface conduction electrons excited by electromagnetic radiation.

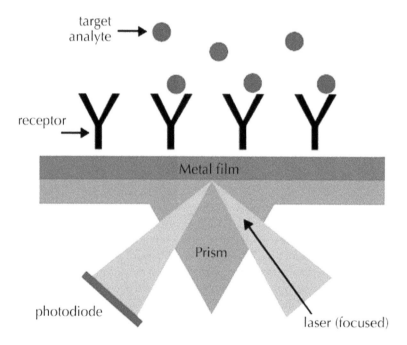

FIGURE 4   Surface Plasmon Resonance detection unit (Daniel and Astruc 2004).

Due to their unique size, AuNPs selectively absorb and reflect certain wavelengths in the visible range of light. This range depends on the size of the particles, with rough-ly spherical AuNPs less than ~ 40 nm in diameter appearing red and shifting in color from pink to purple as the size of the particles increases. The same color-shifting effect can be achieved by bringing two smaller AuNPs together so that their absorbance properties behave as if the smaller particles were a larger single particle. This effect lasts only as long as the particles are in sufficient proximity to each other, enabling the creation of a sensing mechanism (Kreibig and Genzel 1985; W.H.Yang 1995). This sensing mechanism has previously been demonstrated by Guarise et. al., (Guarise, Pasquato et al. 2005) by first stabilizing 12 nm AuNPs with a monothiol and then binding them together with a dithiol cleavable by hydrazine. The bound particles shift in absorbance from near 520 nm (appearing red) to a new peak near 600 nm (appearing purple). A colorimetric adenosine biosensor based on the aptazyme-directed assembly of AuNPs is reported in literature (Liu and Lu 2006). Very recently, AuNPs have been used for the detection of sequence-specific DNA binding using colorimetric biosensing strategy (Ou, Jin et al. 2010). The AuNP based highly sensitive and colorimetric detection of the temporal evolution of superoxide dismutase (SOD1) aggregates, and have been carried out by Hong et. al., (Hong, Choi et al. 2009). Combination of unlabeled DNAzyme and AuNPs have been employed for colorimetric detection of lead (Pb^{+2}) (Wang, Lee et al. 2008). The binding of specific molecules onto the surface of metallic films can induce a variation in the dielectric constant, which can cause a change in the reflection of laser light from a metal-liquid surface (Figure 5) and have been studied intensively in SPR to provide better analytical characteristics.

The signal amplification mechanism of AuNPs can be generally summarized into two points:

(i) The electronic coupling between the localized surface plasmon of AuNPs and the propagating plasmon on the SPR gold surface

(ii) The high density and high molecular weight of AuNPs increase the apparent mass of the analytes immobilized on them.

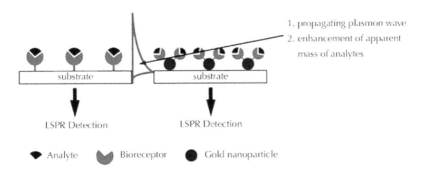

FIGURE 5 Signal amplification mechanism of AuNPs in SPR biosensor (Li, Schluesener et al. 2010).

The AuNPs have also been used for optical virus detection using spatial arrangement of Au nanoparticles on the virus-like particles (VLP) surface.(Niikura, Nagakawa et al. 2009) This structure produces a red shift in the absorption spectrum due to Plasmon coupling between adjacent Au particles, leading to the construction of an optical virus detection system. A fiber-based biosensor has been developed for organophosphorous pesticide determination using LSPR effect of AuNPs (Lin, Huang et al. 2006). An acetylcholinesterase (AChE) layer has been self-assembled by covalent coupling onto the GNP layer. When suitable pesticides presented, the activity of AChE to hydrolyze acetylcholine chloride would be inhibited leading to the change of the light attenuation due to a local increase of the refractive index. The comparative study of the fiber sensor with and without AuNPs suggested that the AuNPs coated on optical fiber can substantially enhance the sensitivity of the sensor. Since GNP-based SPR biosensors can be fabricated into an array format as well, Matsui *et. al.,* (Matsui, Akamatsu et al. 2005) demonstrated a SPR sensor chip for detection of dopamine using dopamine-imprinted polymer gel with embedded AuNPs .It was observed in the studies that SPR signal of the chip is much higher than a chip immobilizing a lower density of AuNPs or no AuNPs. Chang *et. al.,* (Chang, Lin et al. 2011) used AuNPs to develop DNA based biosensor for the detection of mercury using an amplified surface plasmon resonance as "turn-on" indicator. Inorganic mercury ion (Hg^{2+}) has been shown to coordinate to DNA duplexes that feature thymine–thymine (T–T) base pair mismatches. The general concept used in this approach is that the "turn-on" reaction of a hairpin probe via coordination of Hg^{2+} by the T–T base pair results in a substantial increase in the SPR response, followed by specific hybridization with a AuNP probe to amplify the sensor performance.

Surface enhanced SERS is another optic transduction mode that can greatly benefit from the use of GNPs Cao et al. reported a multiplexed detection method of oligonucleotide targets using oligonucleotides functionalized nanoparticles and Raman-active dyes (Hossain, Huang et al. 2009). The AuNPs facilitated the formation of a silver coating that acted as a promoter for the SERS of the dyes. High sensitivity down to the 20 fM DNA level has been reported. The AuNPs can also be used to enhance the fluorescence signal of a labeled antibody. Simonian *et. al.,* (Simonian, Good et al. 2005) used AuNPs for organophorous analytes using fluorophore (7-hydroxy-9H-(1, 3-dichloro-9,9-dimethylacridin-2-one) phosphate, which binds weekly to the active site of an enzyme, that in turn binds covalently to the gold nanoparticles. The fluorescence intensity of the fluorophore is significantly enhanced through the strong local electric field of the gold nanoparticle. The sensor showed linear response in μm when exposed to paraoxon solutions. Sun *et al.,*(Sun, Liu et al. 2009) used AuNPs for fabrication of biosensor for BSA using self- assembly of butyl rhodamine B fluorescent dye. A recent report presents a method of preparing well-ordered nanoporous gold arrays using a porous silicon (PSi) template to fabricate an optical DNA biosensor (Feng, Zhao et al. 2011) . The mechanism of the optical response caused by DNA hybridization on the Au–Psi surface was qualitatively explained by the electromagnetic theory and electrochemical impedance spectroscopy (EIS).

Current AuNP solution based methods of biosensing are not limited strictly to this one type of nanoparticle but can incorporate other particles as well. Peptide linked AuNP – quantum dot biosensors have been created by Chang *et al.,* that rely on the ability of the AuNPs to quench the photoluminescence of the quantum dots when in their close proximity bound state (Chang, Miller et al. 2005). The method of sensing is also considered an "on sensor" since the default state of the particles is "off" (no luminescence), and it is converted to "on" (luminescence) once sensing takes place.

AuNPs BASED ELECTROCHEMICAL BIOSENSOR

In the last few years, electrochemical biosensors created by coupling biological elements with electrochemical transducers based on or modified with AuNPs have played an increasingly important role in biosensor research. AuNPs can be very usefully and provide a stable surface for the immobilization of biomolecules, such that the molecules retain their biological activities. Modification of an electrode surface with AuNPs provides a microenvironment similar to that of the redox proteins in native systems and offers the protein molecules more freedom in orientation. This results in weakening of insulating property of the protein shell for direct electron transfer and facilitate the electron transfer through the conducting tunnels of colloidal gold (Liu, Leech et al. 2003)· (Yáñez-Sedeño and Pingarrón 2005). They can act as an electron-conducting pathway between prosthetic groups and electrode surface.

Gold nanoparticle-modified electrode surfaces can be prepared in three ways:

(a) By binding AuNPs with functional groups of self-assembled monolayers (SAMs);
(b) By direct deposition of nanoparticles onto the bulk electrode surface;
(c) By incorporating colloidal gold into the electrode by mixing the gold with the other components in the composite electrode matrix.

The effect of AuNPs (50–130 nm in diameter) on the response of tyrosinase based biosensor for phenol detection was investigated (Sharina, Lee et al. 2008). The addition of AuNPs to the biosensor membrane led to improvement in the response time by a reduction of approximately 5 folds to give response times of 5–10s. The linear response range of the phenol biosensor was also extended from 24 to 90 mM of phenol. The AuNPs were electrodeposited onto a glassy carbon (GC) electrode to increase the sensitivity of the tyrosinase (TYR) electrode for pesticide detection.(Kim, Shim et al. 2008) The quantitative relationships between the inhibition percentage and pesticide concentration in various water samples were measured at the TYR-AuNP-GC electrode, showing an enhanced performance attributed by the use of AuNPs. Ahirwal and Mitra recently developed an electrochemical immunosensor using AuNPs and an attached antibody for investigating their sensing capabilities with cyclic voltammetry and electrochemical impedance spectroscopy. (Ahirwal 2010). The resulting data indicated that this sandwich-type immunosensor allowed for antibody stability and adequate sensitivity, suggesting its potential application for immunoassays. Wang *et. al.*, (Wang 2011) fabricated gold nanostructure modified electrodes by a simple one-step electrodeposition method for electrochemical DNA biosensor. The DNA immobilization and hybridization on gold nano-flower modified electrode was studied with the use of $[Ru(NH3)_6]^{3+}$ as a hybridization indicator. The double-stranded DNA complex was chemi-absorbed to a gold electrode to produce an electrochemical biosensor for L-histidine (L-histidine-dependent DNAzymes) (Li, Chen et al. 2011).

Based on surface plasmon resonance of AuNPs, electrochemical biosensor (EC-SPR configuration) has also been demonstrated for hydrogen peroxide (H_2O_2) (Zhu, Luo et al. 2009). One advantage of the EC-SPR configuration is the ability to simultaneously obtain information about the electrochemical and optical properties of films with thicknesses in the nanometer range. Cytochrome c has been stably immobilized onto the Au/TiO$_2$ film and enhanced analytical performance that is 4-fold larger than obtained at the Au/TiO$_2$ film was observed. The enhanced photocurrents generated are from surface plasmon resonance of AuNPs, which was confirmed by the good match between action spectrum for photocurrent changes and ultraviolet-visible light (UV-Vis) adsorption spectrum of AuNPs. Besides this characteristic, the present biosensor for H$_2$O$_2$ has also exhibited a low detection limit, a wide dynamic linear range, and good stability.

Composite based electrodes have shown improved selectivity by inhibiting the interference reaction at the electrode. The large surface area, electrochemical properties, catalytic abilities, and inherent biocompatibility make composites suitable for use in amperometric biosensors. Composites of gold nanoparticles; carbon based materials, and Prussian blue nanoparticles have been utilized for the fabrication of electrochemical biosensors (Song, Yuan et al. 2010; Yang, Qu et al. 2010; Zhong, Wu et al. 2010; Huo, Zhou et al. 2011). Gold nanoparticle-coated multiwall carbon nanotube-modified electrodes have been used for electrochemical determination of methyl parathion and hybridization of oligonucleotides.(Ma and Zhang 2011) In recent approaches, the utilization of DNA sequences attached on AuNPs and carbon nanotubes (CNTs) have been reported to improve the recognition power of genosensors (Chu, Yan et al. 2011). A simple and sensitive sandwich-type electro-chemiluminescence immunosensor for α-1-fetoprotein (AFP) on a (nano-Au) modified glassy carbon electrode (GCE) has been developed using silica doped Ru(bpy)$_3{}^{2+}$ and nano-Au composite as labels (Yuan, Yuan et al. 2010). The prepared Ru-silica@Au composite nanoparticles own the large surface area, good biocompatibility, and highly effective electrochemiluminescence properties. The immunosensor performed high sensitivity and wide liner for detection The AFP in the range of 0.05–50 ng/mL with limit of detection of 0.03 ng/mL. Meanwhile, Liu *et. al.,*(Liu, Yuan et al. 2006) have reported a new electrode interface by using l-cysteine–gold particle composite immobilized in the network of a Nafion membrane on a glassy carbon electrode. This HRP biosensor exhibited good response to H$_2$O$_2$ and displayed the remarkable sensitivity and repeatability. An amperometric uric acid biosensor was fabricated using gold nanoparticle (AuNP)/multiwalled carbon nanotube (MWCNT) composite, which exhibited linearity from 0.01–0.8 mM with limit of detection (LOD) of 0.01 mM. The sensor measured uric acid levels in serum of healthy individuals and persons suffering from gout (Chauhan and Pundir 2011). AuNP–CaCO$_3$ composite has been prepared and applied for HRP biosensor (Cai, Xu et al. 2006). AuNP–CaCO$_3$ can retain the porous structure and inherits the advantages from its parent materials, such as satisfying biocompatibility and good solubility and dispersibility in water.

Utilization of hybrids formed by AuNPs and dendrimers has been reported as efficient systems to improve charge transfer on electrode surfaces and created the concept of electroactive nanostructured membranes (ENM) (Crespilho, Emilia Ghica et al. 2006). This strategy involved the utilization of dendrimer-polyamido-amine generation 4 (PAMAM G4) containing AuNPs and PVS as polyelectrolyte matrices for bilayer fabrication on ITO (indium tin oxide) electrodes. This modified electrode was utilized with redox mediator (Me) around AuNPs to improve the electrochemical performance on electrode/electrolyte. Composite of gold and poly(propyleneimine) dendrimer have been used for electrochemical DNA biosensor (Arotiba, Owino et al. 2008). The DNA probe was effectively wired onto the GCE/PPI-AuNP via Au-S linkage and electrostatic interactions. AuNPs were also assembled on the surface of AgCl@PANI core-shell to fabricate AuNPs–AgCl/PANI hybrid material (Yan, Feng et al. 2008). This hybrid material has been used to develop an amperometric glucose biosensor. The composite could provide a biocompatible surface for high enzyme loading and due to size effect, AuNPs could act as a good catalyst for both oxidation and reduction of H_2O_2. Feng *et. al.,* (Feng, Yang et al. 2008) employed a AuNP/PANI nanotube membrane on a glassy carbon electrode for the impedimetric sensing of the immobilization and hybridization of non-labeled DNA, thus obtaining a much wider dynamic detection range and lower detection limit for the DNA analysis.

CONCLUSION

The combination of low toxicity, high surface area, biocompatibility, and colloidal stability allow AuNPs to be safely integrated into the biosensors. Many research groups have demonstrated that they can be used for a wide range of sensing applications, ranging from chemical to biological sample. Major advancements in biosensors revolve around immobilization and interface capabilities of biological material with electrode surface. The use of nanomaterials and their composite results not only on stable immobilization matrix but also act as catalyst for many reactions, thus resulting in enhanced signal response. Looking into many newly explored properties of AuNPs, it is believed there will be a tremendous growth in coming era towards the development of AuNPs based bio-sensing devices for therapeutic and diagnostic applications, like in cancer treatment for laser ablation, chemotherapy, and so on.

This chapter is a part of PhD work of Suman Singh.

KEYWORDS

- **Biosensor**
- **Nanoparticles**
- **Colloids**
- **Nanomaterials**

REFERENCES

1. Adlim, M. M. and Abu Bakar, et. al., Synthesis of chitosan-stabilized platinum and palladium nanoparticles and their hydrogenation activity. *Journal of Molecular Catalysis A: Chemical*, **212**(1–2) 141–149 (2004).
2. Ahirwal, G. K. M., and C. K. AuNPs based sandwich electrochemical immunosensor. *Biosensors and Bioelectronics* **25**(9), 2016–2020 (2010).
3. Alcock, S. and A. Turner. "Continuous analyte monitoring to aid clinical practice. *Engineering in Medicine and Biology Magazine IEEE,* **13**(3), 319–325 (1994).

4. Andrievskii, R. A. "Directions in Current Nanoparticle Research. *Powder Metallurgy and Metal Ceramics*, **42**(11), 624–629 (2003).

5. Arotiba, O., and J. Owino, et al. (2008). An Electrochemical DNA Biosensor Developed on a Nanocomposite Platform of Gold and Poly (propyleneimine) Dendrimer. *Sensors*, **8**(11), 6791–6809.

6. Bagotsky, V. S. *Fundamentals of Electrochemistry*, John Wiley & Sons, Hoboken, New Jersey. (2006).

7. Bhat, S. and U. Maitra. Facially amphiphilic thiol capped gold and silver nanoparticles. *Journal of Chemical Sciences*, **120**(6), 507–513 (2008).

8. Boisselier, E., and A. K. Diallo, et al. Encapsulation and Stabilization of AuNPs with "Click" Polyethyleneglycol Dendrimers. *Journal of the American Chemical Society* **132**(8), 2729–2742 (2010).

9. Bönnemann, H. and Ryan M. Richards. Nanoscopic Metal Particles – Synthetic Methods and Potential Applications. *European Journal of Inorganic Chemistry*, **2001**(10), 2455–2480 (2001).

10. Boopathi, S., and S. Senthilkumar, et. al., Facile and One Pot Synthesis of AuNPs Using Tetraphenylborate and Polyvinylpyrrolidone for Selective Colorimetric Detection of Mercury Ions in Aqueous Medium. *Journal of Analytical Methods in Chemistry* 2012.

11. Brust, M., J. Fink, et. al., Synthesis and reactions of functionalised gold nanoparticles. *Journal of the Chemical Society, Chemical Communications* (16), 1655–1656 (1995).

12. Brust, M. and C. J. Kiely. Some recent advances in nanostructure preparation from gold and silver particles: a short topical review. *Colloids and Surfaces A: Physicochemical and Engineering Aspects* **202**(2–3), 175–186 (2002).

13. Brust, M., M. Walker, et. al., Synthesis of thiol-derivatised AuNPs in a two-phase Liquid-Liquid system. *Journal of the Chemical Society, Chemical Communications*, (7), 801–802 (1994).

14. Bunz, U. H. F. and V. M. Rotello. Gold Nanoparticle–Fluorophore Complexes: Sensitive and Discerning "Noses" for Biosystems Sensing. *Angewandte Chemie International Edition*, **49**(19), 3268–3279 (2010).

15. Busbee, B. D., S. O. Obare, et. al., An Improved Synthesis of High-Aspect-Ratio Gold Nanorods. *Advanced Materials*, **15**(5), 414–416 (2003).

16. C F Bohren, and D. R. H. A. *Absorption of Light by Small Particles*, New York: Wiley (1983).

17. C.J. Murphy, A. M. G., S.E. Hunyadi, J.W. Stone, P.N. Sisco, A. Alkilany, B.E. Kinard, and P. Hankins. Chemical sensing and imaging with metallic nanorods. *Chem Commun (Camb)* **5** (2008).

18. Cai, W. Y., Q. Xu, et. al., Porous gold-nanoparticle-$CaCO_3$ hybrid material: Preparation, characterization, and application for horseradish peroxidase assembly and direct electrochemistry. *Chemistry of materials*, **18**(2), 279–284 (2006).

19. Campion, A. and P. Kambhampati Surface-enhanced Raman scattering. *Chem. Soc. Rev.*, **27**(4), 241–250 (1998).

20. Carotenuto, G. and L. Nicolais. Size-controlled synthesis of thiol-derivatized gold clusters. *Journal of Materials Chemistry* **13**(5), 1038–1041 (2003).

21. Carrot, G., J. C. Valmalette, et. al., AuNP synthesis in graft copolymer micelles. *Colloid & Polymer Science*, **276**(10), 853-859 (1998).

22. Cassagneau, T. and J. H. Fendler. Preparation and Layer-by-Layer Self-Assembly of Silver Nanoparticles Capped by Graphite Oxide Nanosheets. *The Journal of Physical Chemistry B*, **103**(11), 1789–1793 (1999).

23. Chai, F., C. Wang, et. al., Colorimetric detection of Pb2+ using glutathione functionalized gold nanoparticles. *ACS Applied Materials & Interfaces*, **2**(5), 1466–1470 (2010).

24. Chandrasekharan, N., P. V. Kamat, et. al., Dye-Capped Gold Nanoclusters: Photoinduced Morphological Changes in Gold/Rhodamine 6G Nanoassemblies. *The Journal of Physical Chemistry B*, **104**(47), 11103–11109 (2000).

25. Chang, C. C., S. Lin, et. al., An amplified surface plasmon resonance "turn-on" sensor for mercury ion using gold nanoparticles. *Biosensors and Bioelectronics*, **30**(1), 235–240 (2011).

26. Chang, E., J. S. Miller, et. al., Protease-activated quantum dot probes. *Biochemical and Biophysical Research Communications*, **334**(4), 1317–1321 (2005).

27. Chauhan, N. and C. S. Pundir. An amperometric uric acid biosensor based on multiwalled carbon nanotube–AuNP composite. *Analytical Biochemistry* **413**(2), 97–103 (2011).

28. Choi, W. K., T. H. Liew, et. al., A Combined Top-Down and Bottom-Up Approach for Precise Placement of Metal Nanoparticles on Silicon. *Small*, **4**(3), 330–333 (2008).

29. Chu, H., J. Yan, et. al., Electrochemiluminescent detection of the hybridization of oligo-nucleotides using an electrode modified with nanocomposite of carbon nanotubes and gold nanoparticles. *Microchimica Acta*, **175**(3), 209–216 (2011).

30. Cigang, X., Z. Harm van, et. al., A combined top-down bottom-up approach for introduc-ing nanoparticle networks into nanoelectrode gaps. *Nanotechnology*, **17**(14), 3333 (2006).

31. Clark, L. C. Monitor and control of blood and tissue oxygen tensions. *Transactions American, Society for Artificial Internal Organs*, **2** (1956).

32. Clark, L. C. and E. W. Clark. A personalized history of the Clark oxygen electrode. *International Anesthesiology Clinics*, **25**(3), 1 (1987).

33. Clark, L. C. and C. Lyons. Electrode Systems For Continuous Monitoring In Cardiovascu-lar Surgery. *Annals of the New York Academy of Sciences*, **102**(1), 29–45 (1962).

34. Corbierre, M. K., N. S. Cameron, et. al., Polymer-Stabilized AuNPs with High Grafting Densities. *Langmuir*, **20**(7), 2867–2873 (2004).

35. Crespilho, F. N., M. Emilia Ghica, et. al., A strategy for enzyme immobilization on lay-er-by-layer dendrimer-AuNP electrocatalytic membrane incorporating redox mediator. *Electrochemistry communications*, **8**(10), 1665–1670 (2006).

36. Daniel, M. C. and D. Astruc. Gold Nanoparticles: Assembly, Supramolecular Chemis-try, Quantum-Size-Related Properties, and Applications Toward Biology, Catalysis, and Nanotechnology. *ChemInform*, **35**(16), no-no. (2004).

37. Daniel, M. C. and D. Astruc. Gold nanoparticles: assembly, supramolecular chemistry, quantum-size-related properties, and applications toward biology, catalysis, and nanotech-nology. *Chemical reviews*, **104**(1), 293–346 (2004).

38. Dennison, M. and A. Turner. Biosensors for environmental monitoring. *Biotechnology advances* **13**(1), 1–12 (1995).

39. Ding, Y., Y. Hu, et al. Polymer-assisted nanoparticulate contrast-enhancing materials. *Science China Chemistry* **53**(3), 479–486 (2010).

40. Eggins, B. R. *Chemical sensors and biosensors*, Wiley (2002).

41. Ershov, B. Metal Nanoparticles in Aqueous Solutions: Electronic, Optical, and Catalytic Properties *Ross. Khim. Zh.*, **45**(3), 20 (2001).

42. Faraday, M. The Bakerian Lecture: Experimental Relations of Gold (and Other Metals) to Light. *Philosophical Transactions of the Royal Society of London* **147**, 145–181 (1857).

43. Feng, J., W. Zhao, et. al., A label-free optical sensor based on nanoporous gold arrays for the detection of oligodeoxynucleotides. *Biosensors and Bioelectronics* **30**(1), 21–27 (2011).

44. Feng, Y., T. Yang, et. al., Enhanced sensitivity for deoxyribonucleic acid electrochemical impedance sensor: Gold nanoparticle/polyaniline nanotube membranes. *Analytica chimica acta* **616**(2), 144–151 (2008).

45. Fleming, D. A. and M. E. Williams. Size-Controlled Synthesis of AuNPs via High-Temperature Reduction. *Langmuir*, **20**(8), 3021–3023 (2004).

46. Freestone, I., N. Meeks, et. al., The Lycurgus Cup — A Roman nanotechnology. *Gold Bulletin*, **40**(4), 270–277 (2007).

47. Fu, X., Y. Wang, et. al., Shape-Selective Preparation and Properties of Oxalate-Stabilized Pt Colloid. *Langmuir*, **18**(12), 4619–4624 (2002).

48. Gibson, M. I., M. Danial, et. al., Sequentially Modified, Polymer-Stabilized AuNP Libraries: Convergent Synthesis and Aggregation Behavior. *ACS Combinatorial Science*, **13**(3), 286–297 (2011).

49. Giersig, M. and P. Mulvaney. Preparation of ordered colloid monolayers by electrophoretic deposition. *Langmuir*, **9**(12), 3408–3413 (1993).

50. Guarise, C., L. Pasquato, et. al., Reversible aggregation/deaggregation of AuNPs induced by a cleavable dithiol linker. *Langmuir*, **21**(12), 5537–5541 (2005).

51. Henglein, A. Reactions of organic free radicals at colloidal silver in aqueous solution. Electron pool effect and water decomposition. *The Journal of Physical Chemistry*, **83**(17), 2209–2216 (1979).

52. Hermes, J. P., F. Sander, et. al., Nanoparticles to Hybrid Organic-Inorganic Superstructures. *CHIMIA International Journal for Chemistry*, **65**(4), 219–222 (2011).

53. Hernández-Santos, D., M. B. González-García, et. al., Metal-nanoparticles based electroanalysis. *Electroanalysis*, **14**(18), 1225–1235 (2002).

54. Hong, S., I. Choi, et. al., Sensitive and Colorimetric Detection of the Structural Evolution of Superoxide Dismutase with Gold Nanoparticles. *Analytical Chemistry* **81**(4), 1378-1382 (2009).

55. Hossain, M. K., G. G. Huang, et. al., Characteristics of surface-enhanced Raman scattering and surface-enhanced fluorescence using a single and a double layer gold nanostructure. *Physical Chemistry Chemical Physics*, **11**(34),7484–7490 (2009).

56. Hostetler, M. J., J. E. Wingate, et. al., Alkanethiolate Gold Cluster Molecules with Core Diameters from 1.5 to 5.2 nm: Core and Monolayer Properties as a Function of Core Size. *Langmuir*, **14**(1), 17–30 (1998).

57. Huang, T. and R. W. Murray. Visible Luminescence of Water-Soluble Monolayer-Protected Gold Clusters. *The Journal of Physical Chemistry B*, **105**(50), 12498–12502 (2001).

58. Huo, Z., Y. Zhou, et. al., Sensitive simultaneous determination of catechol and hydroquinone using a gold electrode modified with carbon nanofibers and gold nanoparticles. *Microchimica Acta*, **173**(1), 119–125 (2011).

59. Hwang, Y.-N., D. H. Jeong. et. al., Femtosecond Emission Studies on Gold Nanoparticles. *The Journal of Physical Chemistry B*, **106**(31), 7581–7584 (2002).

60. J. P. Wilcoxon, J. E. M., F. Parsapour, B. Wiedenman, and D. F. Kelley. Photoluminescence from nanosize gold clusters. *J. Chem. Phys*, **108**, 9137 (1998).

61. Jain, K. Current status of molecular biosensors. *Medical device technology* **14**(4), 10 (2003).

62. Jana, N. R., L. Gearheart, et al. Evidence for Seed-Mediated Nucleation in the Chemical Reduction of Gold Salts to Gold Nanoparticles. *Chemistry of materials* **13**(7), 2313–2322 (2001).

63. Jiang, W., E. Papa. et. al., Semiconductor quantum dots as contrast agents for whole animal imaging. *Trends in Biotechnology*, **22**(12), 607–609 (2004).

64. Kamat, P. V. Photophysical, Photochemical and Photocatalytic Aspects of Metal Nanoparticles. *The Journal of Physical Chemistry B*, **106**(32), 7729–7744 (2002).

65. Karhanek, M., J. T. Kemp, et. al., Single DNA Molecule Detection Using Nanopipettes and Nanoparticles. *Nano Letters*, **5**(2), 403–407 (2005).

66. Katz, E., I. Willner, et. al., Electroanalytical and bioelectroanalytical systems based on metal and semiconductor nanoparticles. *Electroanalysi*, **16**(1–2), 19–44 (2004).

67. Kim, G. Y., J. Shim, et. al., Optimized coverage of AuNPs at tyrosinase electrode for measurement of a pesticide in various water samples. *Journal of Hazardous Materials*, **156**(1–3), 141–147 (2008).

68. Klabunde, K. J. C. T., G. In Active Metals: Preparation,Characterization, and Applications. *VCH*, New York (1996)

69. Kohler, N., C. Sun, et. al., Methotrexate-Modified Superparamagnetic Nanoparticles and Their Intracellular Uptake into Human Cancer Cells. *Langmuir*, **21**(19), 8858–8864 (2005).

70. Kostoff, R., R. Koytcheff, et. al., Structure of the nanoscience and nanotechnology applications literature. *The Journal of Technology Transfer*, **33**(5), 472–484 (2008).

71. Kreibig, U. and L. Genzel. Optical absorption of small metallic particles. *Surface Science*, **156,** Part 2(0), 678–700 (1985).

72. Kress-Rogers, E. *Handbook of Biosensors and Electronic Noses: Medicine, Food and the Environment*, CRC Press, Boca Raton, USA (1996).

73. Kwon, K., K. Y. Lee, et. al., Controlled Synthesis of Icosahedral AuNPs and Their Surface-Enhanced Raman Scattering Property. *The Journal of Physical Chemistry C*, **111**(3), 1161–1165 (2006).

74. Lahav, M., A. N. Shipway, et. al., An enlarged bis-bipyridinium cyclophane-Au nanoparticle superstructure for selective electrochemical sensing applications. *Journal of Electroanalytical Chemistry*, **482**(2), 217–221 (2000).

75. Li, L. D., Z. B. Chen, et. al., Electrochemical real-time detection of l-histidine via self-cleavage of DNAzymes. *Biosensors and Bioelectronics*, **26**(5), 2781–2785 (2011).

76. Li, Y., E. Boone, et. al., Size Effects of PVP−Pd Nanoparticles on the Catalytic Suzuki Reactions in Aqueous Solution. *Langmuir*, **18**(12), 4921–4925 (2002).

77. Li, Y., H. Schluesener, et. al., Gold nanoparticle-based biosensors. *Gold Bulletin*, **43**(1), 29–41 (2010).

78. Lin, C. S., M. R. Khan, et. al., Platinum states in citrate sols by EXAFS. *Journal of Colloid and Interface Science*, **287**(1), 366–369 (2005).

79. Lin, T.-J., K.-T. Huang, et. al., Determination of organophosphorous pesticides by a novel biosensor based on localized surface plasmon resonance. *Biosensors and Bioelectronics*, **22**(4), 513–518 (2006).

80. Link, S., A. Beeby, et. al., Visible to Infrared Luminescence from a 28-Atom Gold Cluster. *The Journal of Physical Chemistry B*, **106**(13), 3410–3415 (2002).

81. Link, S. and M. A. El-Sayed. Size and Temperature Dependence of the Plasmon Absorption of Colloidal Gold Nanoparticles. *The Journal of Physical Chemistry B*, **103**(21), 4212–4217 (1999).

82. Link, S. and M. A. El-Sayed. Shape and size dependence of radiative, non-radiative and photothermal properties of gold nanocrystals. *International Reviews in Physical Chemistry* **19**(3), 409–453 (2000).

83. Liu, J. and Y. Lu. Fast colorimetric sensing of adenosine and cocaine based on a general sensor design involving aptamers and nanoparticles." *Angewandte Chemie*, **118**(1), 96–100 (2006).

84. Liu, S., D. Leech, et. al., Application of colloidal gold in protein immobilization, electron transfer, and biosensing. *Analytical letters*, **36**(1), 1–19 (2003).

85. Liu, Y., R. Yuan, et. al., Direct electrochemistry of horseradish peroxidase immobilized on gold colloid/cysteine/nafion-modified platinum disk electrode. *Sensors and Actuators B: Chemical*, **115**(1), 109–115 (2006).

86. Luo, X. L., J. J. Xu, et. al., A novel glucose ENFET based on the special reactivity of MnO2 nanoparticles. *Biosensors and Bioelectronics*, **19**(10), 1295–1300 (2004).

87. Ma, H., B. Yin, et. al., Synthesis of Silver and AuNPs by a Novel Electrochemical Method. *ChemPhysChem*, **5**(1), 68–75 (2004).

88. Ma, J. C. and W. D. Zhang. Gold nanoparticle-coated multiwall carbon nanotube-modified electrode for electrochemical determination of methyl parathion. *Microchimica Acta*, **175**(3), 309–314 (2011).

89. Mallick, K., M. J. Witcomb, et. al., Polymer-stabilized colloidal gold: a convenient method for the synthesis of nanoparticles by a UV-irradiation approach. *Applied Physics A: Materials Science & Processing*, **80**(2), 395–398 (2005).

90. Martin, C. R. Membrane-Based Synthesis of Nanomaterials. *Chemistry of materials*, **8**(8), 1739–1746 (1996).

91. Masala, O. and R. Seshadri. Synthesis Routes for Large Volumes of Nanoparticles. *Annual Review of Materials Research*, **34**(1), 41–81 (2004).

92. Matsui, J., K. Akamatsu, et. al., SPR Sensor Chip for Detection of Small Molecules Using Molecularly Imprinted Polymer with Embedded Gold Nanoparticles. *Analytical Chemistry*, **77**(13), 4282-4285 (2005).

93. Mayer, A. B. R. and J. E. Mark Colloidal AuNPs protected by water-soluble homopolymers and random copolymers. *European Polymer Journal*, **34**(1), 103–108 (1998).

94. Meli, L. and P. F. Green. Aggregation and Coarsening of Ligand-Stabilized AuNPs in Poly(methyl methacrylate) Thin Films. *ACS Nano*, **2**(6), 1305–1312 (2008).

95. Meltzer, S., R. Resch, et al. "Fabrication of Nanostructures by Hydroxylamine Seeding of AuNP Templates. *Langmuir*, **17**(5), 1713–1718 (2001).

96. Menard, L. D., S. P. Gao, et. al., Sub-Nanometer Au Monolayer-Protected Clusters Exhibiting Molecule-like Electronic Behavior: Quantitative High-Angle Annular Dark-Field Scanning Transmission Electron Microscopy and Electrochemical Characterization of Clusters with Precise Atomic Stoichiometry. *The Journal of Physical Chemistry B*, **110**(26), 12874–12883 (2006).

97. Mie, G. Beiträge zur Optik trüber Medien, speziell kolloidaler Metallösungen. *Annalen der Physik*, **330**(3), 377–445 (1908).

98. Murray, R. W. Nanoelectrochemistry: Metal Nanoparticles, Nanoelectrodes, and Nanopores. *Chemical reviews*, **108**(7), 2688–2720 (2008).

99. Narayanan, R. and M. A. El-Sayed. Effect of Catalytic Activity on the Metallic Nanoparticle Size Distribution: Electron-Transfer Reaction between Fe(CN)$_6$ and Thiosulfate Ions Catalyzed by PVP–Platinum Nanoparticles. *The Journal of Physical Chemistry B*, **107**(45), 12416–12424 (2003).

100. Narayanan, R. and M. A. El-Sayed. Shape-Dependent Catalytic Activity of Platinum Nanoparticles in Colloidal Solution. *Nano Letters*, **4**(7), 1343–1348 (2004).

101. Niikura, K., K. Nagakawa, et al. AuNP Arrangement on Viral Particles through Carbohydrate Recognition: A Non-Cross-Linking Approach to Optical Virus Detection. *Bioconjugate Chemistry*, **20**(10), 1848–1852 (2009).

102. Nijhuis, C. A., N. Oncel, et. al., Room-Temperature Single-Electron Tunneling in Dendrimer-Stabilized AuNPs Anchored at a Molecular Printboard." *Small* **2**(12), 1422–1426 (2006).

103. Norman, T. J., C. D. Grant, et. al., Near Infrared Optical Absorption of AuNP Aggregates. *The Journal of Physical Chemistry B*, **106**(28), 7005–7012 (2002).

104. Ou, L. J., P. Y. Jin, et. al., Sensitive and Visual Detection of Sequence-Specific DNA-Binding Protein via a Gold Nanoparticle-Based Colorimetric Biosensor. *Analytical Chemistry*, **82**(14), 6015–6024 (2010).

105. Pal, A., K. Esumi, et. al., Preparation of nanosized gold particles in a biopolymer using UV photoactivation. *Journal of Colloid and Interface Science* **288**(2), 396–401 (2005).

106. Park, J. E., M. Atobe, et. al., Synthesis of multiple shapes of AuNPs with controlled sizes in aqueous solution using ultrasound. *Ultrasonics Sonochemistry*, **13**(3), 237–241 (2006).

107. Pillai, Z. S. and P. V. Kamat. What Factors Control the Size and Shape of Silver Nanoparticles in the Citrate Ion Reduction Method? *The Journal of Physical Chemistry B*, **108**(3), 945–951 (2003).

108. Pittelkow, M., K. Moth-Poulsen, et al. Poly(amidoamine)-Dendrimer-Stabilized Pd(0) Nanoparticles as a Catalyst for the Suzuki Reaction. *Langmuir*, **19**(18), 7682–7684 (2003).

109. Prasad, B. L. V., S. I. Stoeva, et. al., AuNPs as Catalysts for Polymerization of Alkylsilanes to Siloxane Nanowires, Filaments, and Tubes. *Journal of the American Chemical Society*, **125**(35), 10488–10489 (2003).

110. R Zsigmondy, L., Akademishe Verlags-gesellschaft, m.b.H.), 44,45,46. *Das Kolloide Gold.* (1925).

111. Rahme, K., P. Vicendo, et al.. A Simple Protocol to Stabilize AuNPs using Amphiphilic Block Copolymers: Stability Studies and Viable Cellular Uptake. *Chemistry – A European Journal*, **15**(42), 11151–11159 (2009).

112. Raveendran, P., J. Fu, et. al., Completely "Green" Synthesis and Stabilization of Metal Nanoparticles. *Journal of the American Chemical Society*, **125**(46), 13940–13941 (2003).

113. Richardson, M. J., J. H. Johnston, et. al., Monomeric and Polymeric Amines as Dual Reductants/Stabilisers for the Synthesis of Gold Nanocrystals: A Mechanistic Study. *European Journal of Inorganic Chemistry*, **2006**(13), 2618–2623 (2006).

114. Richter, J., R. Seidel, et al. Nanoscale palladium metallization of DNA. *Advanced Materials* **12**(7), 507–507 (2000).

115. Saha, K., S. S. Agasti, et. al., AuNPs in Chemical and Biological Sensing. *Chemical reviews* **112**(5), 2739–2779 (2012).

116. Sau, T. K. and C. J. Murphy. Room Temperature, High-Yield Synthesis of Multiple Shapes of AuNPs in Aqueous Solution. *Journal of the American Chemical Society,* **126**(28), 8648–8649 (2004).

117. Sau, T. K., A. Pal, et. al., Size Controlled Synthesis of AuNPs using Photochemically Prepared Seed Particles. *Journal of Nanoparticle Research*, **3**(4): 257–261 (2001).

118. Schierhorn, M., S. J. Lee, et. al., Metal–Silica Hybrid Nanostructures for Surface-Enhanced Raman Spectroscopy. *Advanced Materials*, **18**(21), 2829–2832 (2006).

119. Schmid, G. and B. Corain. Nanoparticulated Gold: Syntheses, Structures, Electronics, and Reactivities. *ChemInform*, **34**(44), no-no (2003).

120. Shang, L., J. Yin, et. al., Gold nanoparticle-based near-infrared fluorescent detection of biological thiols in human plasma. *Biosensors and Bioelectronics*, **25**(2), 269–274 (2009).

121. Sharina, A., Y. Lee, et. al., Effects of AuNPs on the Response of Phenol Biosensor Containing Photocurable Membrane with Tyrosinase. *Sensors* **8**(10), 6407–6416 (2008).

122. Sharma, J., S. Mahima, et. al., (2004). Solvent-Assisted One-Pot Synthesis and Self-Assembly of 4-Aminothiophenol-Capped Gold Nanoparticles. *The Journal of Physical Chemistry B*, **108**(35): 13280-13286.

123. Shen, X. C., L. F. Jiang, et. al., Determination of 6-mercaptopurine based on the fluorescence enhancement of Au nanoparticles. *Talanta*, **69**(2), 456–462 (2006).

124. Shimizu, T., T. Teranishi, et. al., Size Evolution of Alkanethiol-Protected AuNPs by Heat Treatment in the Solid State. *The Journal of Physical Chemistry*, **107**(12), 2719–2724 (2003).

125. Shipway, A. N., M. Lahav, et. al., Nanostructured Gold Colloid Electrodes. *Advanced Materials*, **12**(13), 993–998 (2000).

126. Shon, Y. S., S. M. Gross, et. al., Alkanethiolate-Protected Gold Clusters Generated from Sodium S-Dodecylthiosulfate (Bunte Salts). *Langmuir*, **16**(16), 6555–6561 (2000).

127. Shou, Q., C. Guo, et. al., Effect of pH on the single-step synthesis of AuNPs using PEO–PPO–PEO triblock copolymers in aqueous media. *Journal of Colloid and Interface Science*, **363**(2), 481–489 (2011).

128. Simonian, A. L., T. A. Good, et. al., Nanoparticle-based optical biosensors for the direct detection of organophosphate chemical warfare agents and pesticides. *Analytica chimica acta*, **534**(1), 69–77 (2005).

129. Song, Z., and R. Yuan, et. al., Multilayer structured amperometric immunosensor based on AuNPs and Prussian blue nanoparticles/nanocomposite functionalized interface. *Electrochimica Acta*, **55**(5), 1778–1784 (2010).

130. Sun, X., and B. Liu, et. al., A novel biosensor for bovine serum albumin based on fluorescent self-assembled sandwich bilayers. *Luminescence*, **24**(1), 62–66 (2009).

131. Svedberg, T. *Colloid Chemistry ACS Monography 16*, New York:Chem Catalog Co. (1924).

132. Swihart, M. T. Vapor-phase synthesis of nanoparticles. *Current Opinion in Colloid & Interface Science*, **8**(1), 127–133 (2003).

133. Tabuani, D., O. Monticelli, et. al., Palladium Nanoparticles Supported on Hyperbranched Aramids: Synthesis, Characterization, and Some Applications in the Hydrogenation of Unsaturated Substrates. *Macromolecules*, **36**(12), 4294–4301 (2003).

134. Taton, T. A., C. A. Mirkin, et. al., Scanometric DNA Array Detection with Nanoparticle Probes. *Science*, **289**(5485), 1757–1760 (2000).

135. Templeton, A. C., J. J. Pietron, et. al., Solvent Refractive Index and Core Charge Influences on the Surface Plasmon Absorbance of Alkanethiolate Monolayer-Protected Gold Clusters. *The Journal of Physical Chemistry B*, **104**(3), 564–570 (2000).

136. Templeton, A. C., W. P. Wuelfing, et. al., Monolayer-Protected Cluster Molecules. Accounts of Chemical Research, **33**(1), 27–36 (1999).

137. Thanh, N. T. K. and Z. Rosenzweig. Development of an Aggregation-Based Immunoassay for Anti-Protein A Using Gold Nanoparticles. *Analytical Chemistry*, **74**(7), 1624–1628 (2002).

138. Thomas, K. G. and P. V. Kamat. Chromophore-Functionalized Gold Nanoparticles. *Accounts of Chemical Research*, **36**(12), 888–898 (2003).

139. Thomas, K. G., J. Zajicek, et. al., Surface Binding Properties of Tetraoctylammonium Bromide-Capped Gold Nanoparticles. *Langmuir*, **18**(9), 3722–3727 (2002).

140. Turkevich, J., P. C. Stevenson, et. al., The Formation of Colloidal Gold. *The Journal of Physical Chemistry*, **57**(7), 670–673 (1953).

141. Turner, A. P. F. Biosensors: Past, present and future. *Last accessed October*, **6**, 2005 (1996).

142. Ung, T., M. Giersig, et. al., Spectroelectrochemistry of Colloidal Silver. *Langmuir*, **13**(6), 1773–1782 (1997).

143. Updike, S. J. and G. P. Hicks. The enzyme electrode. *Nature*, **214**(5092), 986–988 (1967).

144. V.A Fabrikanos, S. A., K. H. Lieser. Dastelung stabiler hydrosole von gold und silber durch reduktion mitathylen diamintetraessingsaure. *Z. Naturforsch B*, **18**, 612 (1963).

145. W.H.Yang, G. C. S., and R. P. Van Duyne. Discrete dipole approximation for calculating extinction and Raman intensities for small particles with arbitrary shapes. *J. Chem. Phys.*, **103**, 7 (1995).

146. Wang, J., G. Liu, et. al., Electrochemical stripping detection of DNA hybridization based on cadmium sulfide nanoparticle tags. *Electrochemistry communications*, **4**(9), 722–726 (2002).

147. Wang, J., R. Polsky, et. al., Silver-enhanced colloidal gold electrochemical stripping detection of DNA hybridization. *Langmuir*, **17**(19), 5739–5741 (2001).

148. Wang, J., D. Xu, et. al., Magnetically-induced solid-state electrochemical detection of DNA hybridization. *Journal of the American Chemical Society*, **124**(16), 4208–4209 (2002).

149. Wang, L., Chen, Xiaohong, Wang, Xiaoli, Han, Xiaoping, Liu, Shufeng, and Zhao, Changzhi. Electrochemical synthesis of gold nanostructure modified electrode and its development in electrochemical DNA biosensor. *Biosensors and Bioelectronics*, **30**(1), 151–157 (2011).

150. Wang, X., H. Kawanami, et. al., Amphiphilic block copolymer-stabilized AuNPs for aerobic oxidation of alcohols in aqueous solution. *Chemical Communications*, (37), 4442–4444 (2008)

151. Wang, Z., J. H. Lee, et. al., Label-Free Colorimetric Detection of Lead Ions with a Nanomolar Detection Limit and Tunable Dynamic Range by using AuNPs and DNAzyme." *Advanced Materials*, **20**(17), 3263–3267 (2008).

152. Wang, Z. L. FUNCTIONAL OXIDE NANOBELTS: Materials, Properties and Potential Applications in Nanosystems and Biotechnology. *Annual Review of Physical Chemistry*, **55**(1), 159–196 (2004).

153. Warner, M. G., S. M. Reed, et. al., Small, Water-Soluble, Ligand-Stabilized AuNPs Synthesized by Interfacial Ligand Exchange Reactions. *Chemistry of materials*, **12**(11), 3316-3320 (2000).

154. Whetten, R. L., M. N. Shafigullin, et. al., Crystal Structures of Molecular Gold Nanocrystal Arrays. *Accounts of Chemical Research*, **32**(5), 397-406 (1999).

155. Wilson, O. M., X. Hu, et. al., Colloidal metal particles as probes of nanoscale thermal transport in fluids. *Physical Review B*, **66**(22), 224301 (2002).

156. Xiao, Y., F. Patolsky, et. al., Plugging into Enzymes: Nanowiring of Redox Enzymes by a Gold Nanoparticle. *Science*, **299**(5614), 1877 (2003).

157. Xiulan, S., Z. Xiaolian, et. al., Preparation of gold-labeled antibody probe and its use in immunochromatography assay for detection of aflatoxin B1. *International Journal of Food Microbiology*, **99**(2), 185–194 (2005).

158. Yan, W., X. Feng, et. al., A super highly sensitive glucose biosensor based on Au nanoparticles-AgCl@ polyaniline hybrid material. *Biosensors and Bioelectronics*, **23**(7), 925–931 (2008).

159. Yáñez-Sedeño, P. and J. M. Pingarrón. Gold nanoparticle-based electrochemical biosensors. *Analytical and Bioanalytical Chemistry*, **382**(4), 884–886 (2005).

160. Yang, P. H., X. Sun, et al. Transferrin-Mediated AuNP Cellular Uptake. *Bioconjugate Chemistry*, **16**(3), 494–496 (2005).

161. Yang, S., L. Qu, et. al., Gold nanoparticles/ethylenediamine/carbon nanotube modified glassy carbon electrode as the voltammetric sensor for selective determination of rutin in the presence of ascorbic acid. *Journal of Electroanalytical Chemistry*, **645**(2), 115–122 (2010).

162. Yuan, S., R. Yuan, et. al., Sandwich-type electrochemiluminescence immunosensor based on Ru-silica@Au composite nanoparticles labeled anti-AFP. *Talanta*, **82**(4), 1468–1471 (2010).

163. Zeng, S., K. T. Yong, et. al., A Review on Functionalized AuNPs for Biosensing Applications. *Plasmonics*, **6**(3), 491–506 (2011).

164. Zhai, S., H. Hong, et. al., Synthesis of cationic hyperbranched multiarm copolymer and its application in self-reducing and stabilizing gold nanoparticles. *Science China Chemistry*, **53**(5), 1114–1121 (2010).

165. Zhang, J., J. Du, et. al., Sonochemical Formation of Single-Crystalline Gold Nanobelts." *Angewandte Chemie*, **118**(7), 1134–1137 (2006).

166. Zhang, Y. X. and H. C. Zeng. Surfactant-Mediated Self-Assembly of Au Nanoparticles and Their Related Conversion to Complex Mesoporous Structures. *Langmuir*, **24**(8), 3740–3746 (2008).

167. Zheng, J., J. T. Petty, et. al., High Quantum Yield Blue Emission from Water-Soluble Au8 Nanodots. *Journal of the American Chemical Society*, **125**(26), 7780–7781 (2003).

168. Zhong, Z., S. Patskovskyy, et. al., The Surface Chemistry of Au Colloids and Their Interactions with Functional Amino Acids. *The Journal of Physical Chemistry B*, **108**(13), 4046-4052 (2004).

169. Zhong, Z., W. Wu, et. al., Nanogold-enwrapped graphene nanocomposites as trace labels for sensitivity enhancement of electrochemical immunosensors in clinical immunoassays: Carcinoembryonic antigen as a model. *Biosensors and Bioelectronics*, **25**(10), 2379–2383 (2010).

170. Zhu, A., Y. Luo, et. al., Plasmon-Induced Enhancement in Analytical Performance Based on AuNPs Deposited on TiO2 Film. *Analytical chemistry*, **81**(17), 7243–7247 (2009).

171. Zhu, T., K. Vasilev, et. al., Surface Modification of Citrate-Reduced Colloidal AuNPs with 2-Mercaptosuccinic Acid. *Langmuir*, **19**(22), 9518–9525 (2003).

172. Zsigmondy, R. Die chemische Natur des Cassiusschen Goldpurpurs. *Justus Liebigs Annalen der Chemie*, **301**(2–3), 361–387 (1898).

CHAPTER 4

INDUSTRIALLY RELEVANT NANOPARTICLES—HEMATITE: ITS SYNTHESIS, FUNCTIONALIZATION, AND APPLICATIONS

MARIMUTHU NIDHIN, KALARICAL JANARDHANAN SREERAM, and BALACHANDRAN UNNI NAIR

CONTENTS

4.1 INCREASING RELEVANCE OF NANOTECHNOLOGY IN SCIENCE AND INDUSTRY

Innovations in modern technology are directed towards smaller, cheaper, faster and smarter products than that is currently available in the market. These properties are interconnected with size reduction. For instance, size reduction to 10^{-9} meters, considerably changes the physical and chemical properties of the material. "Metals" or "Metal oxides" are old materials. However, their fabrication in new forms, more specifically in the nano-scale has aroused tremendous interest. (Kline et. al., 2006) Nanotechnology is a broad interdisciplinary area of R&D and industrial activity. It involves manufacturing, processing, and application of materials with size below 100 nm. Research areas in nanotechnology can be broadly grouped under nanomedicine, nanofabrication, nanometrology, nanomaterials, and nanoparticles. (Bakunin et. al., 2004; Aitken et. al., 2006) Through the years, we have seen increasing relevance for nanoparticles and nanomaterials in fundamental and applied science. These nanoparticles and nanostructures exhibit different physical and chemical properties compared to bulk. They have been of keen interest to science and technology, especially in miniaturized devices. Nanotechnology based industries are expected to have a market of USD 1 trillion by 2015. Aitken et. al.,, has provided a detailed review on the potential of nanotechnology. (Aitken et. al., 2006) He opines that an impediment to the quantum jump in growth of this industry is the issue of scale-up from lab scale to industrial level production. There is a need to develop or reorient the synthesis and functionalization strategies of industrially relevant metal or metal oxide nanoparticles.

Synthesis of nanoparticles of iron oxides have been reported in recent times (especially α-Fe_2O_3) by using different chemical methods (Tartaj et. al., 2005). Iron oxide nanoparticles are usually synthesized by the coprecipitation of ferrous (Fe^{2+}) and ferric (Fe^{3+}) ions by a base. Other synthetic methods include the thermal decomposition of an alkaline solution of an Fe^{3+}chelate in the presence of hydrazine and the sonochemical decomposition of an Fe^{2+}salt followed by thermal treatment (Perez et. al., 2003). Uniformly sized iron oxide nanoparticles were synthesized by the high-temperature reaction of Fe(acac) 3 in octyl ether and oleic acid or lauric acid or a mixture of four solvents and ligands, namely, 1,2-hexadecanediol, phenyl ether, oleic acid, and oleylamine. But the major drawbacks of these synthetic procedures are the low dispersion in solvents and wide particle size distribution (Lian et. al., 2009). However, the uniformity of the size and shape of the particles of these nanomaterials was rather poor. These methods usually involve synthesizing a precursor gel of iron, followed by decomposing the gel or precursor into the designed crystalline iron oxide phase at an elevated temperature (Liang et. al., 2006). They require expensive and often toxic reagents, complicated synthetic steps, and are not biocompatible. To understand the environmental implications of these nanoparticles and facilitate their potential applications, it is important to develop a simple, green, and generic method for the preparation of Fe_2O_3 nanoparticles (Lin X. -M. and Samia A. C. S. 2006a). Green synthesis of nanosized Fe_2O_3 with uniform size is one of the important issues of the present research for understanding the fundamental properties of nanomaterials and utilizing their nano scaled properties for various applications.

One of the industrially most relevant and fascinating properties of iron oxide is its magnetism. To a large extent, magnetism is a nano phenomenon (Lin X. -M. and Samia A. C. S. 2006a). When particle dimensions are reduced to nanosize and shape, new properties such as exchange magnetic moments, (Bucher et. al., 1991) exchanged coupled dynamics,(Fullerton et. al., 1998) quantization of spin waves(Jung et. al., 2002) and giant magneto-resistance (Berkowitz et. al., 1992) have been reported. These new properties lead to applications in permanent magnets, data storage devices, magnetic refrigeration, catalysis, targeted drug delivery and MRI. Superparamagnetic iron oxide nanoparticles (SPIONs) possess high transverse (spin-spin) relaxivity and are therefore good T2 contrast agents, enabling them to produce darker images when accumulated in certain areas of the body(Rudzka et. al., 2012).

4.2 HEMATITE: ABUNDANT TUNABLE INDUSTRIAL APPLICATIONS

Almost all processes or environment have iron oxide in it. This includes the depths of Earth, the surfaces of Mars, factories, high-tech magnetic recorders, brains of birds and magnetotactic bacteria. Nanotechnology has made the applications of iron oxide more interesting and increasing. Iron oxides are technologically important transition metal oxides. Sixteen pure phases of iron oxides, viz., oxides, hydroxides or oxy-hydroxides are known till date (Cornell R M and Schwertmann U. 2007). These include $FeOH^{-3}$, $Fe(OH)_2$, $Fe_5HO_8.4H_2O$, Fe_3O_4, FeO, five polymorphs of FeOOH and four polymorphs of Fe_2O_3 (α, β, γ, ϵ), and so on etc. This complexity has brought about a limitation in the knowledge of structural details, thermodynamics and reactivity of iron oxides (Navrotsky et. al., 2008). The physical and chemical properties associated with these particles are dependent on size and degree of hydration and therefore there is a larger emphasis on the synthesis strategies employed, more so in the case of α-Fe_2O_3. α-Fe_2O_3 (Hematite), is an n-type semiconductor, and the most prevalent metal oxide on Earth. It is stable, corrosion resistant and non-toxic under ambient conditions. Historically, one of the first uses of α-Fe_2O_3 was in the production of red pigments (Cornell R M and Schwertmann U. 2007). Relatively small bandgap (2.2 eV; 564 nm) and related visible light absorption that (d $\rightarrow$ d type) (Quinn et. al., 1976) are not strongly absorptive, low cost and stability under deleterious chemical conditions make α-Fe_2O_3 an ideal candidate for several industrial applications such as in catalysis, gas sensors, solar cells, anode material for lithium ion batteries, field emission devices, magnetic materials, hydrogen storage, pigments and water splitting. (Park et. al., 2000; Zhou H. and Wong S. S. 2008; Chen et. al., 2010; Lin et. al., 2012; Wheeler et. al., 2012) However, challenges such as poor conductivity and short hole-diffusion length (2 -4 nm) are some of the draw-backs which need to be overcome prior to practical usage (Cherepy et. al., 1998; Sivula et. al., 2010; Van et. al., 2012).

α-Fe_2O_3 has a hexagonal unit cell and entirely octahedrallyoctahedral coordinated Fe^{3+} atoms (corundum structure). (Cornell R M and Schwertmann U. 2007) Crystal structure consists of alternating iron and oxygen layers stacked along the [001] axis of the hexagonal unit. At temperatures below the Neel temperature (TN–955 K), the magnetization directions of neighboring Fe layers become antiparallel due to antiferromagnetic ordering of the layers. A first-order spin-reorientation transition (Morin transition) occurs at 263K. Below the Morin transition temperature (TM¬), moments of any two magnetic sublattices are antiparallel and aligned along the rhombohedra [111] axis and above morin transition temperature (TM), moments lie in the basal plane, resulting in a weak net magnetic moment(Lu, H. M. and Meng, X. K. 2010). The polymorphs of Fe_2O_3 (Figure 1) and their nanoparticles can in general, be synthesized through natural and synthetic thermal transformations of iron-bearing materials in an oxidizing atmosphere (Zboril et. al., 2002). For instance, iron bacteria coated with ferrihydrite is known to precipitate iron as α-Fe_2O_3(Robbins et. al., 1991). A large host of literature exists on synthetic α-Fe_2O_3 with crystallographic surfaces, tunable through synthesis parameters such as temperature, pressure, additives, pH, and so on etc (Schwertmann U. and Cornell R. M.2000). Many of these variables are also known to affect the sorption reactivity, surface area, particle size, point of zero charge and color (Cornell R M and Schwertmann U. 2007). These variables also affect size and shape of the particles (Liang et. al., 2006). For instance, a surfactant mediated hydrothermal synthesis results in uniform nanocrystalline α-Fe_2O_3(Liu et. al., 2007).

The iron oxide polymorphs, more particularly, their nanoparticles, have unique magnetic, catalytic, optical, sorption, and other properties (Machala et. al., 2007). Decreasing size to less than 20 nm, makes α-Fe_2O_3 nanoparticles smaller than a single magnetic domain and shows superparamagnetic behavior (Parker et. al., 1993; del Monte et. al., 1997). Decrease in size also brings down the TM and TN, with TM completely vanishing between 8–20 nm (Xue et. al., 2003). Above 950 K (TN of α-Fe_2O_3), α-Fe_2O_3 loses its magnetic ordering and becomes a paramagnet (Machala et. al., 2007). Nanoscale particles of α-Fe_2O_3 have been reported to trigger different toxicological reaction pathways rather than microscale particles (Bhattacharya et. al., 2012).

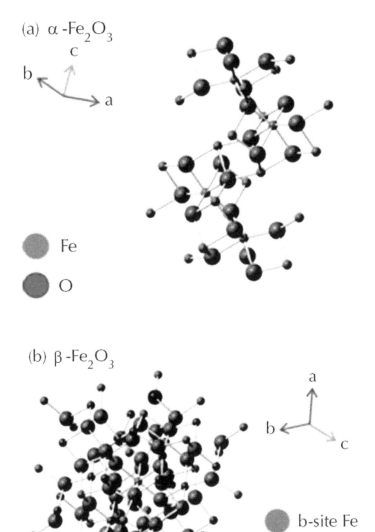

(a) α-Fe_2O_3

c
b
a

Fe

O

(b) β-Fe_2O_3

a
b
c

b-site Fe

d-site Fe

O

FIGURE 1 *(Continued)*

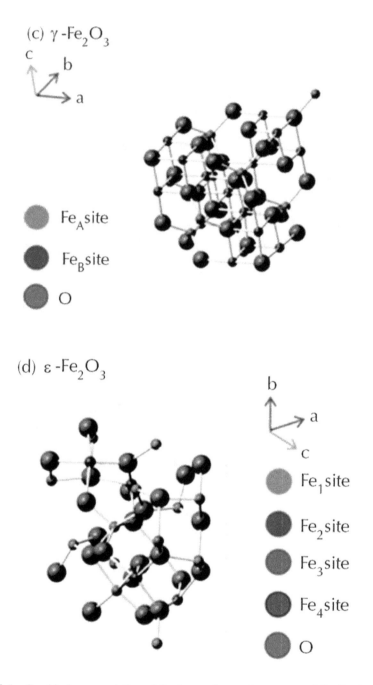

FIGURE 1 Graphical representation of fundamental crystal structures of Fe_2O_3 (Reprinted with permission from Polymorphous Transformations of Nanometric Iron (III) Oxide: A Review, Libor Machala et. al., *Chemistry of Materials*, 2011, 23, 3255. Copyright (2011) American Chemical Society)

4.3 SIZE TUNING IN HEMATITE

It is a big challenge to develop simple and reliable synthesis strategies for nano-structures with controlled morphologies, which strongly influences the application and specific properties of the metal oxide (Huber D. L. 2005). α-Fe_2O_3 nanostructures compared to the bulk, have been used in heterogeneous catalysis, alkylation of phenols, selective catalysis of cyclohexane oxidation, dye solar cells, electrolysis of water, gas sensors, photo catalysts, transistors, controlled drug delivery, and so forth (Agarwala et. al., 2012). α-Fe_2O_3 nanoparticles coated with gold nanoparticles are used as catalysts in CO oxidation (Zhong et. al., 2007). Until now, only a few reports are concerned with the synthesis of α-Fe_2O_3 nanoparticles, especially through methodologies with a potential for scale up. Colloidal chemistry has offered scope to create nanostructures with high mono-dispersity, well-controlled sizes, unique shapes, and complex structures. Compared to top-down approaches like lithography, bottom up approaches such as those based on colloidal chemistry have advantages to achieve small particle size and unmatched structural complexity (Lin X.-M. and Samia A. C. S. 2006b). Based on bottom up approach, α-Fe_2O_3 nanostructures have been synthesized in various shapes such as nanowires, nanorods, nanotubes, hollow fibres, nanorings, and cubes (Woo et. al., 2003; Wen et. al., 2004; Jia et. al., 2005; Jones et. al., 2005; Zhan et. al., 2007; Agarwala et. al., 2012).

Methods of synthesis includes, but not limited to:

a) Chemical precipitation such as mixing of ferric or ferrous salts with alkali to form goethite nanoparticles and subsequent conversion into α-Fe_2O_3 nanoparticles through thermal transformation (Fan et. al., 2005),

b) Forced hydrolysis of $FeCl_3$ using HCl, resulting in direct transformation of amorphous iron oxides to crystalline α-Fe_2O_3 (Wang et. al., 2008)

c) Microemulsion method based on ferrihydride and Fe(II) catalyst (Han et. al., 2011), and

d) Through sol-gel process, such as with ethylene oxide and $FeCl_3$ as starting material (Dong W. and Zhu C. 2002).

However, the uniformity of size of the nanoparticles obtained by most of these methods was relatively poor (Wang et. al., 2007). Modification to these methods includes:

a) Microwave irradiation based hydrolysis of iron salt using urea(Daichuan et. al., 1995) or ethylenediamine (Ni et. al., 2012)

b) Hydrothermal treatment followed by calcinations (Zhang et. al., 2012),

c) Alcoholysis of ferric ion under solvothermal condition(Lian et. al., 2012),

d) Reverse microemulsion technique employing water, chloroform, 1-butanol, and a surfactant (Housaindokht M. R. and Pour A. N. 2011),

e) Employing hydrothermal methods with amino acids as morphology control agent (Wang et. al., 2011), and so on .

These modifications in the synthesis routes can provide size reduction as well as improved physiochemical properties of the nanoparticles.

4.4   SHAPE TUNING IN HEMATITE

Materials with directional properties are opening new avenues in nanomaterial research. Synthesis of nanomaterial's with anisotropic morphologies so as to incorporate them directly into devices has enthused researchers worldwide (Hall S R. 2009). Particle anisotropy produces great changes in properties that are difficult to obtain simply by size tuning of spherical nanoparticles. While 0D or spherical particles cannot be easily tuned owing to the confinement of electrons to the same extent in all dimensions, 1, 2, or, 3D nanostructures where the electron motions are possible in more than one dimension are easily tunable (Sajanlal et. al., 2011). Understanding the interactions between solids, interfacial reactions and kinetics, and solution or, vapor chemistry is essential in developing methodologies for particle shape control. Shape controllable synthesis of nanomaterial's is of great interest and is actively pursued. Efforts on precise control of shape is poised towards potential scale-dependent applications in catalysis, drug delivery, active material encapsulation, ionic intercalation, light weight fillers, surface functionalization, energy storage, and so on etc (Lian et. al., 2009). Shape controlled synthesis involves processes such as seed-mediated (Murphy et. al., 2005) (synthesis of seed → growth of seed in the presence of shaping agent), polyol synthesis (Lee et. al., 2011) (reduction of inorganic salt by polyol at elevated temperature), biological (Huang et. al., 2009) (growth under constrained environments such as peptides), hydro/solvothermal synthesis (Kumar S. and Nann T. 2006) (synthesis in hot water/solvent in autoclave under high pressure), galvanic replacement reactions (Dement'eva O. V. and Rudoy V. M. 2011) (spontaneous reduction in the absence of an electric field), photochemical reactions (Khomutov G. B. 2004) (use of radiolytic or photochemical methods for reduction), electrochemical methods (Huang et. al., 2010) (bulk metal as sacrificial anode is oxidized and the metal cations migrate to cathode and get reduced in the presence of a stabilizing agent), and template mediated synthesis (Hall S R. 2009) (easy fabrication, low cost, high through-put, and adaptability). Significantly, large numbers of these methods are best suited for noble metal nanoparticles, with limited scope for metal oxide nanoparticles, resulting in lower options for their morphology control.

Based on the above methods, researchers have developed many approaches for the preparation of α-Fe_2O_3 with different morphologies.

These include:

a)   Hierarchical mesoporous microspheres,
b)   Acicular nanoparticles,
c)   Nanocubes,
d)   Tube-in-tube structures,
e)   Hollow spheres,
f)   Cigar shapes,
g)   Dendric forms,
h)   Urchin like,

i) Flower like,
j) Nanorods,
k) Nanowires,
l) Nanobelts,
m) Nanochains,
n) Spindles, and so on etc (Matijevic E. 1993; Suber et. al., 1999; Ngo A. T. and Pileni M. P. 2003; Cao et. al., 2005; Zhu et. al., 2006; Jia et. al., 2007; Li et. al., 2007; Zeng et. al., 2008; Lv et. al., 2010; Lv et. al., 2011; Van et. al., 2012).

Most of these approaches advantageously utilize the phase transfer process that occurs when akaganeite (β-FeOOH) synthesized by hydrolysis of FeCl$_3$ is transformed to α-Fe$_2$O$_3$ by heat treatment. Introduction of an organic or inorganic additive during the phase transfer process enables the confinement of growth of the products in specific directions through selective adsorption on specific crystal surfaces (Jia B. and Gao L. 2008). For instance, sodium dodecyl benzene sulfonate can cooperate with Fe^{3+} through electrostatic interactions to form a yellow Fe^{3+}-DBS complex in water solution, which then decomposes during hydrothermal process. The free DBS groups adsorb preferentially on certain surface planes of the freshly formed nanocrystals to confine the growth progress, thus modifying the geometry and size (Yan-yan et. al., 2011). However, as in the case of size, it is still a challenge to produce large homogeneous quantities of various shapes of nanoparticles with similar morphologies. Ionic liquid assisted hydrothermal synthesis of α-Fe$_2$O$_3$ in various morphologies has been reported by, (Lian et. al., 2009) recently. In this method, an ionic liquid, 1-n-butyl-3-methylimidazolium chloride was used as soft template, in a two part reaction, whose mechanism was illustrated as:

$$CH_3COO^- + H_2O \Leftrightarrow CH_3COOH + OH^- \tag{1}$$

$$Fe^{3+} + 3OH^- \rightarrow Fe(OH)_3 / FeOOH \xrightarrow[\text{Ionic liquid template}]{\text{hydrothermal condition}} \alpha - Fe_2O_3 \tag{2}$$

$$\alpha - Fe_2O_3 + 6CH_3COOH \Leftrightarrow 2Fe^{3+} 6CH_3COO^- + 6H_2O \tag{3}$$

While in the second step, α-Fe$_2$O$_3$ nanoparticles were formed, in the third step, to reduce the total surface energy, the core dissolves in the presence of excess CH-3COOH, while a shell of α-Fe$_2$O$_3$ nanoparticles keeps forming on the dissolving core, resulting in hollow spheres. The schematic diagram of the reaction (Figure 2) is a clear indication of the relevance of template in the synthesis of anisotropic nanomaterials.

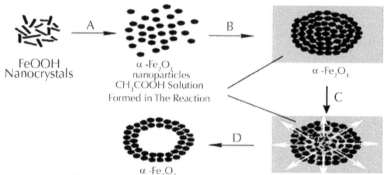

α -Fe$_2$O$_3$
Mesoporous Hollow Microspheres
A: Hydrothermal Condition; B: Self-assembly in the presence of [bmim]Cl;
C: Dissolution-Diffusion in the presence of CH3COOH; D: Ostwald Ripening

FIGURE 2 Schematic illustration for the growth process of the hematite (α-Fe$_2$O$_3$) mesoporous hollow microspheres by ionic liquid based template (Reprinted with permission from hHematite (α-Fe$_2$O$_3$) with various morphologies: Ionic liquid-assisted synthesis, formation mechanism, and properties, *ACS Nano*, 2009, 3, 3749. Copyright (2009) American Chemical Society).

4.4.1 TEMPLATE LLESS SYNTHESIS

During the past few years, template-free methods such as hydrothermal/solvothermal synthesis, based on mechanisms such as oriented attachments(Liu B. and Zeng H. C. 2004b), Kirkendall effects (Liu B. and Zeng H. C. 2004a) and Ostwald ripening have been developed to synthesize nanoparticles. Hydrothermal synthesis based on FeCl$_3$ and CH3COONH4 has been reported. In this method by appropriate choice of reagent concentration, reaction temperature and time, nanocubes of Fe$_2$O$_3$ could be generated as per the following mechanism.

$$3CH_3COONH_4 + FeCl_3 \rightarrow Fe(CH_3COO)_3 + 3NH_4Cl$$

$$Fe(CH_3COO)_3 + 2H_2O \rightarrow Fe(OH)_2(CH_3COO)\downarrow +2(CH_3COOH)$$

$$Fe(OH)_2(CH_3COO) \rightarrow \beta - FeOOH + CH_3COOH$$

$$2\beta - FeOOH \rightarrow \alpha - Fe_2O_3 + H_2O$$

Overall reaction can be summarized as

$$Fe^{3+}(l) \rightarrow Fe(OH)_2CH_3COO(s) \rightarrow \beta - FeOOH(s) \rightarrow \alpha - Fe_2O_3(s)$$

Solvothermal reaction of ethanolic solution of iron acetate in the presence of L-Lycine to produce magnetite, which can subsequently be converted by heat treatment to α-Fe_2O_3 hollow spheres, has been reported. Such hollow spheres have applications in catalysis, gas sensors, and lithium ion batteries (Kim et. al., 2011). However, in many cases the synthesis of nanoparticles in the absence of template not provided good control over the morphology of nanoparticles.

4.4.2 TEMPLATE SYNTHESIS

Substrates that have their surfaces modified to have active sites which can induce nanoparticle deposition are generally called as templates. Examples include porous alumina, polymer nanotubes, and patterned catalysts to control the oriented growth (Routkevitch et. al., 1996; Cao et. al., 2001; Huang et. al., 2001). They can be smaller or larger in size than the nanoparticle deposited. It can also be considered as a scaffold wherein the particles can be arranged into a structure with a morphology that is complementary to that of the template (Grzelczak et. al., 2010). Examples of templates include single molecules, microstructures, or block copolymers. They are classified as soft and hard (Nie et. al., 2010). In a soft-template method, growth occurs by chemical or electrochemical reduction, usually in the presence of a surfactant or structure directing molecule. The template governs growth of certain faces or structures. Polymer-surfactant complexes are promising templates. At first, the surfactant associated with the polymer chain form micellar aggregates at a critical association concentration (much lower than the critical micelle concentration of the surfactant in water). Secondly, both polymer and surfactant are good crystal stabilizers, preventing their aggregation and at the same time connecting them together into loose flowcs through a bridging mechanism (Leontidis et. al., 2001). Finally, through electrostatic, hydrophobic and van der Waals interactions between polymer and the surfactant, stable structures are generated, which serve as ideal soft templates, providing monodisperse nanoparticles. Soft template methods include electrochemical reduction, seeding followed by chemical reduction, redox reactions, selective etching, and so on etc. Soft-template technique offers advantage of scalability (Kline et. al., 2006). In hard-template method, a porous membrane of inorganic or polymeric material serves as a rigid mold for chemical or electrochemical replication of structure. This method provides an easy manner for production of 1-D nanostructures, but with difficulties of scale up. Hard templates such as silica or carbon spheres are also ideal for synthesis of hollow structures (Chen et. al., 2003). Classical examples where the template enables the control of morphology of α-Fe_2O_3 nanoparticles can be found in literature (Table 1).

TABLE 1   Examples of template enabled morphology control in α-Fe_2O_3 nanoparticles

TEMPLATE	FEATURES
Carbon nanotubes (Lu et. al., 2008)	Nanochainsweakly ferromagnetic in room temperature, antiferromagnetic below TM
Amino acids (Cao et. al., 2008)	Nanocubesdouble hydrophilic functional groups -NH2 and –COOH to control growth
Carbon spheres (Jagadeesan et. al., 2008)	Hollow spheres to nanocupsvariable magnetic properties – calcination before gel stage results in spheres and calcination after thixotropic gel formation in cups.
Sulfonatedpolystyrene microspheres (Zhang et. al., 2007)	large scale monodisperse urchin like hollow microspheres and high remanent magnetization
Anionic surfactant as a rod-like template (Mandal S. and Muller A. H. E. 2008)	Porous structureweak ferromagnetic behavior
TiO_2 nanotubes	1D Fe_2O_3 TiO_2nanorod-nanotube arrays
Polycarbonate/Alumina (Zhou H. and Wong S. S. 2008)	Nanotubes to amorphous hydroxides
Ferritin protein cage (Klem et. al., 2010)	Boiling aqueous solution by refluxing the ferritin protein cage/ferrihydrite composite
PVA (Mahmoudi et. al., 2010)	Oriented growth
PEG (Sreeram et. al., 2011)	Necklace shapedsacrificial template
PEG + CTAB (Liu et. al., 2005)	Shuttle like structures
ZnO nanowire arrays (Liu et. al., 2010)	Iron oxide-based nanotube arrayssacrificial template
MCM-48 type silica support and wide-pore silica gel (Surowiec et. al., 2011)	Ordered porous structure

TABLE 1 *(Continued)*

FDU-1 type cubic ordered mesoporous silica (Martins et. al., 2010)	Sacrificial template
DNA (Sarkar et. al., 2011)	Chain like structures ntiferro to ferromagnetic transition at a temperature similar to 240 K
Polymethyl methacrylate as imprint template (Zhang et. al., 2011)	Hierarchically mesoporous and macroporous hematite Fe_2O_3
Protonated triethylenetetramine (Zhou et. al., 2012)	Sacrificial templatenanoribbons
Greek wood and mill scale wastes (Bantsis et. al., 2012)	Iron oxide ceramics
Cellulose films (Liu et. al., 2012)	Sacrificial, looped structures, and replicates template

However, there are also reports that the template synthesis results in nanoparticles with low crystallinity. From Table 1 it can be seen that the template is likely to be present as an impurity or as a part of a composite with α-Fe_2O_3, other than in cases where template is selectively sacrificed. Even when sacrificed, it would leave a residue on top of the nanoparticle.

4.5 GREEN SYNTHESIS

Green nanotechnology transforms existing processes and products to enhance environmental quality, reduce pollution and conserve natural and non-renewable resources. Nano -synthesis involving:

a) Cost effective and non-toxic precursors,
b) Negligible quantities of carcinogenic reagents or solvents,
c) Few number of reagents,
d) Lesser number of reaction steps and hence lesser waste generation, reagent use and power consumption,
e) Little or no byproducts, and
f) Room temperature synthesis under ambient conditions are considered ideal under green nanotechnology principles.

Such processes also need to be efficient in terms of scale-up (Mao et. al., 2007). Molten-salt synthesis method is one of the most versatile and cost effective green synthesis approaches. With NaCl as the reaction medium, large-scale production of single-crystalline α-Fe$_2$O$_3$ rhombohedra have been obtained from relatively polydisperse, polycrystalline and/or amorphous, commercially available starting precursor materials (Park T. -J. and Wong S. S. 2006). Reactor conditions where the autogenous pressure far exceeds ambient pressure, allows solvents to be brought to temperatures much above their boiling points. When chemical reactions are performed under such conditions, they are referred to as solvothermal synthesis. Reaction temperature, time, pH, solvent choice and concentration, additives, autoclave geometries, and so on etc. influence the size and shape of the synthesized nanoparticles (Michailovski A. and Patzke G. R.2006). Template assisted green synthesis is one of the environmentally progressive methodologies for nanoparticle synthesis. Most of these reactions can be run under ambient conditions, with reliable control over shape and dimensionality. A potential to develop generalized protocol for synthesis, capping and functionalization exists with template synthesis, which is more biologically relevant when biopolymers are employed as templates.

4.5.1   POLYSACCHARIDE TEMPLATES

Polysaccharides are a class of biopolymers with repeating units of mono- or disaccharides linked by glycosidic bonds. They can be linear or branched, have high variability of building block composition and physico-chemical properties (Dias et. al., 2011). They have predominant applications in the area of biomaterials (Rinaudo M. 2008). For a nanotechnologist, they are attractive candidates as stabilizing agents, functionalization moieties, drug delivery vehicles, and reducing cum capping agents in the synthesis of noble metal nanoparticles (Liu et. al., 2008; Sreeram et. al., 2008). Simple monosaccharide such as sucrose is a ready source of carbon in the synthesis of carbide nanoparticles (Yang et. al., 2004). Fibre like morphology of cellulose (1→4, glycoside linked β-glucose) is advantageously employed for replicating the fibre morphology on to the nanoparticles (Zheng et. al., 2007). Similarly, the branched structure of dextran enables formation of nanowires (Kong et. al., 2006). Starch, by itself, is a complex glucose polymer with inherent molecular anisotropy and the same is transferrable to other materials with ease (Shi et. al., 2007). Alginate is composed of blocks of poly-guluronate and poly-mannuronate and their ratio and distribution varies with source of seaweed (Schnepp et. al., 2008). By introducing metal cations to a solution of alginate, metallic species are preferentially taken up by poly-G species leading to controlled nucleation and growth. The cocooning effect of alginate prevents nanoparticle coalescence leading to uniform and homogeneous dispersion of nanoparticles (Ozin et. al., 2009). Finally, chitosan being cationic, forms stable complexes with anionic inorganics, which on calcination forms networked structures (Ogawa et. al., 1993).

In a series of works the ability of polysaccharide to aid nanoparticle synthesis when employed as templates were reported. Polysaccharides with varying charge such as cationic chitosan, anionic pectin/alginate, and neutral starch were chosen to form iron–polysaccharide complexes which were calcined to generate iron oxide nanoparticles (Figure 3).

The crystallographic phases matched well with that of hematite (α-Fe$_2$O$_3$). While Dynamic light scattering measurements indicated an average intensity average diameter of around 270 nm, the SEM images suggested spherical morphology for the nanoparticles generated on chitosan, alginate and starch (Nidhin et. al., 2008). In the case of pectin, a linear aggregation of spherical nanoparticles resulting in a rod like morphology was observed. Narrow particle size distribution was observed with starch. At the concentration range investigated in that study, the spatial separation of iron(II) centrescenters during the complexation process is expected to have provided the monodispersity to the nanoparticles.

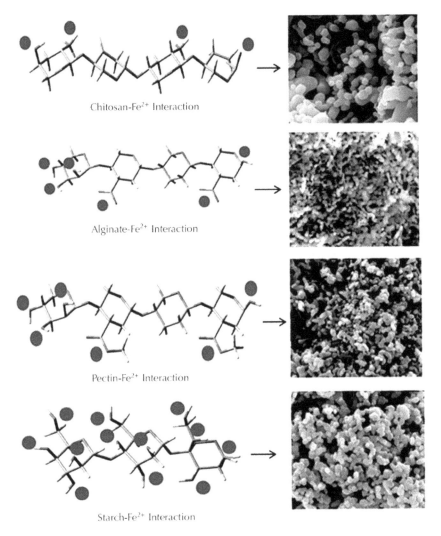

Chitosan-Fe^{2+} Interaction

Alginate-Fe^{2+} Interaction

Pectin-Fe^{2+} Interaction

Starch-Fe^{2+} Interaction

FIGURE 3 Fe (II)-Polysaccharide interaction and the formation of hematite (α-Fe$_2$O$_3$) nanoparticles.

Thermogravimetric analysis suggested that the iron-polysaccharide complexes were stable to degradation at high temperature, when compared to polysaccharides alone and that they were not completely removed even at temperatures of 800°C, leading to the presence of a residual carbon shell around the nanoparticle core. The authors by employing the Peniche method estimated the percentage of nanoparticle content in the core-shell as 5.4%.

The low size and monodispersity of starch templatedtemplate nanoparticles, conferred a higher level of water dispersible character when compared to other nanoparticles. The nanoparticles were not cytotoxic in the concentration range of 50 to 200 μM, possibly due to the carbonaceous shell. The nanoparticles were easily dispersed into medium employed in surface coating, leather finishing and plastic coloration, where a uniform distribution of the color at low concentrations of the nanoparticles was observed. The ability of the nanoparticles to disperse well into polymeric matrices, thus conferring thermal stability to the same was found advantageous in automobile coatings. Dispersibility was attributed to the carbon coating of polysaccharide template over the nanoparticle.

The study was also extended to CoO nanoparticles (Nidhin et. al., 2012a) employing a sacrificial starch template. This method provided uniform shape and size, and was easy to perform at bulk levels. The size distribution of the nanoparticles was in the range of 15–30 nm, with a crystallite size of 18 nm. The low size and aggregation free character of the nanoparticles resulted in a NIR reflectance of above 75%, in spite of a black color (where NIR reflectance values are expected to be less than 30%). This provided feasibility for employing CoO nanoparticles as components of cool coatings on leather, automobile surfaces, and so on etc.

With an aim to further reduce the size of the nanoparticles and to prove that the spatial distribution between the iron(II) centrescenters was responsible for the aggregation free synthesis of nanoparticles, the nanoparticles are synthesized on green templates such as chitosan polyion and blended films.

Through electrostatic interaction between cationic chitosan and anionic alginate/ pectin, self-assembled films were prepared. Similarly chitosan supported starch films were also obtained. Iron(II) in solution interacted with the functional groups in the film, following their adsorption on to polysaccharide films, which were subsequently calcined. Based on pores and the pore distribution, aligned α-Fe_2O_3 nanostructures were observed in the case of chitosan-alginate film template, rhombohedra shaped nanoparticle in the case of chitosan-pectin and spherical nanoparticles in the case of chitosan-starch film template (Figure 4). The weight percentage of the nanoparticles in core-shell dropped to 2.0%, indicating a reduction in size to as low as 2 nm (in the case alginate film). Here, again the biocompatibility of the nanoparticles was established using MTT assay.

Interaction of Fe(II)

with Chitosan-alginate
Polyion film

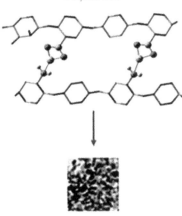

with Chitosan-pectin
Polyion film

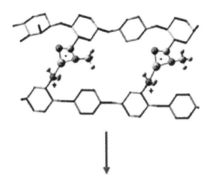

FIGURE 4   *(Continued)*

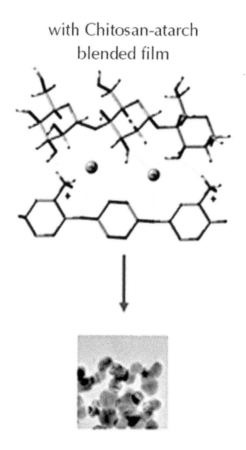

FIGURE 4   Schematic representation of Fe (II)-Polysaccharide films interaction (Source: Nidhin et. al., 2012 b Polysaccharide films as templates in the synthesis of hematite nanostructures with special properties, Applied Surface Science, 258(12), 5179 -5184.).

     The above two methods of synthesis of hematite nanoparticles on polysaccharide template alone and chitosan polyion and blended films were high energy consuming calcination reactions, where after the template removal, potential for aggregation cannot be ruled out. The synthesized nanoparticles by the above methods had also demonstrated super paramagnetic behavior ideal for biological applications (chitosan-alginate film template). The synthesis methodology was modified to a reflux process as against calcination so as not to remove the polysaccharide template by calcination. Nanoparticles were found to grow into rhombohedra with reflux duration. Possibly based on the ions adsorbed from the medium, the nanoparticles demonstrated fluorescence properties and were found ideal for stabilizing collagen in solution. Starch mediated interaction with amino acids in collagen, resulting in H-bonded networks provided a 3.1°C increase in denaturation temperature, over the native collagen (37.2°C).

The study was extended to cobalt ferrite, where incorporation of a surfactant such as CTAB, provided an opportunity for the assembly of the nanoparticles into flower shaped nanostructures with a particle size of around 25 ± 3 nm. This method also provided an opportunity to systematically study its morphology, crystallinity, particle size, magnetic properties, biocompatibility, cytotoxicity, and their biomedical applications such a contrast agents for MRI. The salient observation from this study was that the assembling of the nanoparticles into nano flowers was essential to utilize them as T2 contrast for MR imaging, while the nanoparticles in the absence of assembling process were neither a T1 nor T2 contrast agents. The reflux method of synthesis of anisotropic ferrite nanoparticles with uniform size and shape was found to be simple, low in reaction temperature, high yielding and low in cost of inorganic precursors.

4.6 SURFACE MODIFICATION OF HEMATITE NANOPARTICLES

Unique and advanced properties observed in nanoparticles are usually associated with bare (uncoated) nanoparticles. However, bare nanoparticles possess excessive surface energy and should be protected by way of capping or surface functionalization. In liquid media, unprotected metal and metal oxide nanoparticles are thermodynamically unstable and tend to spontaneously coalesce. In nanostructured hematite materials, long range magnetic dipole interactions can have strong influence on magnetic dynamics. When in close proximity, exchange bias which manifests as shift in hysteresis curves after field cooling, is generally observed. In α-Fe_2O_3, suppression of superparamagnetic relaxation is observed when particles are in close proximity, which is more in the case of uncoated particles than coated particles (Morup et. al., 2010).

A large number of synthesis methods have been developed to prevent the aggregation of nanoparticles. These include sonication by ultrasound, capping with polymers and surfactants, and so on etc. In several instances, equilibrium is established between nanoparticles, aggregates, dispersed particles and re-aggregated particles, resulting in a polydispersed system of particles. One of the drawbacks of these methods is their limited stability in aqueous solutions. Particles with a polymer coating, more so polymeric shell have more stability against aggregation because of large decrease in surface energy. Functionalization of hematite core with polymeric materials can result in three types of structures , viz, core-shell, matrix, and shell$_a$-Core-shell$_b$, as depicted in Figure 5. The choice of solvent for functionalization plays a crucial role in achieving sufficient repulsive interactions to prevent agglomeration. For instance, when functionalized with surfactants, the functionalized nanoparticles can be divided as oil-soluble, water-soluble, and amphiphilic. This division is based on the nature of interaction of the surfactant with the solvent employed.

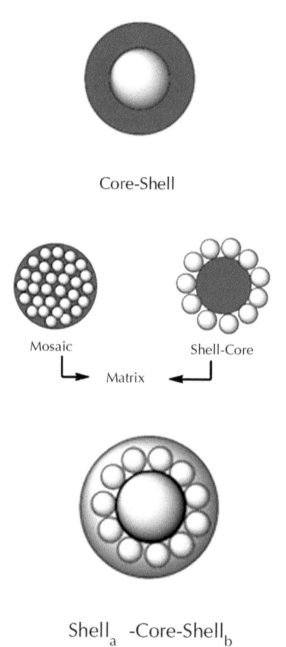

FIGURE 5 Representative structure of organic materials functionalized hematite nanoparticles (if iron oxide nanoparticles always assumed as the core). Reprinted with permission from Magnetic iron oxide nanoparticles: synthesis and surface functionalization strategies, Wei Wu et. al.,, *Nanoscale Research Letters*, 2008, 3, 397. Copyright (2008) Springer).

Accordingly with water as solvent, fatty acid, or alkyl phenol functionalized nanoparticles are oil-soluble, polyol, or lysine coated particles are water-soluble and sulfuric lysine coated particles are amphiphilic. Stabilization and functionalization of nanoparticles in an environmentally benign manner are among one of the goals of green nanotechnology. Such agents, which are from a natural/renewable source, can function as reducing/capping/stabilizing agents, can be effectively used for functionalization. Interaction of nanoparticles with variety of compounds found in nature which provides good control over the size and morphology are highly desirable for various applications of nanotechnology (Virkutyte J. and Varma R. S. 2011).

One of the best ways by which aggregation can be avoided is surface passivation (Bakunin et. al., 2004). The largest class of compounds explored in nano synthesis for passivation is biopolymers. They are produced by living organisms and consist of simple biological compounds. They are generally renewable, non-toxic and biodegradable (Satyanarayana et. al., 2009; Sundar et. al., 2010). Metal and metal oxide nanoparticles when dispersed into biopolymers can overcome many of the short comings of bare nanoparticles such as aggregation and reduction in surface area to volume ratio, without affecting the properties of parent nanoparticles. Appropriate choice of biopolymers would also provide functional groups and enhancement of properties (Sarkar et. al., 2012). Donor or acceptor species that are bound to the hematite nanoparticle surface such as polysaccharides are finding increasing relevance in surface passivation. A detailed biotechnological perspective on the applications of iron oxide nanoparticles modified with polysaccharides can be found in the work of Dias et. al., 2011. Polysaccharides, can be neutral (agarose, dextran, pullalan and starch), negative (alginate, carrageenans, gum Arabic, and heparin), or positively charged (chitosan). They carry functional groups such as OH, COO^-, OSO^{3-}, COO^-, and NH^{3+}. They act as recognition markers in several biological processes (Park et. al., 2008). Chemical co-precipitation is the preferred method of in-situ coating, while encapsulation, microemulsion, hydrothermal treatment, covalent bonding, adsorption or sonication is the preferred post-synthesis coating methods. The biomedical applications of such coated iron oxide nanoparticles include cell tracking, drug delivery, detection of emboli, liver tumor, liver lesion, and so on by MRI, cardiovascular applications and hyperthermia.

CONCLUSION

Iron oxide nanoparticles are undoubtedly one of the most investigated nanoparticles owing to their importance in industrial and medical applications. Hematite nanoparticles are comparatively more stable and therefore have wider applications as well. While several methods are available for the synthesis of nanoparticles, the green synthesis routes are preferred, more so in biological applications. It has been demonstrated that natural products such as polysaccharides can be employed as effective templates, both through sacrificial and otherwise routes to generate monodispersed nanoparticles of below 20 nm. Such methods are easy to adopt for other ferrite nanoparticles as well. Incidentally, some of the polysaccharide methods are facile and possibly simple for replication at industrial scales.

KEYWORDS

- **Green nanotechnology transforms**
- **Iron oxide nanoparticles**
- **Nanotechnology**
- **Polysaccharides**
- **Thermogravimetric analysis**

REFERENCES

1. Agarwala, S., Lim, Z. H., Nicholson, E., and Ho, G. W. Probing the morphology-device relation of Fe2O3 nanostructures towards photovoltaic and sensing applications. *Nanoscale*, **4**(1), 194–205 (2012).

2. Aitken, R. J., Chaudhry, M. Q., Boxall, A. B. A., and Hull, M. Manufacture and use of nanomaterials: current status in the UK and global trends. *Occupational Medicine-Oxford*, **56**(5), 300–306 (2006).

3. Bakunin, V. N., Suslov, A. Y., Kuzmina, G. N., and Parenago, O. P. Synthesis and application of inorganic nanoparticles as lubricant components - a review. *Journal of Nanoparticle Research*, **6**(2–3), 273–284 (2004).

4. Bantsis G., Betsiou M., Bourliva A., Yioultsis T. and Sikalidis C. Synthesis of porous iron oxide ceramics using Greek wooden templates and mill scale waste for EMI applications. *Ceramics International*, **38**(1), 721–729 (2012).

5. Berkowitz, A. E., Mitchell, J. R., Carey, M. J., Young, A. P., Zhang, S., Spada, F. E., Parker, F. T., Hutten, A. and Thomas, G. Giant magnetoresistance in heterogeneous Cu-Co alloys. *Physical Review Letters*, **68**(25), 3745–3748 (1992).

6. Bhattacharya, K., Hoffmann, E., Schins, R. F. P., Boertz, J., Prantl, E. M., Alink, G. M., Byrne, H. J., Kuhlbusch, T. A. J., Rahman, Q., Wiggers, H., Schulz, C., and Dopp, E. Comparison of Micro- and Nanoscale Fe3+-Containing (Hematite) Particles for Their Toxicological Properties in Human Lung Cells in Vitro. *Toxicological Sciences*, **126**(1), 173–182 (2012).

7. Bucher, J. P., Douglass, D. C., and Bloomfield, L. A. Magnetic properties of free cobalt clusters. *Physical Review Letters*, **66**(23), 3052–3055 (1991).

8. Cao, H. Q., Wang, G. Z., Warner, J. H., and Watt, A. A. R. Amino-acid-assisted synthesis and size-dependent magnetic behaviors of hematite nanocubes. *Applied Physics Letters*, **92**(1), 3 (2008).

9. Cao, H. Q., Xu, Z., Sang, H., Sheng, D., and Tie, C. Y. Template Synthesis and Magnetic Behavior of an Array of Cobalt Nanowires Encapsulated in Polyaniline Nanotubules. *Advanced Materials*, **13**(2), 121–123 (2001).

10. Cao, M., Liu, T., Gao, S., Sun, G., Wu, X., Hu, C., and Wang, Z. L. Single-Crystal Dendritic Micro-Pines of Magnetic α-Fe2O3: Large-Scale Synthesis, Formation Mechanism, and Properties. *Angewandte Chemie International Edition*, **44**(27), 4197–4201 (2005).

11. Chen, D., Chen, D., Jiao, X., and Zhao, Y. Hollow-structured hematite particles derived from layered iron (hydro)oxyhydroxide-surfactant composites. *Journal of Materials Chemistry*, **13**(9), 2266–2270 (2003).

12. Chen, J. S., Zhu, T., Yang, X. H., Yang, H. G., and Lou, X. W. Top-Down Fabrication of a-Fe2O3 Single-Crystal Nanodiscs and Microparticles with Tunable Porosity for Largely Improved Lithium Storage Properties. *Journal of the American Chemical Society*, **132**(38), 13162–13164 (2010).

13. Cherepy, N. J., Liston, D. L., Lovejoy, J. A., Deng, H., and Zhang, J. Z. Ultrafast Studies of Photoexcited Electron Dynamics in α- and ☒-Fe2O3 Semiconductor Nanoparticles. *The Journal of Physical Chemistry B*, **102**(5), 770–776 (1998).

14. Cornell, R. M. and Schwertmann, U. *The Iron Oxides: Structure, Properties, Reactions, Occurrences and Uses*. John Wiley & Sons (2007).

15. Daichuan, D., Pinjie, H., and Shushan, D. Preparation of uniform a-FeO(OH) colloidal particles by hydrolysis of ferric salts under microwave irradiation. *Materials Research Bulletin*, **30**(5), 537–541 (1995).

16. del Monte, F., Morales, M. P., Levy, D., Fernandez, A., Oca, M., Roig, A., Molins, E., O'Grady, K., and Serna, C. J. Formation of a-Fe2O3 Isolated Nanoparticles in a Silica Matrix. *Langmuir*, **13**(14), 3627–3634 (1997).

17. Dement'eva, O. V. and Rudoy, V. M. Colloidal synthesis of new silver-based nanostructures with tailored localized surface plasmon resonance. *Colloid Journal*, **73**(6), 724–742 (2011).

18. Dias, A. M. G. C., Hussain, A., Marcos, A. S., and Roque, A. C. A. A biotechnological perspective on the application of iron oxide magnetic colloids modified with polysaccharides. *Biotechnology Advances*, **29**(1), 142–155 (2011).

19. Dong, W. and Zhu, C. Use of ethylene oxide in the sol-gel synthesis of α-Fe2O3 nanoparticles from Fe(iii) salts. *Journal of Materials Chemistry*, **12**(6), 1676–1683 (2002).

20. Fan, H., Song, B., Liu, J., Yang, Z., and Li, Q. Thermal formation mechanism and size control of spherical hematite nanoparticles. *Materials Chemistry and Physics*, **89**(2–3), 321–325(2005).

21. Fullerton, E. E., Jiang, J. S., Sowers, C. H., Pearson, J. E., and Bader, S. D. Structure and magnetic properties of exchange-spring Sm-Co/Co superlattices. *Applied Physics Letters*, **72**(3), 380–382 (1998).

22. Grzelczak, M., Vermant, J., Furst, E. M., and Liz-Marzaìn, L. M. Directed Self-Assembly of Nanoparticles. *ACS Nano*, **4**(7), 3591–3605 (2010).

23. Hall, S. R. Biotemplated syntheses of anisotropic nanoparticles. *Proceedings of the Royal Society a-Mathematical Physical and Engineering Sciences*, **465**(2102), 335–366 (2009).

24. Han, L. H., Liu, H., and Wei, Y. In situ synthesis of hematite nanoparticles using a low-temperature microemulsion method. *Powder Technology*, **207**(1â€"3), 42–46 (2011).

25. Housaindokht, M. R. and Pour, A. N. Precipitation of hematite nanoparticles via reverse microemulsion process. *Journal of Natural Gas Chemistry*, **20**(6), 687–692 (2011).

26. Huang, M. H., Mao, S., Feick, H., Yan, H. Q., Wu, Y. Y., Kind, H., Weber, E., Russo, R., and Yang, P. D. Room-temperature ultraviolet nanowire nanolasers. *Science*, **292**(5523), 1897–1899 (2001).

27. Huang, X., Neretina, S., and El-Sayed, M. A. Gold Nanorods: From Synthesis and Properties to Biological and Biomedical Applications. *Advanced Materials*, **21**(48), 4880–4910 (2009).

28. Huang, X., Qi, X., Huang, Y., Li, S., Xue, C., Gan, C. L., Boey, F., and Zhang, H. Photochemically Controlled Synthesis of Anisotropic Au Nanostructures: Platelet-like Au Nanorods and Six-Star Au Nanoparticles. *ACS Nano*, **4**(10), 6196–6202 (2010).

29. Huber, D. L. Synthesis, Properties, and Applications of Iron Nanoparticles. *Small*, **1**(5), 482–501 (2005).

30. Jagadeesan, D., Mansoori, U., Mandal, P., Sundaresan, A., and Eswaramoorthy, M. Hollow Spheres to Nanocups: Tuning the Morphology and Magnetic Properties of Single-Crystalline α-Fe2O3 Nanostructures. *Angewandte Chemie International Edition*, **47**(40), 7685–7688 (2008).

31. Jia, B. and Gao, L. Growth of Well-Defined Cubic Hematite Single Crystals: Oriented Aggregation and Ostwald Ripening. *Crystal Growth & Design*, **8**(4), 1372–1376 (2008).

32. Jia, C. J., Sun, L. D., Yan, Z. G., Pang, Y. C., You, L. P., and Yan, C. H. Iron Oxide Tube-in-Tube Nanostructures. *The Journal of Physical Chemistry C*, **111**(35), 13022–13027 (2007).

33. Jia, C. J., Sun, L. D., Yan, Z. G., You, L. P., Luo, F., Han, X. D., Pang, Y. C., Zhang, Z., and Yan, C. H. Single-Crystalline Iron Oxide Nanotubes. *Angewandte Chemie International Edition*, **44**(28), 4328–4333(2005).

34. Jones, N. O., Reddy, B. V., Rasouli, F., and Khanna, S. N. Structural growth in iron oxide clusters: Rings, towers, and hollow drums. *Physical Review B*, **72**(16), 165411 (2005).

35. Jung, S., Watkins, B., DeLong, L., Ketterson, J. B., and Chandrasekhar, V. Ferromagnetic resonance in periodic particle arrays. *Physical Review B*, **66**(13), 132401 (2002).

36. Khomutov, G. B. Interfacially formed organized planar inorganic, polymeric and composite nanostructures. *Advances in Colloid and Interface Science*, **111**(1–2), 79–116 (2004).

37. Kim, H. J., Choi, K. I., Pan, A., Kim, I. D., Kim, H. R., Kim, K. M., Na, C. W., Cao, G., and Lee, J. H. Template-free solvothermal synthesis of hollow hematite spheres and their applications in gas sensors and Li-ion batteries. *Journal of Materials Chemistry*, **21**(18), 6549–6555 (2011).

38. Klem, M. T., Young, M., and Douglas, T. Biomimetic synthesis of photoactive alpha-Fe2O3 templated by the hyperthermophilic ferritin from Pyrococus furiosus. *Journal of Materials Chemistry*, **20**(1), 65–67 (2010).

39. Kline, T. R., Tian, M., Wang, J., Sen, A., Chan, M. W. H., and Mallouk, T. E. Template-Grown Metal Nanowires. *Inorganic Chemistry*, **45**(19), 7555–7565 (2006).

40. Kong, R., Yang, Q., and Tang, K. A Facile Route to Silver Nanowires. *Chemistry Letters*, **35**(4), 402–403 (2006).

41. Kumar, S. and Nann, T. Shape control of II-VI semiconductor nanomateriats. *Small*, **2**(3), 316–329 (2006).

42. Lee, G., Cho, Y. S., Park, S., and Yi, G. R. Synthesis and assembly of anisotropic nanoparticles. *Korean Journal of Chemical Engineering*, **28**(8), 1641–1650 (2011).

43. Leontidis, E., Kyprianidou-Leodidou, T., Caseri, W., Robyr, P., Krumeich, F., and Kyriacou, K. C. From Colloidal Aggregates to Layered Nanosized Structures in Polymer-Surfactant Systems. 1. Basic Phenomena. *The Journal of Physical Chemistry B*, **105**(19), 4133–4144(2001).

44. Li, L., Chu, Y., Liu, Y., and Dong, L. Template-Free Synthesis and Photocatalytic Properties of Novel Fe2O3 Hollow Spheres. *The Journal of Physical Chemistry C*, **111**(5), 2123–2127 (2007).

45. Lian, J., Duan, X., Ma, J., Peng, P., Kim, T., and Zheng, W. Hematite (α-Fe2O3) with Various Morphologies: Ionic Liquid-Assisted Synthesis, Formation Mechanism, and Properties. *ACS Nano*, **3**(11), 3749–3761 (2009).

46. Lian, S. Y., Li, H. T., He, X. D., Kang, Z. H., Liu, Y., and Lee, S. T. Hematite homogeneous core/shell hierarchical spheres: Surfactant-free solvothermal preparation and their improved catalytic property of selective oxidation. *Journal of Solid State Chemistry*, **185**, 117–123 (2012).

47. Liang, X., Wang, X., Zhuang, J., Chen, Y. T., Wang, D. S., and Li, Y. D. Synthesis of nearly monodisperse iron oxide and oxyhydroxide nanocrystals. *Advanced Functional Materials*, **16**(14), 1805–1813 (2006).

48. Lin, X. M. and Samia, A. C. S. Synthesis, assembly and physical properties of magnetic nanoparticles. *Journal of Magnetism and Magnetic Materials*, **305**(1), 100–109 (2006a).

49. Lin, X. M. and Samia, A. C. S. Synthesis, assembly and physical properties of magnetic nanoparticles. *Journal of Magnetism and Magnetic Materials*, **305**(1), 100–109 (2006b).

50. Lin, Y. J., Xu, Y., Mayer, M. T., Simpson, Z. I., McMahon, G., Zhou, S., and Wang, D. W. Growth of p-Type Hematite by Atomic Layer Deposition and Its Utilization for Improved Solar Water Splitting. *Journal of the American Chemical Society*, **134**(12), 5508–5511 (2012).

51. Liu, B. and Zeng, H. C. Fabrication of ZnO Dandelions via a Modified Kirkendall Process. *Journal of the American Chemical Society*, **126**(51), 16744–16746 (2004a).

52. Liu, B. and Zeng, H. C. Mesoscale Organization of CuO Nanoribbons. *Journal of the American Chemical Society*, **126**(26), 8124–8125 (2004b).

53. Liu, J. P., Li, Y. Y., Fan, H. J., Zhu, Z. H., Jiang, J., Ding, R. M., Hu, Y. Y., and Huang, X. T. Iron Oxide-Based Nanotube Arrays Derived from Sacrificial Template-Accelerated Hydrolysis: Large-Area Design and Reversible Lithium Storage. *Chemistry of Materials*, **22**(1), 212–217 (2010).

54. Liu, S., Tao, D., and Zhang, L. Cellulose scaffold: A green template for the controlling synthesis of magnetic inorganic nanoparticles. *Powder Technology*, **217**(0), 502–509 (2012).

55. Liu, X. H., Qiu, G. Z., Yan, A. G., Wang, Z., and Li, X. G. Hydrothermal synthesis and characterization of α-FeOOH and α-Fe2O3 uniform nanocrystallines. *Journal of Alloys and Compounds*, **433**(1–2), 216–220 (2007).

56. Liu, Z. H., Jiao, Y. P., Wang, Y. F., Zhou, C. R., and Zhang, Z. Y. Polysaccharides-based nanoparticles as drug delivery systems. *Advanced Drug Delivery Reviews*, **60**(15), 1650–1662 (2008).

57. Lu, H. B., Liao, L., Li, J. C., Shuai, M., and Liu, Y. L. Hematite nanochain networks: Simple synthesis, magnetic properties, and surface wettability. *Applied Physics Letters*, **92**(9), 3 (2008).

58. Lu, H. M. and Meng, X. K. Morin Temperature and Neĺ☐el Temperature of Hematite Nanocrystals. *The Journal of Physical Chemistry C*, **114**(49), 21291–21295 (2010).

59. Lv, B., Liu, Z., Tian, H., Xu, Y., Wu, D., and Sun, Y. Single-Crystalline Dodecahedral and Octodecahedralα-Fe2O3 Particles Synthesized by a Fluoride Anion–Assisted Hydrothermal Method. *Advanced Functional Materials*, **20**(22), 3987–3996 (2010).

60. Lv, B., Xu, Y., Wu, D., and Sun, Y. Single-crystal [small alpha]-Fe2O3 hexagonal nanorings: stepwise influence of different anionic ligands (F- and SCN- anions). *Chemical Communications*, **47**(3), 967–969 (2011).

61. Machala, L., Zboril, R., and Gedanken, A. Amorphous Iron(III) OxideA Review. *The Journal of Physical Chemistry B*, **111**(16), 4003–4018 (2007).

62. Mahmoudi, M., Simchi, A., Imani, M., Stroeve, P., and Sohrabi, A. Templated growth of superparamagnetic iron oxide nanoparticles by temperature programming in the presence of poly(vinyl alcohol). *Thin Solid Films*, **518**(15), 4281–4289 (2010).

63. Mandal, S. and Muller, A. H. E. Facile route to the synthesis of porous α-Fe2O3 nanorods. *Materials Chemistry and Physics*, 111(2–3), 438–443 (2008).

64. Mao, Y., Park, T. J., Zhang, F., Zhou, H., and Wong, S. S. Environmentally Friendly Methodologies of Nanostructure Synthesis. *Small*, **3**(7), 1122–1139(2007).

65. Martins, T. S., Mahmoud, A., da Silva, L. C. C., Cosentino, I. C., Tabacniks, M. H., Matos, J. R., Freire, R. S., and Fantini, M. C. A. Synthesis, characterization and catalytic evaluation of cubic ordered mesoporous iron-silicon oxides. *Materials Chemistry and Physics*, **124**(1), 713–719 (2010).

66. Matijevic, E. Preparation and properties of uniform size colloids. *Chemistry of Materials*, **5**(4), 412–426(1993).

67. Michailovski, A. and Patzke, G. R. Hydrothermal Synthesis of Molybdenum Oxide Based Materials: Strategy and Structural Chemistry. *Chemistry – A European Journal*, **12**(36), 9122–9134 (2006).

68. Morup, S., Hansen, M. F., and Frandsen, C. Magnetic interactions between nanoparticles. *Beilstein Journal of Nanotechnology*, **1**, 182–190 (2010).

69. Murphy, C. J., San, T. K., Gole, A. M., Orendorff, C. J., Gao, J. X., Gou, L., Hunyadi, S. E., and Li, T. Anisotropic metal nanoparticles: Synthesis, assembly, and optical applications. *Journal of Physical Chemistry B*, **109**(29), 13857–13870 (2005).

70. Navrotsky, A., Mazeina, L., and Majzlan, J. Size-driven structural and thermodynamic complexity in iron oxides. *Science*, **319**(5870), 1635–1638(2008).

71. Ngo, A. T. and Pileni, M. P. Assemblies of cigar-shaped ferrite nanocrystals: orientation of the easy magnetization axes. *Colloids and Surfaces a-Physicochemical and Engineering Aspects*, **228**(1–3), 107–117(2003).

72. Ni, H., Ni, Y. H., Zhou, Y. Y., and Hong, J. M. Microwave-hydrothermal synthesis, characterization and properties of rice-like alpha-Fe2O3 nanorods. *Materials Letters*, **73**, 206–208(2012).

73. Nidhin, M., Sreeram, K. J., and Nair, B. U. Green synthesis of rock salt CoO nanoparticles for coating applications by complexation and surface passivation with starch. *Chemical Engineering Journal*, **185–186**(0), 352–357 (2012a).

74. Nidhin, M., Sreeram, K. J., and Nair, B. U. Polysaccharide films as templates in the synthesis of hematite nanostructures with special properties. *Applied Surface Science*, **258**(12), 5179–5184 (2012b).

75. Nidhin, M., Indumathy, R., Sreeram, K. J., and Nair, B. U. Synthesis of iron oxide nanoparticles of narrow size distribution on polysaccharide templates. *Bulletin of Material Science*, **31**, 93–96 (2008).

76. Nie, Z., Petukhova, A., and Kumacheva, E. Properties and emerging applications of self-assembled structures made from inorganic nanoparticles. *Nat Nano*, **5**(1), 15–25 (2010).

77. Ogawa, K., Oka, K., and Yui, T. X-ray study of chitosan-transition metal complexes. *Chemistry of Materials*, **5**(5), 726–728 (1993).

78. Ozin, G. A., Arsenault, A. C., Cademartiri, L., and Chemistry, R. S. o. Nanochemistry: a chemical approach to nanomaterials. *Royal Society of Chemistry* (2009).

79. Park, S. J., Kim, S., Lee, S., Khim, Z. G., Char, K., and Hyeon, T. Synthesis and Magnetic Studies of Uniform Iron Nanorods and Nanospheres. *Journal of the American Chemical Society*, **122**(35), 8581–8582 (2000).

80. Park, S., Lee, M. R., and Shin, I. Chemical tools for functional studies of glycans. *Chem Soc Rev*, **37**(8), 1579–91 (2008).

81. Park, T. J. and Wong, S. S. As-Prepared Single-Crystalline Hematite Rhombohedra and Subsequent Conversion into Monodisperse Aggregates of Magnetic Nanocomposites of Iron and Magnetite. *Chemistry of Materials*, **18**(22), 5289–5295(2006).

82. Parker, F. T., Foster, M. W., Margulies, D. T., and Berkowitz, A. E. Spin canting, surface magnetization, and finite-size effects in a-Fe2O3 particles. *Physical Review B*, **47**(13), 7885–7891 (1993).

83. Perez, J. M., Simeone, F. J., Saeki, Y., Josephson, L., and Weissleder, R. Viral-Induced Self-Assembly of Magnetic Nanoparticles Allows the Detection of Viral Particles in Biological Media. *Journal of the American Chemical Society*, **125**(34), 10192–10193 (2003).

84. Quinn, R. K., Nasby, R. D., and Baughman, R. J. Photoassisted electrolysis of water using single crystal a-Fe2O3 anodes. *Materials Research Bulletin*, **11**(8), 1011–1017 (1976).

85. Rinaudo, M. Main properties and current applications of some polysaccharides as biomaterials. *Polymer International*, **57**(3), 397–430 (2008).

86. Robbins, E. I. and Iberall, A. S. Mineral remains of early life on earth - on mars. *Geomicrobiology Journal*, **9**(1), 51–66 (1991).

87. Routkevitch, D., Bigioni, T., Moskovits, M., and Xu, J. M. Electrochemical Fabrication of CdS Nanowire Arrays in Porous Anodic Aluminum Oxide Templates. *The Journal of Physical Chemistry*, **100**(33), 14037–14047 (1996).

88. Rudzka, K., Delgado, Ã. N. V., and Viota, J. N. L. Maghemite Functionalization for Antitumor Drug Vehiculization. *Molecular Pharmaceutics* (2012).

89. Sajanlal, P. R., Sreeprasad, T. S., Samal, A. K., and Pradeep, T. *Anisotropic nanomaterials: structure, growth, assembly, and functions* (2011).

90. Sarkar, D., Mandal, K., and Mandal, M. Synthesis of Chainlike alpha-Fe2O3 Nanoparticles in DNA Template and Their Characterization. *Nanoscience and Nanotechnology Letters*, **3**(2), 170–174 (2011).

91. Sarkar, S., Guibal, E., Quignard, F., and SenGupta, A. K. Polymer-supported metals and metal oxide nanoparticles: synthesis, characterization, and applications. *Journal of Nanoparticle Research*, **14**(2), 24 (2012).

92. Satyanarayana, K. G., Arizaga, G. G. C., and Wypych, F. Biodegradable composites based on lignocellulosic fibers-An overview. *Progress in Polymer Science*, **34**(9), 982–1021 (2009).

93. Schnepp, Z. A. C., Wimbush, S. C., Mann, S., and Hall S. R. Structural Evolution of Superconductor Nanowires in Biopolymer Gels. *Advanced Materials*, **20**(9), 1782–1786 (2008).

94. Schwertmann, U. and Cornell, R. M. *Iron Oxides in the Laboratory: Preparation and Characterization*. Wiley-VCH (2000).

95. Shi, W., Liang, P., Ge D., Wang, J., and Zhang, Q. Starch-assisted synthesis of polypyrrole nanowires by a simple electrochemical approach. *Chemical Communications*, **23**, 2414–2416 (2007).

96. Sivula, K., Zboril, R., Le Formal, F., Robert, R., Weidenkaff, A., Tucek, J., Frydrych, J., and Graìtzel, M. Photoelectrochemical Water Splitting with Mesoporous Hematite Prepared by a Solution-Based Colloidal Approach. *Journal of the American Chemical Society*, **132**(21), 7436–7444 (2010).

97. Sreeram, K. J., Nidhin, M., and Nair, B. U. Microwave assisted template synthesis of silver nanoparticles. *Bulletin of Materials Science*, **31**(7), 937–942 (2008).

98. Sreeram, K. J., Nidhin, M., and Nair, B. U. Synthesis of aligned hematite nanoparticles on chitosan-alginate films. *Colloids and Surfaces B: Biointerfaces*, **71**(2), 260–267 (2009a).

99. Sreeram, K. J., Nidhin, M., and Unni, Nair, B. Formation of necklace-shaped haematite nanoconstructs through polyethylene glycol sacrificial template technique. *Journal of Experimental Nanoscience*, 1–13 (2011).

100. Suber, L., Fiorani, D., Imperatori, P., Foglia, S., Montone, A., and Zysler, R. Effects of thermal treatments on structural and magnetic properties of acicular α-Fe2O3 nanoparticles. *Nanostructured Materials*, **11**(6), 797–803(1999).

101. Sundar, S., Kundu, J., and Kundu, S. C. Biopolymeric nanoparticles. *Science and Technology of Advanced Materials*, **11**(1) (2010).

102. Surowiec, Z., Gac, W., and Wiertel, M. The Synthesis and Properties of High Surface Area Fe2O3 Materials. *Acta Physica Polonica A*, **119**(1), 18–20(2011).

103. Tartaj, P., Morales, M. P., Gonzalez-Carreno, T., Veintemillas-Verdaguer, S., and Serna, C. J. Advances in magnetic nanoparticles for biotechnology applications. *Journal of Magnetism and Magnetic Materials*, **290–291**, Part 1(0), 28–34 (2005).

104. Virkutyte, J. and Varma, R. S. Green synthesis of metal nanoparticles: Biodegradable polymers and enzymes in stabilization and surface functionalization. *Chemical Science*, **2**(5), 837–846 (2011).

105. Wang, G. H., Li, W. C., Jia, K. M., and Lu, A. H. A Facile Synthesis of Shape and SizeControlled α-Fe2O3 Nanoparticles Through Hydrothermal Method. *Nano*, **6**(5), 469–479 (2011).

106. Wang, S. B., Min, Y. L., and Yu, S. H. Synthesis and Magnetic Properties of Uniform Hematite Nanocubes. *The Journal of Physical Chemistry C*, **111**(9), 3551–3554 (2007).

107. Wen, X., Wang, S., Ding, Y., Wang, Z. L., and Yang, S. Controlled Growth of Large-Area, Uniform, Vertically Aligned Arrays of Î±-Fe2O3 Nanobelts and Nanowires. *The Journal of Physical Chemistry B*, **109**(1), 215–220 (2004).

108. Wheeler, D. A., Wang, G., Ling, Y., Li, Y., and Zhang, J. Z. Nanostructured hematite: synthesis, characterization, charge carrier dynamics, and photoelectrochemical properties. *Energy & Environmental Science*, **5**(5), 6682–6702 (2012).

109. Woo, K., Lee, H. J., Ahn, J. P., and Park, Y. S. Sol–Gel Mediated Synthesis of Fe2O3 Nanorods. *Advanced Materials*, **15**(20), 1761–1764(2003).

110. Xue, D. S., Gao, C. X., Liu, Q. F., and Zhang, L. Y. Preparation and characterization of haematite nanowire arrays. *Journal of Physics: Condensed Matter*, **15**(9), 1455 (2003).

111. Yang, Z., Xia, Y., and Mokaya, R. High Surface Area Silicon Carbide Whiskers and Nanotubes Nanocast Using Mesoporous Silica. *Chemistry of Materials*, **16**(20), 3877–3884 (2004).

112. Yanyan, X., Shuang, Y., Guoying, Z., Yaqiu, S., Dongzhao, G., and Yuxiu, S. Uniform hematite a-Fe2O3 nanoparticles: Morphology, size-controlled hydrothermal synthesis and formation mechanism. *Mater Lett*, **65**(12), 1911–1914 (2011).

113. Zboril, R., Mashlan, M., and Petridis, D. *Iron(III) Oxides from Thermal Processes Synthesis, Structural and Magnetic Properties, Mossbauer Spectroscopy Characterization, and Applications Chemistry of Materials*, **14**(3), 969–982 (2002).

114. Zeng, S. Y., Tang, K. B., Li, T. W., Liang, Z. H., Wang, D., Wang, Y. K., Qi, Y. X., and Zhou, W. W. Facile route for the fabrication of porous hematite nanoflowers: Its synthesis, growth mechanism, application in the lithium ion battery, and magnetic and photocatalytic properties. *Journal of Physical Chemistry C*, **112**(13), 4836–4843 (2008).

115. Zhang, X. J., Hirota, R., Kubota, T., Yoneyama, Y., and Tsubaki, N. Preparation of hierarchically meso-macroporous hematite Fe2O3 using PMMA as imprint template and its reaction performance for Fischer-Tropsch synthesis. *Catalysis Communications*, **13**(1), 44–48 (2011).

116. Zhang, Y. P., Chu, Y., and Dong, L. H. One-step synthesis and properties of urchin-like PS/ alpha-Fe2O3 composite hollow microspheres. *Nanotechnology*, **18**(43), 5 (2007).

117. Zheng, Z., Huang, Ma, Zhang, Liu, LiuLiu Z., Wong, K. W., and Lau, W. M. Biomimetic Growth of Biomorphic CaCO3 with Hierarchically Ordered Cellulosic Structures. *Crystal Growth & Design*, **7**(9), 1912–1917 (2007).

118. Zhong, Z., Ho, J., Teo, J., Shen, S., and Gedanken, A. Synthesis of Porous α-Fe2O3 Nanorods and Deposition of Very Small Gold Particles in the Pores for Catalytic Oxidation of CO. *Chemistry of Materials*, **19**(19), 4776–4782(2007).

119. Zhou, H. and Wong, S. S. A Facile and Mild Synthesis of 1-D ZnO, CuO, and α-Fe2O3 Nanostructures and Nanostructured Arrays. *ACS Nano*, **2**(5), 944–958 (2008).

120. Zhou, Y. X., Yao, H. B., Yao, W. T., Zhu, Z., and Yu, S. H. Sacrificial Templating Synthesis of Hematite Nanochains from [Fe18S25](TETAH)14 Nanoribbons: Their Magnetic, Electrochemical, and Photocatalytic Properties. *Chemistry-a European Journal*, **18**(16), 5073–5079 (2012).

121. Zhu, L. P., Xiao, H. M., Liu, X. M., and Fu, S. Y. Template-free synthesis and characterization of novel 3D urchin-like [small alpha]-Fe2O3 superstructures. Journal of Materials Chemistry, 16(19), 1794–1797 (2006).

CHAPTER 5

NOVEL DIELECTRIC NANOPARTICLES (DNP) DOPED NANO-ENGINEERED GLASS BASED OPTICAL FIBER FOR FIBER LASER

M. C. PAUL, A. V. KIR'YANOV, S. DAS, M. PAL, S. K. BHADRA, YU.O. BARMENKOV, A. A. MARTINEZ-GAMEZ, J. L. LUCIO MARTHNEZ, S. YOO, and J. K. SAHU

CONTENTS

ABSTRACT

We have developed the technology for making of dielectric nano-particles (DNP) doped nano-engineered glass based optical fibers. Two kinds of DNP containing silica glass based Yb_2O_3 doped fibers are made successfully through solution doping (SD) technique. One: Yb_2O_3 doped yttria-rich alumino-silica nano-particles based optical fiber developed during drawing of D-shaped low RI resin coated large mode area (LMA) optical fiber from the modified preform which annealed at 1450–1550°C for 3 hours under heating and cooling rates of 20°C/min and other: Yb_2O_3 doped zirconia-germanium-alumino (ZGA) rich yttria-silica nano-particles based optical fibers developed during drawing of normal RI coated single mode optical fiber from the modified preform which annealed at 1000–1100°C for 3 hr under heating and cooling rates of 20°C/min. Fabrication of Yb_2O_3 doped yttria-rich alumino-silica nano-particles based D–shaped low RI coated large core optical fibers having core diameter around 20.0–30.0 micron was made. The size of DNP nano-particles was maintained within 5–10 nm under doping of 0.20 mole% of fluorine. The start fiber preforms are studied by means of EPMA, EDX, and electron diffraction analyses, revealing phase-separated nano-sized ytterbium-rich areas in their cores. There is a great need to engineer the composition as well as doping levels of different elements within the core glass during the preform making stages to generate phase-separated Yb_2O_3 doped DNP nano particles in the fiber. The matter concentrates on making of Yb_2O_3 doped DNP containing optical fibers along with material characterizations, study of spectroscopic properties, photo-darkening phenomena, and lasing characteristics. Such kind of nano-engineered glass based optical fibers shows good lasing efficiency with improved photodarkening (PD) phenomena compared to the standard silica glass based optical fibers.

5.1 INTRODUCTION

Considerable work was carried out on incorporation of rare-earth (RE) oxide nano-particles into different glass hosts. A number of processes were inspected to solve the task, such as the co-sputtering technique (Fujii et al., 1998), pyrolysis (John et al., 1999), ion implantation (Chryssou et al., 1999), laser ablation (Nichols et al., 2001), sol-gel processes (Yeatman et al., 2000), and direct deposition of nano-particles (Rajala et al., 2003). However, when the target is the fabrication of re doped optical fibers with nano-engineered core glass from the respective optical preforms which annealed under different heating conditions, the SD technique (Bandyopadhyay et al., 2004; Bhadra et al., 2003; Sen et al., 2005; Townsend et al., 1987) in the modified chemical vapor deposition (MCVD) process seems to be the most natural approach. An example of synthesis of Er_2O_3-doped nano-particles based calcium-germano-silicate glass for fiber applications by applying basic principles of phase-separation phenomena was reported earlier (Dacapito et al., 2008; Blanc et al., 2009). Ytterbium-doped silica fiber lasers have been a front runner in the race for higher powers due to its excellent conversion efficiency. As the ytterbium-doped silica fiber laser make a breakthrough in power scaling to kilowatt level (Jeong et al., 2004), the efforts to improve the host material properties become much more important. In particular, PD observed in Yb-doped fibers is host material dependent. The PD in Yb-doped fibers induces permanent excess loss in the pump and signal bands of Yb-doped fibers (Koponen et al., 2006 ; Yoo et al., 2007) which degrades the laser efficiency. The induced loss is much more pronounced when Yb^{3+} ions are incorporated in aluminosilica host than in phosphor-silica host (Shubin et al., 2007 ; Jetschke et al., 2008) and also it is found to be temperature dependent (Jeong et al., 2004 ; Yoo et al., 2010). Thus, the PD appears to be controllable by modifying the host composition. Here, we have investigated the fabrication process of nano-engineered glass based optical fibers with yttrium-rich (Y) Yb_2O_3-doped alumino-silica core glass and zirconia-germano-alumino (ZGA) rich yttria-silica core glass where the dominant portion of ytterbium is within the phase-separated nano-particles. More attention is presently paid to find out the host materials with phase-separated Yb_2O_3 doped nano-particles having low phonon energy in order to improve lasing efficiency and suppress or mitigate the concentration quenching and PD in heavily Yb-doped optical fibers.

5.2   EXPERIMENTAL RESULTS

Nano-engineered glass based Yb_2O_3 fiber preforms were fabricated by employing MCVD process and SD technique. The details about the fabrication process of both type of nano-engineered glass based Yb_2O_3 doped fiber preforms were described earlier (Paul et al., 2010; Kiryanov et al., 2011). To make Yb_2O_3 doped yttria-rich alumino-silica nano-particles based optical fiber, yttria-alumino-silica glass based preforms are annealed at 1450–1550°C for 3.0 hr under heating and cooling rates of 20°C/min and Yb_2O_3 ZGA rich yttria-silica nano-particles based optical fibers zirconia-yttria-germano-alumino-silica glass based preforms are annealed at 1000–1100°C for 3.0 hr under heating and cooling rates of 20°C/min. Yb_2O_3 doped yttria-rich alumino-silica nano-particles based D-shaped low refractive index (RI) resin coated optical fiber with different core diameter 20–30 micron as well as NA 0.06–0.10 and Yb_2O_3 doped ZGA rich yttria-silica nano-particles based single mode optical fibers having core diameter 3.2–3.4 micron with NA 0.25–0.22 were developed from the respective annealed preforms using fiber drawing tower.

High-resolution transmission electron microscope (HTEM) Model: Tecnai G2 30ST (FEI Company, USA) images of annealed preform and fiber samples were used to study the core glass morphology. To evaluate the composition of phase-separated particles the electron beam was focused on the particles and then focused in an area outside of the particles, when the energy dispersive X-ray analyses (EDX) data were taken. The nature of the particles was evaluated from their electron diffraction pattern. The average dopant percentages in samples were measured by an electron probe micro analyser (EPMA) at Electron Microprobe Laboratory, University of Minnesota, USA.

The microscopic pictures of the fiber cross-sections were taken using an optical microscope (Model: Nikon Eclipse LV 100). The Yb-doped fibers were then investigated through a set of characterization procedures that included the measurements of absorption and fluorescence spectra, using an optical spectrum analyzer (OSA) of the Ando type, as well as the measurements of fluorescence lifetimes and analysis of the resultant fibers in the sense of PD phenomena and laser properties.

5.3 DISCUSSIONS AND RESULTS

To make Yb_2O_3 doped yttria-rich alumino-silica nano-particles based optical fiber with varying core diameter from 20–30 micron, yttria-alumino-silica glass based preforms with core diameter 1.50–2.85 mm within 10.0 mm diameter of Yb_2O_3 doped preforms based on yttria-alumino-silica glass hosts were fabricated through modification of several fabrication parameters such as number of deposited porous layers, composition of deposited porous layers, and CSA of the starting deposited tube. In this case the glass modifiers such as Al_2O_3, BaO, Li_2O, Yb_2O_3 and Y_2O_3 are incorporated by the SD technique using an alcoholic-water (1:5) mixture of appropriate strength of $YbCl_3.6H_2O$, $AlCl_3.6H_2O$, $YCl_3.6H_2O$, $LiNO_3$ and $BaCl_2.2H_2O$. The cross-sectional view of Yb_2O_3 doped yttria-rich alumino-silica nano-particles based optical fibers of different diameters were shown in Figure 1. To make Yb_2O_3 doped ZGA rich yttria-silica nano-particles based optical fibers, preforms are fabricated through deposition of single silica-germamo-phosphorous layer at optimum deposition temperature of 1375±25°C with flow of a mixture of $SiCl_4$, $GeCl_4$, $POCl_3$, O_2 and, He followed by soaking of the porous soot layer into a solution using an alcoholic-water (1:5) mixture of suitable strength of $YbCl_3.6H_2O$, $AlCl_3.6H_2O$, $YCl_3.6H_2O$, and $ZrOCl_2.8H_2O$ for one hour.

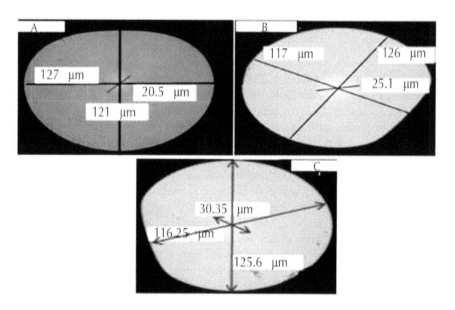

FIGURE 1 Cross-sectional view of nano-engineered yttrium-rich Yb_2O_3-doped alumino-silica core glass based fibers of different diameters (A) 20.0 μm, (B) 25.0 μm, and (C) 30.0 μm

The refractive index profiles of two kinds of nano-engineered glass based optical fibers are shown in Figure 2.

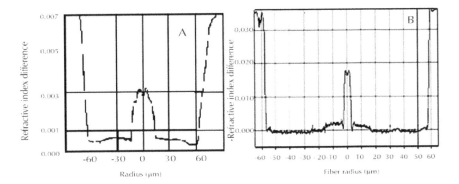

FIGURE 2 RI profile of nano-engineered optical fibers: (A) yttrium-rich Yb$_2$O$_3$-doped alumino-silica core glass and (B) zirconia-germano-alumino (ZGA) rich yttria-silica core glass.

The distribution of different dopants within the whole core region of nano-engineered glass based optical fiber preform sample was shown in Figure 3 measured from EPMA result. The formation of phase-separated Yb$_2$O$_3$ doped yttrium aluminium garnet (YAG) crystalline nano-particles starts when the molar ratio of Al:Y reaches ~1.70–1.75 under suitable doping levels of Y$_2$O$_3$ and Al$_2$O$_3$ after thermal annealing of preform samples within the temperature range of 1450–1550°C given in Figure 3 which shows the crystalline nano-structure of less than 20 nm in size and surrounded by the amorphous SiO$_2$ matrix. The crystalline nature was detected from the electron diffraction pattern. All the particles were observed to be near round shape. The composition of such phase-separated crystalline particles was evaluated by comparing their energy dispersive X-ray analyses (EDX) data, taken from directly on the crystal particles and from the areas of outside the particles shown in Figure 3. The evaluated data consists of 13.0 at % of Al, 0.02 at % of Yb, 8.0 at % of Y, and 34.0 at % of O. The doping level of the formation of Yb:YAG crystals within fiber preform sample is 0.02 at %. Such compositional analyses indicate that the phase-separated crystalline nano-particles were Yb:YAG crystals dispersed into the silica glass matrix. The particles detected Si, Al, and O without Yb and Y. The doping level of the formation of Yb:YAG crystals within fiber preform sample is 0.02 at %. Such compositional analyses indicate that the phase-separated crystalline nano-particles were Yb:YAG crystals dispersed into the silica glass matrix.

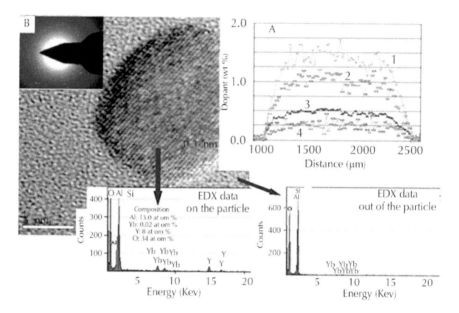

FIGURE 3 Dopant distribution curve of yttrium-rich Yb_2O_3-doped alumino-silica core glass based optical fiber preform having core diameter around 25.0 micron: Al_2O_3 (1), Y_2O_3 (2), Yb_2O_3 (3), and F (4) along with (B) HRTEM picture with electron diffraction pattern and EDX curves on and out the particles.

On the other hand similar phenomena occurs for formation of Yb_2O_3 doped ZGA rich yttria-silica nano-particles based optical fibers where preforms are annealed at 1000–1100°C for 3.0 hr under heating and cooling rates of 20 c/min. The TEM pictures of doped ZGA preform along with fiber samples are shown in Figure 4. The EDX curves shows that the particles are based on zirconia-germano-alumino rich yttria-silica. The average size of the particles was found to be 15–20 nm. The size increases with increasing the ratio of Zr/Yb. Both ZrO_2 (Tomazawa et al., 1979) as well as P serves as nucleating agents for phase-separation. With increasing doping levels of ZrO_2, the numbers of nano-particles will increase at each portion of the core region as well as from the center to the core-clad boundary region.

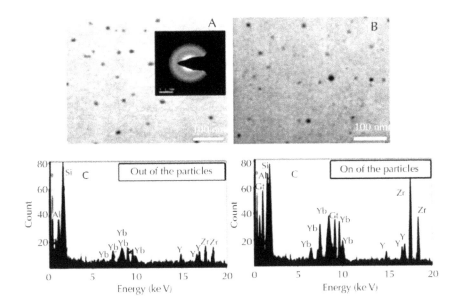

FIGURE 4 TEM pictures of zirconia-germano-alumino (ZGA) rich yttria-silica core glass based preform (A) and optical fiber (B) along with electron diffraction pattern and EDX curves (C) on and out of the particles.

The doping of minor amount of fluorine reduces the size of the nano-particles within 5–10 nm range as shown in Figure 5. The doping of fluorine in yttria-alumino silicate glass composition may have a dramatic effect on the nucleation and crystallization behavior of the glass contributing to the bimodal size distribution.

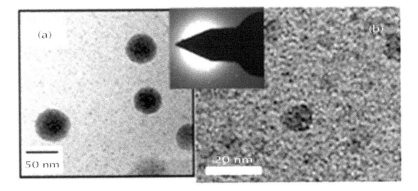

FIGURE 5 TEM pictures of different yttrium-rich Yb_2O_3-doped alumino-silica core glass based optical fiber preforms, (a) without F doping, (b) with fluorine doping along with electron diffraction

This effect was not only a result of the stoichiometric considerations of crystal formation but also of the network disrupting role of fluorine within a glass network. Here fluorine may act as a nucleating agent, promoting crystallization of Yb_2O_3-doped YAG crystals within the silica glass matrix and also serves as a facilitator of the kinetics of crystallization through rearrangement of the glass network. The non-bridging oxygens preferentially bond to the silicon atoms present, thereby preventing the formation of silicon–fluorine bonds and the formation of volatile SiF_4. In such glass, P and F may play different kinds of nucleation as well as crystallization phenomenon for the formation of observed bimodal size distribution. The formation of nano-particles start from the annealing temperature of 1300°C but the particles becomes amorphous nature. On the other hand, the nano-particles also becomes amorphous in nature at thermal annealing above 1600°C. One of the reasons may be that yttria-alumino-silicate (YAS) glass undergoes phase-separating once it "enters" the immiscible region of a ternary diagram of the YAS system. The composition of such kind of glass having silica content around 90 mol% forms both the liquid and clear glass zones.

Generally, the formation of such type of Yb:YAG crystals in bulk glass samples occurs by the following way:

$$5Al_2O_3 + 3(1-x)Y_2O_3 + 3xYb_2O_3 = 2Y_{3(1-x)}Yb_{3x}Al_5O_{12}, \text{ where } x = 0.1$$

Such type of nano-structuration retains within the core glass matrix of an optical fiber. Increasing the P_2O_5 content in glass accelerates growth of formation of phase separated particles upon heating through thermal perturbation, where P_2O_5 serves as a nucleating agent owing to the higher field strength difference (> 0.31) between Si^{4+} and P^{5+} (Vomacka et. al., 1995). The nature of the nano-particles changes from crystalline to dielectric state when fiber was drawn at high temperature around 2000°C from such annealed preforms and keep the sizes of nano-particles within 5–10 nm in case of Yb_2O_3 doped yttria-rich alumino-silica nano-particles based optical fiber.

The background loss of the fibers at 1285 nm wavelength was measured by high-resolution optical time-domain reflectrometer (OTDR). It was observed that background loss of the fibers very much related to the sizes of the Yb_2O_3-doped nano-particles. The nano-particles silica having sizes 5–10 nm shows the minimum background loss around 40–50dB/km at 1285 nm wavelength. Whereas the nano-particles silica fibers containing the larger sized particles show the high background loss. The background loss at 1285 nm was 40–2400 dB/km, depending on the core composition and size of the nano-particles. The fluorescence life-time of both kinds of optical fibers observed to be 1.0 ± 0.01 ms and 0.70 ± 0.01 ms, respectively.

The PD of both types of nano-engineered glass based Yb_2O_3 doped fibers was evaluated by monitoring the transmitted output power through the fiber under 975 nm irradiation (Tomazawa et al., 1979; Yoo et al., 2010). The temporal characteristics of the transmitted probe power is represented in Figure 6 for Yb_2O_3 doped nano-engineered glass based optical fibers. The PD induced loss is significantly reduced in Yb_2O_3 doped yttria-rich alumino-silica nano-particles based optical fibers. When we fitted the measured results with stretched exponential form, we found that the saturated induced loss is reduced by 10–20 times compared to the aluminosilicate counterpart, depending on the host material compositions.

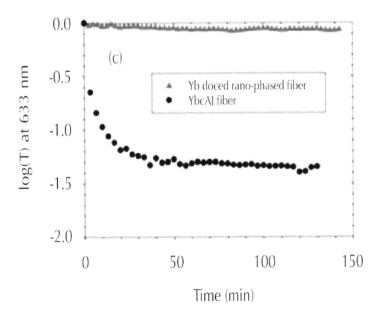

FIGURE 6 Temporal characteristics of transmitted power at 633 nm for Yb-doped nanoparticle fiber and Yb:Al fiber.

The effect of PD in the was also evaluated by monitoring the transmitted probe power at 633 nm wavelength (a He–Ne laser, 5 mW). The probe beam transmission was measured after sequencing doses of doped ZGA rich yttria-silica nano-particles based optical fibers under irradiation at 975 nm wavelength (at the moments when the pump light was temporally switched off). The LD output was spliced to one of the input ports of wavelength division multiplexer (WDM) while one of its output ports was spliced to such nano-engineered glass based fiber; the pump beam propagated through such fiber and photo darkened it. The probe beam from the He–Ne laser was coupled to the active nano-engineered glass based fiber through the second input port of WDM (so, it propagated in the same direction as the pump one). The attenuated output power at 633 nm was detected by a photo detector.

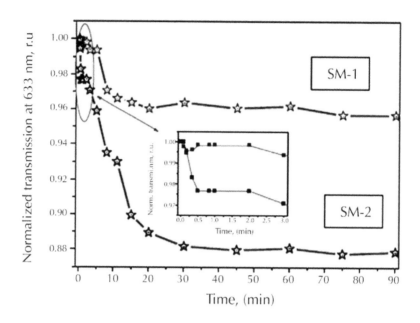

FIGURE 7 Temporal behaviors of normalized transmitted power at 633-nm wavelength at PD accompanying CW pumping (975 nm, 255 mW) of SM-1 and SM-2 YFs.

We used in the experiments short pieces of such active fiber samples: ≈30–35cm (SM-1) and ≈20–25 scm (SM-2) , correspondingly, in order to have a possibility to measure the difference in the attenuation spectra (these spectra were obtained after some of the irradiation doses). Launched pump power at 975 nm was maintained in the experiments at 250 mW to provide a high Yb^{3+} population inversion that was practically at the same level (>45%) throughout the active fiber length. The temporal characteristics of the transmitted power at 633 nm for SM-1 and SM-2 fibers are presented in Figure 7 and the correspondent spectra differences as appeared through the PD experiments are shown in Figure 8.

It is seen that for the less Yb^{3+} doped fiber SM-1 the PD effect is almost negligible as compared to the heavier doped SM-2, an expectable fact. One can also reveal from the data that although the PD effect in new doped ZGA nano-particles based ytterbium fibers (YFs) is relatively weak, its appearance resembles the PD features in other alumino-silicate YFs. Only one typical detail that stems from the PD loss dependences shown in Figure 7 (see inset) should be emphasized: The transmitted power at 633 nm, at the initial PD stage, drops by a stepwise (not smooth) (Shubin et. al., 2007 ; Guzman-Chavez et. al., 2007) manner, probably being the appearance of different particular rates of PD within and outside the nano-sized particles, correspondingly rich and poor in Yb_2O_3 content. However, this peculiarity has to be proved by more detailed studies in the future.

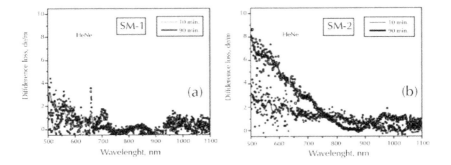

FIGURE 8   The examples of difference (PD loss) spectra obtained for SM-1 (a) and SM-2 (b) YFs after 10 and 90 min of 975-nm PD.

The laser efficiency of Yb_2O_3 doped yttria-rich alumino-silica nano-particles based D-shaped low RI resin coated LMA optical fibers of varying diameter from 20–35 micron having NA around 0.06–0.10 were tested under 976 nm end pumping configuration. The fiber laser was pumped by a multimode fiber-coupled laser diode stack operating at 976-nm wavelength through a fiber combiner. The laser cavity was formed by two fibers Bragg gratings (FBGs) with reflectivity 99% (HR) and 10% (LR) and both centered at 1080 nm, which were spliced directly with an Yb-doped fiber. The output characteristics of the lasers with the active fiber having D- and P- shape are highlighted by Figure 8 (a, b).

It is seen that both lasers release around 18 W of CW power at 1080 nm wavelength (at pump power of 23 W) and remarkably high optical efficiency, in excess of 80%, one of the highest values accessible to-date using Yb-doped fibers. It is of mention that output power of the lasers was almost not subjected to degradation at long-term (hours) tests, revealing a negligible PD effect in these new Yb_2O_3 doped DNPs based optical fibers, which is one of their important advantages. The laser performance of such kind of Yb_2O_3 doped nano-engineered optical fiber is shown in Figure 9, where the lasing efficiency occurs within 80–85%. The lasing efficiency of Yb_2O_3 doped ZGA rich yttria-silica nano-particles based optical fibers is shown in Figure 9. The slope efficiency of such kind of nano-engineered optical fiber is found to be around 60% which becomes lower than that of yttria-rich alumino-silica nano-particles based optical fiber due to presence of larger sized particle.

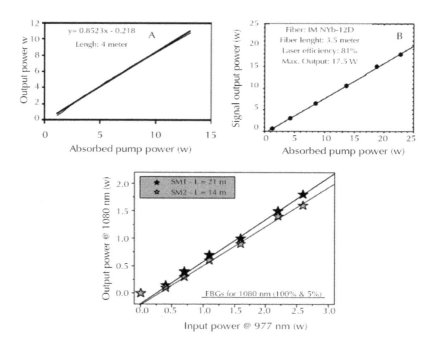

FIGURE 9 Lasing efficiency curves of nano-engineered glass based optical fibers: yttrium-rich Yb_2O_3-doped alumino-silica core glass based optical fibers of different core diameters, (A) 20.0 micron, (B) 30.0 micron, and (C) zirconia-germano-alumino (ZGA) rich yttria-silica core glass based two different optical fibers

5.4 CONCLUSION

Making of large core nano-engineered glass based optical fibers having the range of 20-30 micron diameter with elimination of unwanted small bubbles formation within the core or at the core-clad boundary are reported in this paper. A uniform distribution of Al, Yb, and other ions along the entire core-region of the preform was observed which corresponds to uniform RI of preform/fiber from the core-clad boundary to the center of core without formation of any central dip or core-clad imperfection. We also found that the presence of yttria-rich aluomino-silica and zirconia-germano-alumino rich yttria-silica phase-separated nano-host strongly modifies the surrounding environment of Yb ions which increase the electric dipole moment of the radiative transition for getting the good lasing efficiency. In our experiments, Y_2O_3 serve as an attractive host material for laser applications. Another interesting property allowing radiative transitions between electronic levels is that the dominant phonon energy is 377cm^{-1} which is one of the smallest phonon cut-off among oxides. On the other hand, zirconium oxide possesses of a stretching vibration at about 470 cm^{-1}, which is very low compared with that of Al_2O_3 (870 cm^{-1}) and SiO_2 (1100 cm^{-1}) . The introduction of a codopant cation such as Zr^{4+} allows one to avoid formation of Yb^{3+} clusters in silica and consequently to exacerbate the luminescence. ZrO_2 also serves as a nucleating agent to induce phase-separation as well as reducing the concentration quenching phenomena. Such type of nano-engineered Yb_2O_3 doped glass based optical fibers shows good lasing efficiency (75–80%) with reducing PD phenomena. This type of nanostructuration of doped fiber will be proposed as a new route to 'engineer' the local dopant environment. All these results will benefit to optical fiber components such as lasers, amplifiers and sensors, which can now be realized with silica glass.

KEYWORDS

- **Dielectric nano-particles**
- **Fiber laser**
- **Optical fibers**
- **Porous soot layer**
- **Scanning electron microscopy**
- **Solution doping**
- **Yb_2O_3 doped silica glass**

ACKNOWLEDGMENT

Authors would like to acknowledge DST, Government of India, and CONACYT, Mexico, for provision of financial support under the Programme of Cooperation in Science & Technology between India and Mexico. Authors are thankful to the Director, Prof. Indranil Manna, CSIR-CGCRI for giving permission to publish this paper.

REFERENCES

1. Bandyopadhyay, T., Sen, R., Dasgupta, K., Bhadra, S. K., and Paul, M. C. *Process for making rare earth doped optical fiber*. US patent 6751990 (2004).
2. Bhadra, S. K., Sen, R., Pal, M., Paul, M. C. Naskar, M. K., Chatterjee, S., Chatterjee, M., and Dasgupta, K. Development of rare-earth doped fibres for amplifiers in WDM system. *IEE Proceedings of Circuits, Devices & Systems*, **150**(6), 480–485 (2003).
3. Blanc, W., Dussardier, B. and Paul, M. C. Er-doped oxide nanoparticles in silica-based optical-fibers. *GlassTechnology: European Journal of Glass Science and Technology*, **A50**, 79–81 (2009).
4. Chryssou, C. E., Kenyon, A. J., Iwayama, T. S., Pitt, C. W., and Hole, D. E. *Journal of Applied Physics Letter*, **75**(14), 2011–2013 (1999).
5. D'Acapito, F., Maurizio, C., Mukul, P., Lee, Th. S., Blanc W., and Dussardier B.. Role of CaO addition in the local order around Erbium in SiO_2–GeO_2–P_2O_5 fiber preforms . *Material Science and Engineering B*, **146**, 167–170 (2008).
6. Fujii, M., Yoshida, M., Hayashi, S., and Yamamoto, K. Photoluminescence from SiO_2 films containing Si nanocrystals and Er: Effects of nanocrystalline size on the photoluminescence efficiency of Er^{3+}. *Journal Applied Physics*. 84, 4525–4531 (1998).

7. Guzman-Chavez, A. D., Kir'yanov, A. V, Yu, O. Barmenkov, Yu.O., and Il'ichev, N. N., Reversible photo-darkening and resonant photo bleaching of Ytterbium-doped silica fiber at in-core 977-nm and 543-nm irradiation," *Laser Phys. Lett.* **4**, 734–739 (2007).
8. Jeong, Y., Sahu, J. K., Payne, D. N., and Nilsson, J. Ytterbium-doped large-core fiber laser with 1.36 kW continuous-wave output power *Journal of Optics Express*, **12**(25), 6088–6092 (2004).
9. Jetschke, S., Unger, S., Schwuchow, A., Leich, M., and Kirchhof, J. Efficient Yb laser fibers with low photodarkening by optimization of the core composition, *Optics Express*, **16**(20), 15540–15545 (2008).
10. John, J. S., Coffer, J. L., Chen, Y. D., and Pinizzotto, R. F. Synthesis and characterization of discrete luminescent erbium-doped silicon nanocrystals. *Journal of American Chemical Society*. 121(9), 1888–1892 (1999)
11. Kiryanov, A. V., Paul, M. C., Bamenkov, Yu, O., Das, S., Pal, M., Bhadra, S.K., Zarate, L.E., and Guzman-Chavez, A.D. Fabrication and characterization of new Yb-doped zirconia-germano-alumino silicate phase-separated nano-particles based fibers, *Opticss Express*, **19**(16), pp.14823–14837 (2011).
12. Koponen, J. J., Söderlund, M. J., and Hoffman, H. J. Measuring photodarkening from single-mode ytterbium doped silica fibers. Optics Express, **14**, 11539–11544 (2006).
13. Nichols, W. T., Keto, J. W., Henneke, D. E., Brock, J. R., Malyavanatham, G., and Becker, M. F.. Large-scale production of nanocrystals by laser ablation of microparticles in a flowing aerosol. *Journal of Applied Physics Letter*, **78**(8), 1128–1130 (2001).
14. Paul, M. C., Bysakh, S., Das, S., Bhadra, S. K., Pal, M., Yoo, S., Kalita, M. P., Boyland, A., and Sahu, J. K. Yb_2O_3-doped YAG nano-crystallites in silica-based core glass matrix of optical fiber preform. *Materials Science and Engineering* B, **175**(2), 108–119 (2010).
15. Rajala, M., Janka, K., and Kykkänen, P. "An industrial method for nanoparticle synthesis with a wide range of compositions. *Review on Advanced Materials Science*, **5**, 493–497 (2003).
16. Shubin, A. V., Yashkov, M. V., Melkumov, M. A., Smirnow, S. A., Bufetov, I. A. & Dianov, E. M. Photodarkening of aluminosilicate and phosphosilicate Yb-doped fibers. *Lasers and Electro-Optics and International Quantum Electronics Conference*, CLEO E-IQEC (2007).
17. Tomazawa, M. and Doremus, R. H. (Eds.). *Phase Separation in Glass. In: Treatise on Materials Science and TechnologyAcademic Press*, NY (1979).
18. Townsend, J. E., Poole, S. B., and Payne, D. N. Solution doping technique for fabrication of rare earth doped optical fibers. Electronics Letter, **23**, 329–331 (1987).
19. Vomacka, P., Babushkin, O., and Warren, R.. Zirconia as a nucleating agent in a yttria-alumina-silica glass. *Journal of the European Ceramic Society* , **15**(11), 1111–1117 (1995).
20. Yeatman, E. M., Ahmad, M. M., McCarthy, O., Martucci, A., and Guglielmi M. *Journal of Sol-Gel Science and Technology*, **19**, 231–236 (2000).
21. Yoo S., Boyland A. J., Standish R. J., and Sahu J. K. Measurement of photodarkening in Yb-doped aluminosilicate fibres at elevated temperature. Electronics Letters, **46**(3), 233–234 (2010).
22. Yoo, S., Basu, C., Boyland, A. J., Stone, C., Nilsson, J., Sahu, J. K., and Payne D.. Photodarkening in Yb-doped aluminosilicate fibers induced by 488 nm irradiation. *Opics Letter,* **32**(12), 1626–1628 (2007).
23. Yoo, S., Kalita, M. P., Boyland, A. J., Webb, A. S., Standish, R. J., Sahu, J. K., Paul, M. C., Das, S., Bhadra, S. K. and Pal, M. Ytterbium-doped Y2O3 nanoparticle silica optical fibers for high power fiber lasers with suppressed photodarkening. *Optics Communications*, **283**(18), 3423–3427 (2010).

CHAPTER 6

ZnO NANOSTRUCTURES AND ITS LUMINESCENT PROPERTIES

P. M. ANEESH and M. K. JAYARAJ

CONTENTS

6.1 INTRODUCTION

Nanotechnology deals with materials or structures in nanometer scales (10^{-9} m), typically ranging from sub nanometers to several hundreds of nanometers. Earlier common notion was that material properties can be changed only by varying the chemical composition. But later it has been found that the material properties can be tuned by varying the size of the material without changing the chemical composition. The transition from micron sized particles to nanoparticles leads to a number of changes in their physical properties. The major change is the increase in the surface area to volume ratio, as the size of the particle reduces. These new properties or phenomena will not only satisfy everlasting human curiosity, but also promise a new advancement in technology. Another very important aspect of nanotechnology is the miniaturization of current and new instruments, sensors, and machines that will have great impact on the world we live in. Nanotechnology has an extremely broad range of potential applications from nanoscale electronics and optics to nanobiological systems.

The band gap of the semiconductors increases compared to their bulk counterpart and the density of states will also be modified as the size of the particles reduces. The luminescent properties of these nanostructures are quite interesting. The luminescence emission and absorption edge can be tuned by changing the particle size. Phosphors are the solid material that emits light when it is exposed to some radiation such as ultraviolet light or an electron beam. Efficient phosphors for lighting applications, flat panel displays, and so on have always been a goal for researchers. The particle size of conventional phosphors is in micrometer scale, hence light scattering at grain boundaries is strong and it decreases the light output. Nanophosphors can be synthesized from tens to hundreds of nanometers that are smaller than the visible wavelength and can reduce the scattering, thereby enhancing the luminescence efficiency.

The properties of the material such as optical, magnetic, mechanical, and chemical lies in the states of electrons in the constituting atoms, their configuration in the atomic structures and the way these electrons form bonds with the neighboring atoms in a molecular system. The typical de Broglie wavelength of electrons is in the range on nanometers. The electrons in the material experiences a trapped state when the dimension of the material reduces to nanometer range and it is termed as 'confinement effects'. This reduction in size will lead to many important properties in material like high surface area to volume ratio, enhancement of electric field related effects, chemical activity, and so on.

In a nanostructured semiconductor, the quantum confinement occurs when electrons or holes in the material are trapped in one or more of the three dimensions within nanometer size regime. In a bulk semiconductor, carrier motion is unrestricted along all three spatial directions. However, a nanostructure has one or more of its dimensions reduced to a nanometer length scale and this produces a quantization of the carrier energy corresponding to motion along these directions. If the quantum confinement is along one direction, the carriers can have free motion in other two directions. If the quantum confinement is in two directions, the carriers have free motion only in one direction. The carriers have no free motion if they are confined in all three directions. The confinement of carriers in one, two, and three dimensions are called quantum well, quantum wire, and quantum dot structures respectively. The total energy of electrons (or holes) will be the sum of allowed energies associated with the motion of these carriers along the confined direction and the kinetic energy due to free motion in the remaining unconfined directions. The density of states, the number of energy states present in a unit energy interval per unit volume of the material structure, will also be modified by reducing the dimension of the material into nanometer range. Due to the quantization of energy levels, the relationship between the density of states, and energy values is also dramatically modified as shown in Figure 1.

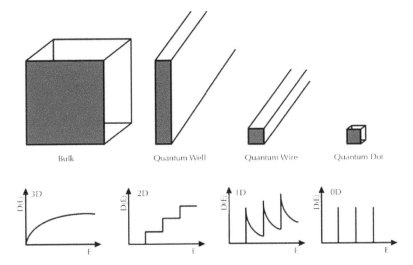

FIGURE 1 Schematic illustration of density of states of a semiconductor as a function of dimension.

Semiconductors with dimensions in the nanometer realm are important because their electrical, optical, and chemical properties can be tuned by changing the size of particles. Optical properties are of great interest for application in optoelectronics, photovoltaics, and biological sensing. Various chemical synthetic methods have been developed to synthesize such nanoparticles.

The ZnO is II-VI metal oxide semiconductor and belongs to the P63mc space group. ZnO is a n-type semiconductor under most growth conditions, p-type conductivity of ZnO has also been achieved by either with codoping or growth under stringent growth conditions (Kim et. al., (2003), Ryuet. al., (2003), Joseph et. al., (2001), Senthil et. al., (2010)). ZnO exhibits a wurtzite structure (hexagonal symmetry) or rock salt structure (cubic symmetry). The ZnO crystals most commonly stabilize with the wurtzite structure (hexagonal symmetry), whereas at high pressure, the crystals exhibit the rock salt phase (cubic symmetry). Even though it is tetrahedrally bonded, the bonds have a partial ionic character. The lattice parameters of ZnO are a = 0.32495 nm and c = 0.52069 nm at 300K, with a c/a ratio of 1.602, which is close to the 1.633

ratio of an ideal hexagonal close-packed structure. In the direction parallel to the c-axis, the Zn-O distance is 0.1992 nm, and it is 0.1973 nm in all other three directions of the tetrahedral arrangement of nearest neighbors. In a unit cell, zinc occupies the (0, 0, 0.3825) and (0.6667, 0.3333, 0.8825) positions and oxygen occupies the (0, 0, 0) and (0.6667, 0.3333, 0.5) positions (Pearson (1967)). The wurtzite structure of ZnO has a direct energy band gap of 3.37 eV at room temperature. The lowest conduction band of ZnO is predominantly s-type and the valance band is p-type (six-fold degenerate). The Zn-O bond is half ionic and half covalent. Doping in ZnO is much easier compared with other covalent-bond wide bandgap semiconductors, such as GaN. By appropriate doping, the electrical conductivity of ZnO can be tailored from semiconducting to semimetal, keeping high optical transparency to the visible and UV spectral regime. ZnO is more resistant to radiation damage than Si, GaAs, and GaN (Look (2001)), which is preferred for the long-term stability of field emission emitters in high electric fields. These make ZnO an ideal candidate among transparent conducting oxides (TCOs) for field emission displays. ZnO nanotips are attractive for field emission due to their low emission barrier, high saturation velocity, and high aspect ratio. ZnO is a wide band gap compound semiconductor that is suitable for short wavelength opto-electronic applications. The high exciton binding energy (60 meV) in ZnO crystal can ensure efficient excitonic emission at room temperature. Room temperature ultraviolet (UV) luminescence has been reported in nanoparticles and thin films of ZnO.

ZnO is a versatile functional material that has diverse growth morphologies such as, nanoparticles (Ajimsha et. al., (2008)), core/shell nanoparticles (Zeng et. al., (2007)) nanowires and nanorods (Baruah et. al., (2008)), nanocombs (Huang et. al., (2006)), nanorings (Hughes (2005)), nanoloops and nanohelices (Kong et. al., (2003)), nanobows (Hughes et. al., (2004)), nanobelts (Sun et. al., (2008)) nanocages (Snure et. al., (2007)), nanocomposites (Bajaj et. al., (2010)), and quantum wells (Misra et. al., (2006)). These structures have been successfully synthesized under specific growth conditions (Wang (2004)). Nanostructures have attracted attention because of their unique physical, optical, and electrical properties resulting from their low dimensionality. ZnO has an effective electron mass of $\sim$0.24 m$_e$, and a large exciton binding energy of 60 meV. Thus bulk ZnO has a small exciton Bohr radius ($\sim$1.8 nm) (Gil et. al., (2002),Wonget. al., (1999)). In ZnO, the quantum confinement effect should be observable when any one of the dimension is at the scale of exciton Bohr radius.

Nanostructured ZnO materials have received broad attention due to their distinguished performance in electronics, optics, and photonics. From 1960s, synthesis of ZnO thin films has been an active field because of their applications as sensors, transducers, and catalysts. In the last few decades, study of one dimensional (1D) materials has become a leading edge in nanoscience and nanotechnology. With reduction in size, novel electrical, mechanical, chemical. and optical properties are introduced, which are largely believed to be the result of increased surface area and quantum confinement effects. Nanowire-like structures are the ideal system for studying the transport process in one-dimensionally (1D) confined objects, which are of benefit not only for understanding the fundamental phenomena in low dimensional systems, but also for developing new generation nanodevices with high performance. ZnO nanostructures have a wide range of technological applications like surface acoustic wave filters

(Emanetoglu et. al., (1999)), photonic crystals (Chen et. al., (2000)), photodetectors (Liang et. al., (2001)), light emitting diodes (Saito et. al., (2002)), photodiodes (Lee et. al., (2002)), gas sensors (Mitra et. al., (1998)), optical modulator waveguides (Koch et. al., (1995)), solar cells (Gratzel (2005)), and varistors (Lin et. al., (1999)). ZnO is also receiving a lot of attention because of its antibacterial property and its bactericidal efficacy has been reported to increase as the particle size decreases (Padmavathy et. al., (2008)).

6.1.1   CONFINEMENT IN THREE DIMENSIONS: QUANTUM DOTS

Quantum dots represent the case of three-dimensional confinement, hence the case of an electron confined in a three-dimensional quantum box, typically of dimensions ranging from nanometers to tens of nanometers. These dimensions are smaller than the de Broglie wavelength of thermal electrons. A quantum dot is often described as an artificial atom because the electron is dimensionally confined just like in an atom (where an electron is confined near the nucleus) and similarly has only discrete energy levels. The electrons in a quantum dot represent a zero-dimensional electron gas (0DEG).

6.2   HYDROTHERMAL METHOD

Most of the ZnO crystals have been synthesized by traditional high temperature solid state method, which is energy consuming and difficult to control the particle properties. ZnO (Hui et. al., (2004), Zhang et. al., (2002), Li et. al., (2001)) nanostructures can be prepared on a large scale at low cost by simple solution-based methods, such as chemical precipitation (Zhong et. al., (1996), Lingna et. al., (1999)), sol-gel synthesis (Bahnemann et. al., (1987)), and solvothermal/hydrothermal reaction (Huiet. al., (2004), Zhang et. al., (2002), Li et. al., (2001)). Hydrothermal technique is a promising alternative synthetic method because of the low process temperature and very easy to control the particle size. The hydrothermal process has several advantage over other growth processes such as use of simple equipment, catalyst-free growth, low cost, large area uniform production, environmental friendliness, and less hazardous. The low reaction temperatures make this method an attractive one for microelectronics and plastic electronics (Lee et. al., (2006)). This method has also been successfully employed to prepare nanoscale ZnO and other luminescent materials. The particle properties such as morphology and size can be controlled *via* the hydrothermal process by adjusting the reaction temperature, time, and concentration of precursors.

FIGURE 2   Hydrothermal furnace and autoclave for the synthesis of nanostructures.

For the growth of ZnO nanoparticles by hydrothermal technique, stock solutions of $Zn(CH_3COO)_2.2H_2O$ (0.1 M) was prepared in 50ml methanol under stirring. To this stock solution 25ml of NaOH (varying from 0.2 M to 0.5 M) solution prepared in methanol was added under continuous stirring in order to get the pH value of reactants between 8 and 11. These solutions was transferred into teflon lined sealed stainless steel autoclaves and maintained at various temperature in the range of 100–200°C for 3–12 hr under autogenous pressure. It was then allowed to cool naturally to room temperature. After the reaction was complete, the resulting white solid products were washed with methanol, filtered, and then dried in air in a laboratory oven at 60°C.

In the precursor solution used in the experiment, the source of Zn is in the forms of $Zn(OH)_2$ precipitates and $Zn(OH)_4^{2-}$ species according to the stoichiometric ratio of Zn^{2+} on OH⁻. The $Zn(OH)_2$ precipitates under the hydrothermal conditions will dissolve to considerable extent to form ions of Zn^{2+} and OH⁻, once the product of (Zn^{2+}) and (OH⁻) exceeds a critical value which is necessary for the formation of ZnO crystals, the ZnO crystals will precipitate from the solution. The solubility of ZnO is significantly smaller than that of $Zn(OH)_2$ under the hydrothermal conditions, consequently, the $Zn(OH)_2$ precipitates strongly tends to be transformed into ZnO crystals during the hydrothermal process, by the following reactions (Ahsanulhaq et. al., (2007), Xiangyang et. al., (2005)).

$$Zn(OH)_2 \quad = \quad Zn^{2+} + 2OH^- \qquad (1)$$

$$Zn^{2+} + 2OH^- = ZnO + H_2O \qquad (2)$$

At the initial stage of the process, the concentrations of Zn^{2+} and OH⁻ were relatively higher so that the crystal growth in different directions was considerable (Ahsanulhaq et. al., (2007), Tok (2006)). When the concentration of Zn^{2+} and OH⁻ reaches the supersaturation degree of ZnO, ZnO begins to nucleate and the crystal growth begins.

Figure 3 shows the TEM image and corresponding selected-area electron diffraction (SAED) pattern of the ZnO nanoparticles synthesized at 100°C for 3 hr from 0.3 M NaOH. The TEM image confirms the formation of ZnO nanoparticle and it has an average size about 8 nm. The diffraction rings of SAED pattern shown in the inset of Figure 3, can be indexed to the reflections from different planes of the ZnO nanocrystals.

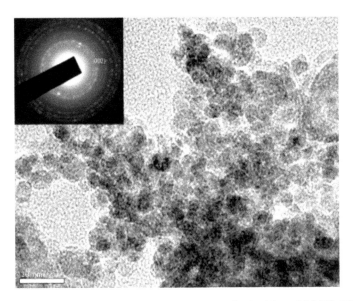

FIGURE 3 The TEM image of the ZnO nanoparticles synthesized from 0.3 M NaOH at 100°C for 3 hr.

The x-ray diffraction data were recorded by using Cu Ka-radiation (1.5406 A°). The intensity data were collected over a 2θ range of 20–80°. X-ray diffraction studies confirmed that the synthesized materials were ZnO with wurtzite phase and all the diffraction peaks agreed with the reported JCPDS data and no characteristic peaks were observed other than ZnO.

The mean grain size (D) of the particles was determined from the XRD line broadening using Scherrer equation (Klug et. al., (1954)).

$$D = 0.89\lambda / (\beta \cos\theta)$$

Where λ is the wavelength (Cu Ka), β is the full width at the half- maximum (FWHM) of the ZnO (101) line and θ is the diffraction angle.

A definite line broadening of the diffraction peaks is an indication that the synthesized materials are in nanometer range. The grain size was found to be in the range of 7–24 nm depending on the growth condition.

Figure 4 shows the variation of FWHM and grain size of ZnO nanoparticles synthesized from 0.3M NaOH at different temperatures for a growth time of 6 hr. The average grain size calculated by Scherrer equation is observed to increases from 7 nm to 16 nm as the temperature increases from 100°C to 200°C. This is due to the change of growth rate between the different crystallographic planes.

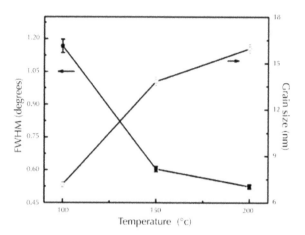

FIGURE 4 Variation of FWHM and grain size of ZnO nanoparticles with growth temperature. The nanoparticles synthesized with 0.3 M NaOH for a growth time of 6 hr.

The ZnO structures with different grain sizes can be obtained by controlling the concentration of the precursors. The ZnO nanoparticles were synthesized by keeping the concentration of $Zn(CH_3COO)_2.2H_2O$ as 0.1 M, reaction temperature at 200°C, for growth duration of 12 hr, but varying the concentration of NaOH from 0.2 M to 0.5 M at 200°C for 12 hr. Figure 5 shows the variation of FWHM and grain size of ZnO nanoparticles with concentration of NaOH precursor for the samples grown at 200°C for 12 hr. The FWHM decreases in a linear fashion on increasing the concentration of NaOH in the precursor solution. The grain size increases from 12 nm to 24 nm as the concentration of NaOH precursors increases from 0.2 M to 0.5 M. These results reveal that the molar ratio of OH^- to Zn^{2+} is a dominant factor for the formation of the ZnO nanoparticles.

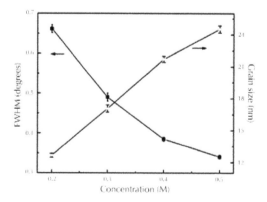

FIGURE 5 Variation of FWHM and grain size of ZnO nanoparticle with various concentration of NaOH grown at 200°C for a duration of 12 hrs.

The grain size of ZnO nanoparticles synthesized from 0.2 M NaOH at 150°C increases from 9 nm to 13 nm as the time of growth increases from 3 to 12 hrs. It shows that grain size has a linear dependence on time of growth (Flor et. al., (2004)).

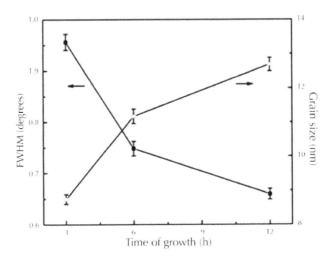

FIGURE 6 Variation of FWHM and grain size of ZnO nanoparticle obtained with different time of growth for 0.2 M NaOH at 150°C.

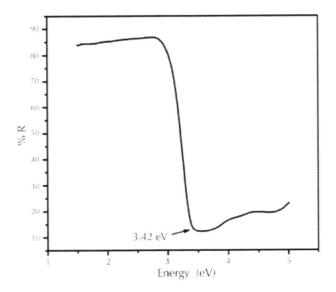

FIGURE 7 The DRS of the typical ZnO nanoparticles synthesized from 0.3 M NaOH at 100°C for 6 hr.

Diffuse reflectance spectral studies in the UV- Vis- NIR region were carried out to estimate the optical band gap of the synthesized nanoparticles. Figure 7 shows the plot for the percentage of reflection as a function of band gap energy (hυ) of the nanoparticles synthesized via hydrothermal method from 0.3M NaOH at 100°C for 6 hr. The band gap estimated for this sample 3.42 eV that is slightly higher than that of bulk ZnO (3.37 eV).

The amount of Na that is incorporated in ZnO nanoparticles was determined using ICP-AES data. The ICP-AES shows that Na incorporated into the nanoparticles is about 0.17% for lower concentration (0.2M NaOH in the precursor solution) and about 1.49% for higher concentration (0.05M) of NaOH in the precursor solution. The Na content in the ZnO nanoparticles increases with NaOH concentration. The theoretical binding energy of dopant atoms on the individual semiconductor nanocrystal surface determines its doping efficiency (Steven et. al., (2005)). The binding energy for wurtzite semiconductor nanocrystal is three times less than that of zinc blend or rock salt structure on (001) faces. The low concentration of incorporated Na is likely to be a consequence of lower doping efficiency of the wurtziteZnO nanocrystal surfaces.

The luminescence of ZnO nanoparticles is one of particular interest from viewpoints of both physical and applied aspects. Figure 8 shows the room temperature photoluminescence spectrum of the ZnO nanoparticles excited at 362 nm.

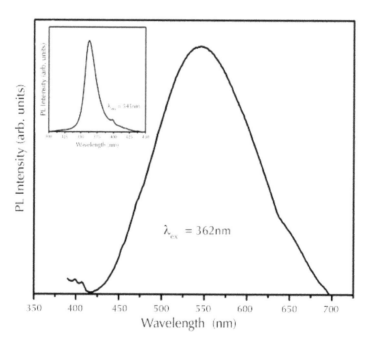

FIGURE 8   Room temperature photoluminescence spectra of ZnO nanoparticle excited at λ_{exc} = 362 nm. The inset shows the corresponding photoluminescent excitation spectra (λ_{em} = 545nm) of ZnO nanoparticles.

Green emission was observed from the hydrothermally synthesized ZnO nanoparticles. It can be attributed to the transition between singly charged oxygen vacancy and photo excited hole or Zn interstitial related defects (Vanheusden et. al., (1996), Peng et. al., (2006)). The inset in the Figure 8 shows the photoluminescent excitation spectra of the ZnO nanoparticles (λ_{em} = 545nm) which indicates that the excitation is at 362 nm. The excitation peak corresponds to the band to band transition which also confirms the blue shift in the band gap of ZnO nanoparticles.

6.3   LIQUID PHASE PULSED LASER ABLATION

It is possible to produce wide variety of compounds and morphology by pulsed laser ablation using different target materials and background gases and varying parameters such as the laser wavelength, fluence, and pulse duration (Henley et. al., (2004)). Liquid phase pulsed laser ablation (LP-PLA) involves the focusing of highly intense laser beam onto the surface of a solid target, which is submerged beneath a liquid. The very first process of LP-PLA is the interaction of laser with the solid target surface and subsequent vaporisation of the solid target as well as a very small amount of surrounding liquid. The ejected species being in highly excited states, there is strong chance for the chemical reactions between the ablated species and molecules in the liquid (Sakka et. al., (2000)). Typically the reaction products are NPs consisting of atoms from both the target and the liquid, which will form a nanoparticle suspension in the liquid. Therefore there is a chance for the prolonged interaction of this nanocolloidal suspension with the laser radiation leading to further changes in the NP's composition, size or morphology (Simakin et. al., (2004)). The major difference between normal pulsed laser ablation (PLA) and LP-PLA is that the expansion of plasma occurs freely either in vacuum or in a gaseous medium in normal PLA where as it is confined by a liquid layer in LP-PLA. The liquid layer causes a covering effect to the ejected species (Berthe et. al., (1997)) and causes delay to the expansion of the plasma leading to a very high plasma pressure and temperature thus allowing the formation of novel materials. In LP-PLA the expansion process happens adiabatically at supersonic velocity generating a shock-wave in front of it which will generate an additional instantaneous pressure that will result in the increase of plasma temperature (Zhu et. al., (2001). Thus high pressure and high density plasma will be formed in LP-PLA compared with PLA plasma formed in gas or vacuum.

One of the advantages of LP-PLA is that, the product will contain atoms from the target material and the liquid. The very special advantage of LP-PLA is that NPs produced by laser ablation of solid targets in a liquid environment are free of any counter-ions or surface-active substances (Wang et. al., (2002)). However, sometimes the surface active agents are also involved in the synthesis process (Zeng et. al., (2005)) the chancesfor whicharevery few.

In the water confined regime (WCR), plasma exerts a much stronger pressure which has significant effects on the mechanical response of solid surfaces immersed in the liquid resulting in a much higher ablation rate. It has been reported that the ablation rate is high in the water confined regime (WCR) as compared to the normal pulsed laser ablation (Yang (2007)). An analytical model was previously introduced for the prediction of laser-induced pressures in the confined ablation mode (Fabbro et. al., (1990)). The plasma pressure on the surface of the substrate will be quite high in the case of the WCR as compared to the value obtained in the direct ablation regime. It is obvious in the WCR that the target surface temperature T_s increases with plasma pressure (Allmen et. al., (1995)) and higher the value of T_s higher the ablation rate (Bauerle et. al., (1996)). The laser ablation in the WCR is a dynamic positive feedback process because the higher ablation rate of material will enhance the confined plasma pressure. Thus ablation rate will be very much enhanced by WCR. Figure 9 shows the setup of laser ablation in the WCR.

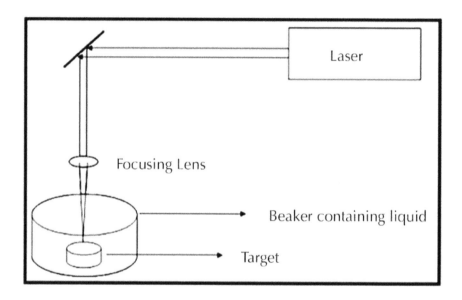

FIGURE 9   The schematic of LP-PLA.

The LP-PLA has been used to produce nanoparticles of many different metal elements including titanium (Simakin et. al., (200)), silicon (Takada et. al., (200)), cobalt (Tsuji et. al., (200)), zinc (Singh et. al., (200)), copper (Yeh et. al., (199)), silver (Shafeev et. al., (200)), and gold (Sylvestre et. al., (200)). This technique can also be used to prepare NPs of compound materials such as TiO_2 (Sugiyama et. al., (200)), Ti C (Dolgaev et. al., (200)), and CoO (Tsuji et. al., (200)) in water, and ZnSe and CdS in various solvents, including water (Ankin et. al., (200)). The use of this method opens up the possibility of studying new materials at the nanoscale range and therefore to envision new applications. Only very recently LP-PLA has gained great attention for its ability to form more complex, higher dimensional nanostructures, and lead to the study of the dynamical process among laser-solid-liquid interactions.

A ZnO (99.99%) mosaic target sintered at 1000°C for 5 hr was used for the synthesis of ZnO nanoparticles. The ZnO target immersed in 15 mL of the liquid media having different pH was ablated at room temperature by thirdharmonic of Nd: YAG laser (355 nm, repetition frequency of 10 Hz, and pulse duration of 9 ns). The laser beam was focused using a lens and the ablation was done at a laser fluence of 15 mJ/ pulse. The spot size of the laser beam is about 1 mm. The duration of ablation was 1 hr for all the experiments. This simple room temperature LP-PLA method produced a highly transparent ZnO nanoparticles well dispersed in the liquid media.

Transmission electron microscopic (TEM) studies confirm the resulting product after laser ablation in different media consisted of particles in the nanoregime. The selected area electron diffraction (SAED) pattern shows concentric rings confirming the hexagonal structure of ZnO. This clearly shows the growth of crystalline ZnO nanoparticles. From these studies, the formation of other species like $Zn(OH)_2$ or ZnO/ Zn core shell structure is not found. Because the ejected molten material from the target normally reacts with medium only at the outer surfac (Nichols et. al., (2006)), the ejected plasma readily cools, thereby forming ZnO itself. Because there are many surface oxygen deficiencies, these nanoparticles will be charged. The ZnO nanoparticles grown by LP-PLA in pure water is usually charged because the isoelectronic points of ZnO (~9.3) is well above the pH = 7.0 of pure wate (Kooli et. al., (1996)). This surface charge provides a shield, preventing further agglomeration and forming self stabilized particles even in the absence of surfactant.

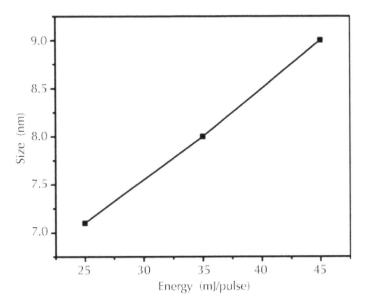

FIGURE 10   Variation of size of the ZnO NPs synthesized by LP-PLA method with laser fluence.

The fluence dependence on the mean size of the particle shows almost a linear increase at lower laser fluences (Figure 10). However at higher laser fluence result in growth of bigger size and wide size distribution of nanoparticles. The longer duration of LP-PLA at lower fluence does not increase the size of the NPs but increase the particle density. The transparency of the ZnO nanoparticle colloid remains as such even for ablation duration of more than 3 hrs at 45 mJ/pulse laser energy. The maximum concentration of ZnO NPs that was achieved while maintaining transparency was 17.5 micro mg/m (Ajimsha et. al., (2008)).

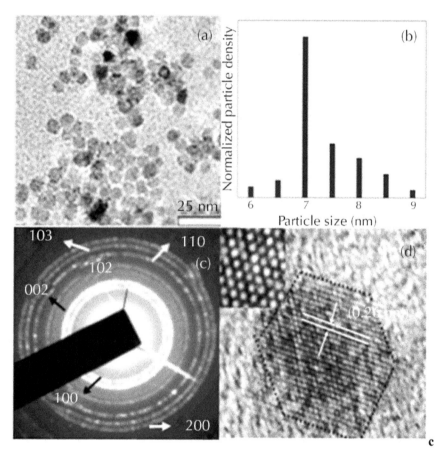

FIGURE 11 (a) TEM image, (b) particle size distributionnd, and (c) SAED pattern of ZnO NPs synthesized in water by LP PLA method with a fluenceof 25 mJ/pulse, and (d) HRTEM image of a single ZnO nanoparticle and inset shows the arrangement in hexagonal close packed mode.

The TEM analysis revealed that ZnO samples after laser ablation with energy 25 mJ/pulse in the neutral media (pH~7) consists of particles in the nano regime as shown in Figure 11a. The particle size distribution shows size distribution is in small range (Figure 11(b)) and majority of the particles have size 7 nm. The d = 0.26 nm shown in the HRTEM image (Figure 11(d)) corresponds to (002) plane of wurtzite ZnO. Almost all particles show uniform size. The selective area electron diffraction (SAED) (Figure 11(c)) exhibit well distinguishable concentric ring pattern representing 100, 002, 102, 110, and 103 plane of hexagonal ZnO. This clearly shows the growth of crystalline ZnO NPs with random orientation. ZnO NPs were arranged in hexagonal shape as observed from high resolution TEM image (Figure 11(d)). The stacking of the 85 hexagonal unit cells make 7 nm.

The TEM image shows that the particles are in spherical shape and it has an average size about 7 nm and the colloid is transparent. The Zn/ZnO composite nanoparticles grown by Zeng et.al (Zeng et. al., (2005)) have an average particle size 18 nm and colored due to turbidity. The size of the particle is found to increase when the experiment is done with oxygen bubbling into the water during laser ablation of ZnO targets. Whilethe size remains the same as that grown in pure water when ZnO NPs were grown in nitrogen atmosphere. Figure 12 (a & b) shows the TEM images of the ZnO NPs prepared in oxygen atmosphere and nitrogen atmosphere. The TEM image of NPs grown in nitrogen atmosphere (Figure 12 (b)) keeping the other parameters of the experiment the same has same size as those grown in neutral demonized water (Figure 12 (c)). The oxygen bubbling during the ablation increases the amount of dissolved oxygen and promotes the growth of ZnO. This leads to bigger ZnO NPs, whereas nitrogen bubbling through the solution does not provide any extra oxygen other than the oxygen in the plasma produced by the laser interaction with the ZnO target. Thus, the size of the particle is same as those obtained by LP-PLA in pure water.

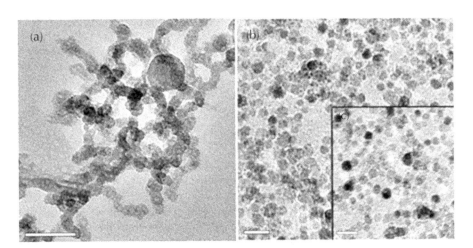

FIGURE 12   The TEM image of the (a) ZnO NPs prepared in oxygen atmosphere, (b) nitrogen atmosphere, and (c) TEM image of ZnO NPs synthesized in water.

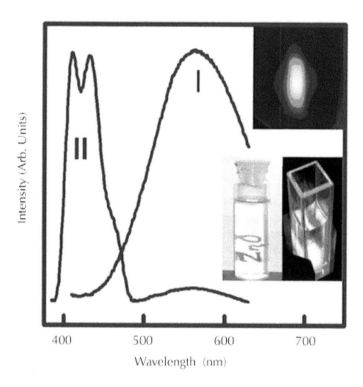

FIGURE 13 The PL emission spectra of ZnO NPs prepared without (curve I) and with (curve II) oxygen atmosphere at an excitation wavelength of 345 nm. Inset bottom shows the photograph of transparent ZnO NPs synthesized in water by LP- PLA method and yellow luminescence emission under UV excitation from ZnO NPs synthesized by LPPLA without oxygen bubbling. Inset top is the bluish- violet luminescence under UV excitation from the NPs grown with oxygen bubbling during LPPLA.

The PL measurement was performed on the NPs dispersed in neutral media at an excitation wavelength of 345 nm. Deep yellow luminescence (Figure 13) was observed from the ZnO NPs dispersed in water. Inset bottom of Figure 13 shows the photograph of highly transparent ZnO NPs dispersed in water and its yellow emission under UV excitation. This yellow luminescence originates from the native oxygen defect (Ajimsha et. al., (2008)) of the prepared ZnO NPs.

The origin of yellow luminescence due to oxygen vacancy was further supported by the experiment carried out with oxygen bubbling into the water during laser ablation of ZnO targets. Figure 13 (curve II) shows PL emissions peaking at 408 nm and 427 nm in the violet blue region, suppressing the yellow emission when oxygen was bubbled through liquid during the ablation. Figure 13 shows the photograph of deep bluish-violet emission. Due to the bubbling of oxygen during ablation, defect density was considerably reduced tending to more stoichiometric ZnO NPs. Whereas the ZnO NPs grown under nitrogen atmosphere has similar size and PL emission characteristic as of those grown in neutral water without oxygen bubbling during the LP-PLA.

This further supports that yellow luminescence originates from oxygen vacancies. Emission at 408 nm is due to the transition of electrons from shallow donor levels to valance band (Zeng et. al., (2006)). According to Lin et. al, (Lin et. al., (2001)) the energy gap between the valance band and energy level of interstitial zinc is 2.9 eV. This is very well consistent with PL emission at 427 nm for the ZnO NPs. The week Raman peak of the solvent corresponding to OH vibration was not detected in the PL spectra mainly because the PL emission intensity was very intense.

The ZnO NPs grown in the present study does not exhibit any green emission. The origin of green emission still remains controversial. However, there is convincing evidence that type of defect responsible for the green emission is located at the surfac (Djurisic et. al., (2006)). The absence of green emission in the LP-PLA grown ZnO NPs suggest the possible presence of $Zn(OH)_2$ on the surface (Zhou et. al., (2002)).

The growth of ZnO NPs by LP-PLA can be modeled as follows. The plasma consisting of ionic and neutral species of Zn and oxygen (Joshy et. al., (2008)) along with water vapor is produced at the solid-liquid interface on interaction between the laser beam and the ZnO target. Due to the high intensity of the laser beam in the nano second scales, high temperature (10^4–10^5K), and pressure of few GP (Berthe et. al., (1997)) in the volume is produced. The adiabatic expansion of the plasma leads to formation of ZnO. The ZnO thus formed interact with the solvent water forming a thin layer of $Zn(OH)_2$ since ZnO is extremely sensitive to H_2O environmen (Nakagawa et. al., (1986)). Thus the ZnO NPs prepared by LP-PLA may have a thin passivation layer of $Zn(OH)_2$. The oxygen bubbling during the ablation increases the amount of dissolved oxygen and promotes the growth of ZnO. This leads to bigger ZnO NPs, where as nitrogen bubbling through the solution does not provide any extra oxygen other than the oxygen in the plasma produced by the laser interaction with the ZnO target. Thus the size of the particle is same as those obtained by LP-PLA in pure water. The ZnO NPs grown by LP-PLA in the acidic medium pH = 5 shows relatively bigger size in comparison with those grown in neutralwater under identical experimental conditions. The very thin passivation layer of $Zn(OH)_2$ during the cooling of laser plasma interacting with the liquid medium may be slower owing to higher dissolution of hydroxide in acidic medium. Hence this favours the growth of bigger ZnO NPs, where as the ablation in alkali medium favours growth of $Zn(OH)_2$ by providing hydroxyl group and hence result in smaller NPs.

Figure 14 (a) shows TEM and Figure 14 (b) is the high resolution transmission electron microscopic (HRTEM) image. The SAED pattern of the ZnO NPs prepared in acid media (pH ~5) keeping all other experimental parameters the same shows the ring pattern corresponding to the 002 plane. The particles have an ellipsoidal shape with 15 nm size in the elongated region and 11 nm along semi minor axis as obtained from the HRTEM image. From the diffraction rings in the SAED pattern, 002 plane of the wurtzite ZnO was identified.

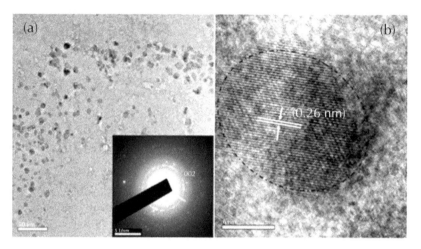

FIGURE 14   (a) TEM image and (b) HRTEM image of ZnO NPs synthesized in acid media by LP- PLA method. Inset of (a) shows the corresponding SAED pattern.

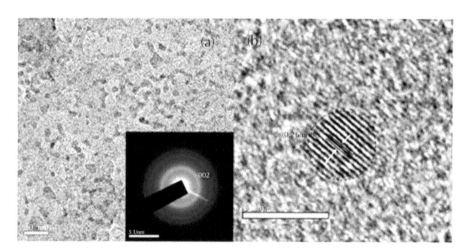

FIGURE 15   The TEM image (a) and HRTEM image (b) ZnO NPs synthesized in basic media by LP- PLA method. Inset of (a) shows the corresponding SAED pattern.

Figure 15 (a) shows the TEM and the SAED pattern (inset) of ZnO NPs prepared by pulsed laser ablation in basic media (pH ~ 9). Spherical particles were observed in the HRTEM image (Figure 15(b)) having a size about 4 nm. The 002 plane of wurtzite ZnO is observed in the SAED pattern. This confirms the formation of crystalline ZnO NPs by pulsed laser ablation in liquid. The ablation in basic medium may favour the faster growth of $Zn(OH)_2$ passivation layer inhibiting the growth of ZnO. This results in smaller size NPs on LPPLA in basic medium.

The thermodynamic conditions created by the laser ablation plume in the liquid are localized to a nano meter scale which is not much influenced by the pH of the solution. The increase of laser energy for the ablation results in increase of size of the NPs due to ablation of more material. The hydroxide passivation layer formation is much influenced by the pH of the aqueous solution which may affect the growth and size of the particles. All the particles grown in acidic, alkali, and neutral medium are well dispersed and no agglomeration of the particles are observed as in the case of ablation of zinc metal targets in aqueous solutio (He et. al., (2007)). In the present study, the particles grown during oxygen bubbling of LP-PLA leads to the formation of bigger particles and agglomeration. This suggests that surface charge of ZnO NPs arise mainly from oxygen deficiency and pH of the medium has less pronounced effect.

6.4 CONFINEMENT IN TWO DIMENSIONS: QUANTUM WIRES

Quantum wires/nanorods represent two-dimensional confinement of electrons and holes. Such confinement permits free-electron behavior in only one direction, along the length of the wire (say the y direction). For this reason, the system of quantum wires describes a one-dimensional electron gas (IDEG) when electrons are present in the conduction band.

The ZnO nanorods were synthesized from $Zn(CH_3COO)_2.2H_2O$ and NH_4OH precursors by hydrothermal method. To the stock solution of $Zn(CH_3COO)_2.2H_2O$, NH_4OH was added drop wise in order to get a pH value of reactants between (8–9). The final solution was transferred into teflon lined sealed stainless steel autoclaves and maintained at 150°C for 3 hr under autogenous pressure. After cooling white solid products of ZnO nanostructures were washed with distilled water, filtered, and then dried in air.

The SEM image of the flower like nanostructure of ZnO is shown in the figure 16. The flower like nanostructure consists of ZnO nanorods that grown at various angles from a single nucleation centre arranging them in a spherical shape exhibiting flower-like morphologies. Each nanorod has hexagonal shape at its tip and these nanorods have an average diameter of 500 nm and a length of 4nmm.

FIGURE 16   The SEM images of ZnO nanostructures grown by hydrothermal method.

The hexagonal nanorods with six defined facets arose to maintain the minimum surface energy as to keep the symmetry of the crystal structure (wurtzite ZnO). It is well known that the radii of newly formed crystal increases linearly with time after nuclei formation. As the new crystals grow, phase boundaries also increases at a given speed and eventually touch each other, forming the base of the structure. Once base is formed, growth rate start to decrease along the transverse direction and growth in the radial direction continues being the top surface an energetically favoured surface. ZnO is a polar crystal, where zinc and oxygen atoms are arranged alternatively along the c-axis and the top surfaces is Zn-terminated (0 0 0 I) while the bottom surfaces are oxygen-terminated (0 0 0 □). The Zn-(0 0 0 1) is catalytically active while the O-(0 0 0 I) is inert (Gao et. al., (2004)). Therefore, the top surface would be energetically active. This will then help in growing in the radial direction once the nuclei is formed.

Raman scattering are studied to further investigate the characteristics of the synthesized ZnO nanorods. The wurtzite ZnO has 12 phonon modes (G) consisting of 9 optical modes and 3 acoustic modes. From the group theory, $A_1 + E_1 + 2E_2$ modes of wurtzite ZnO (space group C_{6v}^4 ($P6_3mc$)) are Raman active. Among these, both E_2 (low) and E_2 (high) modes are Raman active and nonpolar modes which are associated with the vibration of heavy Zn sub-lattice and the motion of oxygen atoms respectively. A_1 and E_1 modes are polar modes and split into transverse optical (TO) and longitudinal optical (LO) phonons. When the incident light is parallel to c-axis of ZnO samples, E_2 modes and LO mode of A_1 are allowed, whereas TO modes of A_1 and E_1 are forbidden according to Raman selection rules. In the Raman measurement, the incident light is exactly perpendicular to the sample surface. Figure 17 shows the Raman spectra of ZnO nanostructures synthesized by hydrothermal method and it confirms the formation of wurtzite ZnO. The broad peak at about 331 cm^{-1} seen in both spectra is attributed to the second-order Raman processe [(Alim et. al., (2005)]). The peak at 379 m^{-1} corresponds to A_1(TO) and 438 m^{-1} corresponds to transverse E line (Tsai et. al., (2012)). A small broad peak at 579 m^{-1} is assigned to the LO mode consisting of A_1(LO) and E_1(LO) modes (Zhang et. al., (2003)).

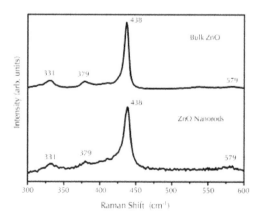

FIGURE 17   Raman spectra of bulk and ZnO nanorods synthesized by hydrothermal method.

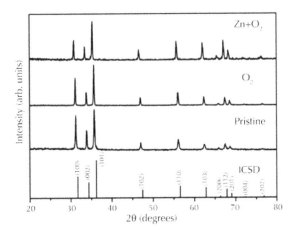

FIGURE 18 The XRD pattern of the ZnO pristine sample and that annealed in oxygen atmosphere and (oxygen + Zn vapor) atmosphere.

In order to determine the defects and mechanism of photoluminescence emission from ZnO nanorods, the hydrothermally synthesized nanorods were annealed at 800°C for 1 hr in oxygen atmosphere and zinc vapors with nitrogen and oxygen as carrier gas. The XRD pattern of the ZnO pristine sample and that annealed in (oxygen+Zn vapor) atmosphere and oxygen atmosphere are shown in the Figure 18. The wurtzite structure of the ZnO was not modified by the annealing in different atmosphere. The room temperature photoluminescence of ZnO nanorods excited with He-Cd laser with an excitation wavelength λ_{ex}) of 325 nm is shown in Figure 19. It shows two peaks in the UV region with a very weak emission in the visible region.

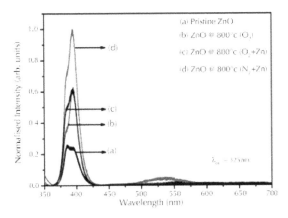

FIGURE 19 Photoluminescence spectra of ZnO nanorods synthesized by hydrothermal method, and then annealed in oxygen, Zn vapor in oxygen and nitrogen atmosphere at 800°C λ_{ex} = 325 nm).

The photoluminescence from ZnO nanorods annealed at 800°C in different atmosphere is also shown in the Figure 19. Pristine ZnO nanorods and the samples annealed at 800°C in nitrogen and oxygen atmosphere in the presence of Zn vapor shows a broad asymmetric PL peak in the UV region corresponding to the recombination of free excitons through an exciton–exciton collision process and defects. The broad UV emission can be deconvoluted into two Gaussian peaks (Figure 20) at 383 and 394 nm. The PL peak at 383 nm is attributed to near band edge emission and the emission at 394 nm is generally attributed to Zn_I related defect (Hanet. al., (2010), Caoet. al., (2006), Chattopadhyay et. al., (2011)). In general, Zn_I are mobile in Zn (Janotti et. al., (2007)). Also, it has been reported that energy state of photo-excited O_V in ZnO may lie close to the conduction band minimu (Lany et. al., (2007)). Theoretically it has been proved that in the presence of O_V, Zn_I may become stable (Kim et. al., (2009)). Stable co-existence of Zn_I and O_V defects have also been identified experimentally in annealed ZnO based ceramic (Cheng et. al., (2008)). From the Figure 19, it is clear that the samples annealed in Zn vapor+ nitrogen atmosphere have higher PL peak intensity at 394 nm than that of the pristine samples and it is due to the Zn_I defects. During the annealing of samples in oxygen atmosphere with the presence of Zn vapors, some of the Zn and oxygen reacts to form stoichiometric ZnO and thus the 394 nm peak have smaller intensity than that annealed in Zn vapor with nitrogen as carrier gas. Relative intensity of near band edge emission was found to decrease for the annealed samples, probably due to the decomposition of ZnO at higher temperatures.

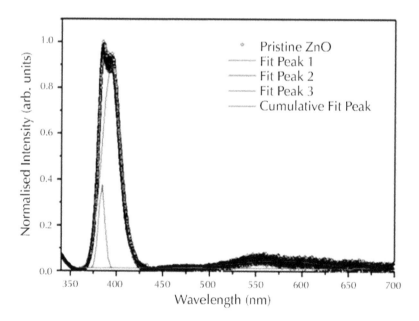

FIGURE 20   Deconvoluted photoluminescence spectra of pristine ZnO nanorods synthesized by hydrothermal method λ_{ex} = 325 nm).

The ZnO nanorods annealed in oxygen atmosphere at 800°C shows a green emission at 535 nm in addition to the peak in UV region. The green luminescence from the ZnO nanostructures is attributed to the presence of the interstitial oxygen defect [(Bayan et. al., (2011)]). Thus, the annealing in oxygen atmosphere leads to appearance of green PL peaks resulting from the interstitial oxygen defects. Samples annealed in the Zn vapors in the oxygen atmosphere shows no PL peak in the visible region because the supplied oxygen and zinc combines to form stoichiometric ZnO. Since, no extra oxygen is introduced in to the system, when annealed in the presence of Zn vapor with nitrogen atmosphere and hence the PL peak intensity of 535 nm remains the same as that of pristine ZnO samples. Thus these studies confirm that the green PL peak at 535 nm is due to oxygen interstitials.

6.5   CONFINEMENT IN ONEDIMENSION: QUANTUM WELL

Quantum wells are structures in which a thin layer (confinement layer) of a smaller band gap semiconductor is sandwiched between two layers (barrier layer) of a wider band gap semiconductor. The heterojunction between the smaller and the wider band gap semiconductors forms a potential well confining the electrons and the holes in the smaller band gap material. The motions of the electrons and the holes are restricted in one dimension (along the thickness direction). This system represents a two-dimensional electron gas (2DEG), when electrons are present in the conduction band.

The optical characteristics of semiconductor bulk materials are mainly determined by the inherent band structure of the material, but utilization of quantum well (QW) structures restricts carrier motion to quasi-two-dimensions and the confinement of carriers at the nanometer scale gives rise to various quantum effects on optical properties. Surface recombination is less important in QW structures as compared to bulk material. Together with high thermal conductivity, high luminous efficiency, and mechanical and chemical robustness, ZnO and its alloys have great prospects in optoelectronic applications in the wavelength range from the ultraviolet to the red (Makino et. al., (2005)). Moreover, excitons in ZnO-based quantum well heterostructures exhibit high stability compared to bulk semiconductors and III-V QWs due to the enhancement of the binding energy (Sun et. al., (2000), (2002)) and the reduction of the exciton-phonon couplin (Ozgur et. al., (2005)) caused by quantum confinement. An important step in order to design high performance ZnO-based optoelectronic devices is the realization of band gap engineering to create barrier layers and quantum wells in heterostructure devices. Band filling is one of the problems associated with single QW structures which lead to luminescence saturation at high current densities. Multiple quantum well (MQW) consists of more than one well so that more carriers can be injected into it.

Recently, much effort has been devoted toward the investigation and fabrication of ZnO/ZnMgO MQWs for UV light emitting application (Gruber et. al., (2004), Makino et. al., (2000), Misra et. al., (2006), Ohtomo et. al., (1999), Zhang et. al., (2005)). Ohtomo et. al., have grown ZnO/ZnMgO MQWs on lattice mismatched sapphire substrates using laser molecular beam epitaxy, but no photoluminescence was observed above 150k (Ohtomo et. al., (1999)a, (2005)). Krishnamoorthy et al (Krishnamoorthy et. al., (2002)) have reported size dependent quantum confinement effects using pulsed PL measurements at 77 K in ZnO/ZnMgO single quantum wells grown on sapphire substrates using the pulsed laser deposition.

Ceramic targets of ZnO and $Zn_{0.9}Mg_{0.1}O$ were used for pulsed laser deposition of $Zn_{0.9}Mg_{0.1}O/ZnO/Zn_{0.9}Mg_{0.1}O$ MQW structure (Figure 21). The $Zn_{0.9}Mg_{0.1}O$ target was prepared by sintering the mixture of 10 at % of high purity MgO and ZnO powder at 1300°C. The ZnO target was prepared by sintering the high purity powder at 1300°C for 5 hrs in air. Q-switched Nd:YAG laser (35 and 266 nm) was used for ablation. The repetition frequency was 10 Hz with a pulse width of 6–7 ns. Laser beam was focused to a spot size 2 mm on the surface of the target and the target was kept in rotation to avoid pitting. The growth chamber was evacuated to a base pressure of 10^{-6} mbar by using turbo molecular pump backed with diaphragm pump, and then O_2 (99.99% purity) was introduced as working gas, maintaining the pressure at 10^{-3} mbar. The target to substrate distance was kept at 6 cm and the substrate temperature was 600°C. The ablation was carried out at constant laser energy of 2 J/cm² for all the depositions. ZnO, and ZnMgO targets were laser ablated individually and the deposited film thickness was measured using stylus profiler. Growth of crystalline and optical quality of ZnO films on sapphire requires a growth temperature of 750°C or highe [(Misra et. al., (2005), Ohkubo et. al., (2000)]) through domain epitax [(Narayan et. al., (2003)]), but the ZnO/ZnMgO hetero interfaces and ZnMgO alloy are chemically stable only beloC 650°C [(Ohtomo (1999)b)]). To accomplish these contradictory growth requirements, a buffer assisted growth methodology has been use [(Misra et. al., (2005)]). To eliminate the lattice mismatch-induced effects, 50 nm ZnO buffer layer was first grown on Al_2O_3 substrate at 700°C by pulsed laser ablation. Then, the ten periods of $Zn_{0.9}Mg_{0.1}O/ZnO/Zn_{0.9}Mg_{0.1}O$ layer was grown on this ZnO template at a substrate temperature at 600°C keeping all the other deposition parameters the same. This ensured a high crystalline quality of the barrier and well layers along with the physically and chemically sharp interfaces. The $Zn_{0.9}Mg_{0.1}O$ barrier layer thickness was 8 nm. The quantum well structures with well layer thickness (ZnO layer) ranging from 2 to 6 nm were grown by pulsed laser deposition. All the layers were deposited sequentially without breaking the vacuum.

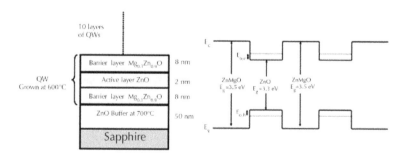

FIGURE 21   Device structure and band diagram of the ZnMgO/ZnO/ZnMgO symmetric MQW structures grown by PLD.

Luminescence emission measurements allow one to determine the energy shift due to the quantum size effects. The band-edge photoluminescence has been studied extensively in ZnO bulk, films, quantum wells, and other nanostructures, and the origin of the PL spectra in the near band-gap region has been propose [(Sun et. al., (2002), Monticone et. al., (1998), Shan et. al., (2005)]). At low temperatures, the PL spectra near the band gap consist of transitions of free and bound excitons, followed by their longitudinal optical phonon replica on the low-energy side as a shoulde [(Sun et. al., (2005)]).

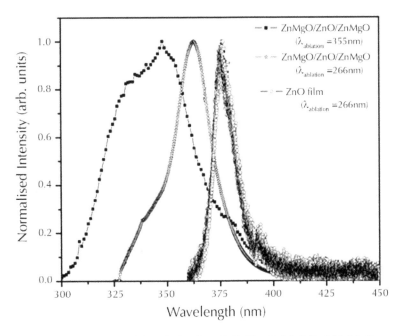

FIGURE 22   Room temperature PL emission from ZnO film, ZnMgO/ZnO/ZnMgO symmetric MQWs at an excitation of λ_{ex} = 266 nm. $\lambda_{ablation}$ is the wavelength used for ablation).

Figure 22 shows the room temperature PL emission at an excitation wave length of 266 nm from ZnO thin film, ZnMgO/ZnO/ZnMgO symmetric grown by third ($\lambda_{ablation}$ = 355 nm) and fourth ($\lambda_{ablation}$ = 266 nm) harmonics of Nd:YAG laser. Room temperature PL emission of ZnMgO/ZnO/ ZnMgO symmetric MQW grown by third harmonic Nd:YAG laser ($\lambda_{ablation}$ = 355 nm) shows a broad peak with small spikes. The large FWHM (~39 nm) and small spikes in the PL spectra is due to the interface roughness and fluctuations in the well layer thickness. The room temperature PL emission from the ZnMgO/ZnO/ ZnMgO symmetric MQW grown by fourth harmonics of Nd:YAG laser with 2 nm as the confinement layer (ZnO layer) thickness shows a sharp PL emission peak centered at 361 nm which corresponds to the band edge of the ZnO. The small peak at 350 nm corresponds to the ZnMgO barrier laye [(Zhu et. al., (2007)]). The PL peak have smaller FWHM (~20 nm) compared to MQW grown by ablating with 355 nm and spikes are absent in the spectrum which indicates the improvement in the layer quality. There is a blue shift in the PL peak position with respect to the bulk which is assigned to the quantum confinement effects. These studies confirms that fourth harmonics of the Nd:YAG laser is better than third harmonics for growing the multiple quantum well structures.

The improved PL emission from the MQW structure indicates the improved quality of barrier and confinement layers and the interface between these layers. Abaltion using fourth harmonic Nd:YAG laser ($\lambda_{ablation}$ = 266 nm) improves the layer properties. The large absorption coefficient of the target material at lower wavelength results in the ablation of a thin surface layer of the material. This can be observed from the lower growth rate of materials at laser energy of $2J/cm^2$ during the ablation with fourth harmonic (laser lambda$_{ablation}$ = 266 nm) of Nd:YAG laser than with third harmonic ($\lambda_{ablation}$ = 355 nm) Nd:YAG laser. The lower laser wave lengthe.that is, at higher incident photon energy results in the highly energetic ablated particles, which improves the crystallinity of the layers.

6.6 CONCLUSION

Various nanostructures of ZnO have been grown by hydrothermal method and pulsed laser ablation technique. The luminescence of these nanostructures can be tuned by controlling the defect chemistry of ZnO nanostructures. The LP-PLA offers the feasibility of growing biocompatible luminescent nanoparticles suitable for biological labeling. The uncapped, non toxic ZnO particles grown in aqueous medium make them attractive for imaging applications. Efficient room temperature photoluminescence emission of the symmetric MQW structures based on ZnO can be utilized for optoelectronic applications.

KEYWORDS

- **Hydrothermal method**
- **Liquid phase pulsed laser ablation**
- **Nanostructures**
- **Nanotechnology**
- **Quantum dots**

ACKNOWLEDGMENT

This work was supported by the Department of Science and Technology under the Nano Science and Technology Initiative. One of the authors (PMA) thanks Kerala State Council for Science, Technology and Environment and Council of Scientific & and Industrial Research for the award of research fellowship.

REFERENCES

1. Ahsanulhaq, Q., Umar, A., and Hahn, Y. B. Growth of aligned ZnOnanorods and nanopencils on ZnO/Si in aqueous solution: growth mechanism and structural and optical properties. *Nanotechnology*, **18**, 115603(1)–115603(7) (2007).
2. Ajimsha, R. S., Anoop, G., Arun, A., and Jayaraj, M. K .Luminescence from Surfactant-Free ZnO Quantum Dots Prepared by Laser Ablation in Liquid. *Electrochem.and Solid-State Lett.*, **11**, K14–K17 (2008).
3. Alim, K. A., Fonoberov, V. A., and Balandin, A. A.. Origin of the optical phonon frequency shifts in ZnO quantum dots. *Appl. Phys. Lett.*, **86**, 053103(1)–053103(3) (2005).
4. Allmen, M. V., and Blatter, A. Laser-Beam Interactions with Material: Physical Principles and Applications, Vol. 2, *Springer Series in Materials Science*, Berlin (1995).
5. Ankin, K. V., Melnik, N. N., Simakin, A. V., Shafeev, G. A., Voronov, V. V., and Vitukhonovsky, A. G. Formation of ZnSe and CdS quantum dots via laser ablation in liquids. Chem. *Phys. Lett.*, **366**, 357–360 (2002).
6. Bajaj, G. and Soni, R. K. Nanocomposite ZnO/Au formation by pulsed laser irradiation. *Appl. Surf. Sci.*, **256**, 6399–6402 (2010).
7. Bahnemann, D. W., Kormann, C., and Hoffmann, M. R. Preparation and Characterization of Quantum Size Zinc Oxide: A Detailed Spectroscopic Study. *J. Phys. Chem.* **91**, 3789–3798 (1987).
8. Baruah, S., Thanachayanont, C., and Dutta, J..Growth of ZnO nanowires on nonwoven polyethylene fibers. *Sci. Technol. Adv. Mater.*, **9**, 025009(1)–025009(8) (2008).
9. Bauerle, D. Laser Processing and Chemistry, *Springer*, Berlin (1996).
10. Bayan, S. and Mohanta, D. Defect mediated optical emission of randomly oriented ZnOnanorods and unusual rectifying behavior of Schottkynanojunctions. *J. Appl. Phys.* **110**, 054316(1)–054316(6) (2011).
11. Berthe, L., Fabbro, R., Peyre, P., Tollier, L., and Bartinicki, E. Shock waves from a water-confined laser-generated plasma. *J. Appl. Phys.* **82**, 2826–2832 (1997).
12. Cao, B., Cai, W., and Zeng, H. Temperature-dependent shifts of three emission bands for ZnOnanoneedlearrays, *Appl. Phys. Lett.* **88**, 161101(1)–161101(3) (2006).
13. Chattopadhyay, S., Dutta, S., Pandit, P., Jana, D., Chattopadhyay, S., Sarkar, A., Kumar, P., Kanjilal, D., Mishra, D. K., and Ray, S. K. Optical property modification of ZnO: Effect of 1.2 MeV Ar irradiation. *Phys. Status Solidi C*, **8**, 512–515 (2011).
14. Chen, Y., Bagnall, D., and Yao, T. ZnO as a novel photonic material for the UV region. Mater. *Sci. Eng. B*, **75**, 190–198 (2000).
15. Cheng, P., Li, S., Zhang, L., and Li, J. Characterization of intrinsic donor defects in ZnO ceramics by dielectric spectroscopy. *Appl. Phys. Lett.* **93**, 012902(1)–012902(3) (2008)
16. Djurisic, A. B. and Leung, Y. H. Optical Properties of ZnO Nanostructures. *Small*, **2**, 944–961 (2006).
17. Dolgaev, S. I., Simakin, A. V., Voronov, V. V., Shafeev, G. A., and Bozon-Verduraz, F.. Nanoparticles produced by laser ablation of solids in liquid environment. *Appl. Surf. Sci.*, **186**, 546–551 (2002).
18. Emanetoglu, N. W., Gorla, C., Liu, Y., Liang, S., and Lu, Y.Epitaxial ZnO piezoelectric thin films for saw filters.*Mater. Sci. Semicond. Process*, **2**, 247–252 (1999)
19. Fabbro, R., Fournier, J., Ballard, P., Devaux, D.,and Virmont, J. Physical study of laser produced plasma in confined geometry. *J. Appl. Phys.* **68**, 775–784 (1990).
20. Flor, J., Marques de. Lima, S. A., and Davolos, M. R. Effect of reaction time on the particle size of ZnO and ZnO:Ce obtained by a sol–gel method. Surface and Colloid Science, **128**, 239–243 (2004).

21. Gao, P.X. and Wang, Z. L. Substrate atomic-termination induced anisotropic growth of ZnO nanowires/nanorods by VLS process. *J. Phys. Chem. B*, **108**, 7534–7537 (2004).

22. Gil, B., and Kavokin, A. V. Giant exciton-light coupling in ZnO quantum dots. *Appl. Phys. Lett.*, **81**, 748–750 (.2002).

23. Gratzel, M. Dye-Sensitized Solid-State Heterojunction Solar Cells. *MRS Bull.*, **30**, 23–27 (2005).

24. Gruber, T., Kirchner, C., Kling, R., and Reuss, F..ZnMgO epilayers and ZnO–ZnMgO quantum wells for optoelectronic applications in the blue and UV spectral region. *Appl. Phys. Lett.*, **84**, 5359–5361 (2004).

25. Han, N. S., Shim, H. S., Seo, J. H., Kim, S. Y., Park, S. M., and Song, J. K. Defect states of ZnO nanoparticles: Discrimination by time-resolved photoluminescence spectroscopy. *J. Appl. Phys.* **107**, 084306(1)–084306(7) (2010).

26. He, C., Saski, T., Usui, H., Shimizu, Y., and Koshizaki, N. Fabrication of ZnO nanoparticles by pulsed laser ablation in aqueous media and pH-dependent particle size: An approach to study the mechanism of enhanced green photoluminescence. *J. Photochem and Photobiology A: Chemistry*, **191**, 66–73 (2007).

27. Henley, S. J., Ashfold, M. N. R., and Cherns, D.The growth of transparent conducting ZnO films by pulsed laser ablation.*Surf.Coat. Tech.*, **177**, 271–276 (2004).

28. Huang, Y., Zhang, Y., Bai, X., He, J., Liu, J., and Zhang X. Bicrystalline Zinc Oxide Nanocombs. *J. Nanosci. Nanotechnol.*, **6**, 2566–2570 (2006).

29. Hughes, W. L. and Wang, Z. L. Formation of Piezoelectric Single-Crystal Nanorings and Nanobows. *J. Am. Chem. Soc.*, **126**, 6703–6309 (2004).

30. Hughes W. L. and Wang Z. L..Controlled synthesis and manipulation of ZnOnanorings and nanobows. *Appl. Phys. Lett.*, **86**, 043106(1)–043106(3) (2005).

31. Hui, Z., Deren, Y., Xiangyang, M., Yujie, J., Jin, X., and Duanlin, Q. Synthesis of flower-like ZnO nanostructures by an organic-free hydrothermal process. *Nanotechnology*, **15**, 622–626 (2004).

32. Janotti, A. and Van de Walle, C. G.. Native point defects in ZnO. *Phys. Rev. B*, **76**, 165202(1)–165202(22) (2007).

33. Joseph, M., Tabata, H., Saeki, H., Ueda, K., and Kawai, T..Fabrication of the low-resistive p-type ZnO by codoping method. *Physica B*, **302**, 140–148 (2001).

34. Joshy, N. V., Saji K. J., andJayaraj, M. K. Spatial and temporal studies of laser ablated ZnO plasma. *J. Appl. Phys*, **104**, 053307(1)–053307(6) (2008).

35. Kim, K. K., Kim, H. S., Hwang, D. K., Lim, J. H., and Park, S. J. Realization of p-type ZnO thin films via phosphorus doping and thermal activation of the dopant. *Appl. Phys. Lett*, **83**, 63–65 (2003).

36. Kim, Y. S. and Park, C. H. Rich Variety of Defects in ZnO via an Attractive Interaction between O Vacancies and Zn Interstitials: Origin of n-Type Doping. *Phys. Rev. Lett.* **102**, 086403(1)–086403(4) (2009).

37. Klug, H. P. and Alexander, L. E. *X-ray diffraction procedures for polycrystalline and amorphous materials*, 1st edn, chapter 9, Wiley, New York (1954).

38. Koch, M. H., Timbrell, P. Y., and Lamb, R. N. The influence of film crystallinity on the coupling efficiency of ZnO optical modulator waveguides. *Semicond. Sci. Technol.*, **10**, 1523–1527 (1995).

39. Kong, X. Y., and Wang, Z. L. Spontaneous Polarization-Induced Nanohelixes, Nanosprings, and Nanorings of Piezoelectric Nanobelts. *Nano Lett.*, **3**, 1625–1631 (2003).

40. Kooli, F., Chsem I,. C., and Vucelic, W. Synthesis and Properties of Terephthalate and Benzoate Intercalates of Mg−Al Layered Double Hydroxides Possessing Varying Layer Charge. *Chem. Mater.*, **8**, 1969–1977 (1996).

41. Krishnamoorthy, S., Iliadis, A. A., Inumpudi, A., Choopun, S., Vispute, R. D., and Venkatesan, T. Observation of resonant tunneling action in $ZnO/Zn_{0.8}Mg_{0.2}O$ devices. *Solid-State Electron.*, **46**, 1633–1637 (2002).

42. Lany, S.and Zunger, A. Dopability, Intrinsic Conductivity, and Nonstoichiometry of Transparent Conducting Oxides. *Phys. Rev. Lett.* **98**, 045501(1)–045501(4) (2007).

43. Lee, J. Y., Choi, Y. S., Kim, J. H., Park, M. O., and Im, S. Optimizing n-ZnO/p-Si heterojunctions for photodiode applications. *Thin Solid Films*, **403**, 553–557 (2002).

44. Lee, C. Y., Tseng, T. Y., Li, S., and Lin, P. Effect of phosphorus dopant on photoluminescence and field-emission characteristics of $Mg_{0.1}Zn_{0.9}O$ nanowires. *J. Appl. Phys*, **99**, 024303(1)–024303(6) (2006).

45. Li, W. J., Shi, E. W., Zheng, Y. Q., and Yin, Z. W. Hydrothermal preparation of nanometer ZnO powders. *J. Mater. Sci. Lett*, **20**, 1381–1383 (2001).

46. Liang, S., Sheng H., Liu, Y., Huo Z., Lu, Y., and Shen, H. ZnO Schottky ultraviolet photodetectors. *J. Crystal Growth*, **225**, 110–113 (2001).

47. Lin, B. X., Fu, Z. X., and Jia, Y. B. Green luminescent center in undoped zinc oxide films deposited on silicon substrates. *Appl. Phys. Lett.*, **79**, 943–945 (2001).

48. Lin, Y., Zhang, Z., Tang, Z., Yuan, F., and Li, J. Characterisation of ZnO-based varistors prepared from nanometre Precursor powders. *Adv. Mater. Opt. Electron.*, **9**, 205–209 (1999).

49. Lingna, W. and Mamoun, M. J. Mater.. Synthesis of zinc oxide nanoparticles with controlled morphology. *J. Mater. Chem*, **9**, 2871–2878 (1999).

50. Look, D. C. Recent advances in ZnO materials and devices. *Mater. Sci. Eng.. B*, **80**, 383–387 (2001).

51. Makino, T., Chia, C. H., Tuan N. T., Sun, H. D., Segawa, Y., Kawasaki, M., Ohtomo, A., Tamura, K., and Koinuma, H. Room-temperature luminescence of excitons in ZnO/(Mg,Zn)O multiple quantum wells on lattice-matched substrates. *Appl. Phys. Lett.*, **77**, 975–977 (2000).

52. Makino, T., Segawa, Y., Kawasaki, M., and Koinuma, H. Optical properties of excitons in ZnO-based quantum well heterostructures. *Semicond. Sci. Technol.*, 20, S78–S91 (2005).

53. Misra, P. and Kukreja, L. M. Buffer-assisted low temperature growth of high crystalline quality ZnO films using Pulsed Laser Deposition. *Thin Solid Films*, **485**, 42–46 (2005).

54. Misra, P., Sharma, T. K., Porwai, S., and Kukreja, L. M. Room temperature photoluminescence from ZnO quantum wells grown on (0001) sapphire using buffer assisted pulsed laser deposition. *Appl. Phys. Lett.*, **89**, 161912(1)–161912(3) (2006).

55. Mitra, A., Chatterjee, A. P., and Maiti, H. S..ZnO thin film sensor.*Mater.Lett.*, **35**, 33–38 (1998).

56. Monticone, S., Tufeu, R., and Kanaev, A. V.Complex Nature of the UV and Visible Fluorescence of Colloidal ZnO Nanoparticles. *J. Phys. Chem.* B, **102**, 2854–2862 (1998).

57. Nakagawa, M. and Mitsudo, H.Anomalous temperature dependence of the electrical conductivity of zinc oxide thin films. *Surf Sci*, **175**, 157–176 (1986).

58. Narayan J., and Larson, B. C. Domain epitaxy: A unified paradigm for thin film growth. *J. Appl.Phys.*, **93**, 278–285 (2003).

59. Nichols, W. T., Sasaki, T., and Koshizaki, N. Laser ablation of a platinum target in water. III. Laser-induced reactions. *J. Appl.*, *Phys.*,**100**, 114913(1)–114913(7) (2006).

60. Ohkubo, I., Matsumoto, Y., Ohtomo, A., Ohnishi, T., Tsukazaki, A., Lippmaa, M., Koinuma H., and Kawasaki M. Investigation of ZnO/sapphire interface and formation of ZnOnanocrystalline by laser MBE. *Appl. Surf. Sci.*, **159**, 514–519 (2000).

61. Ohtomo, A. and Tsukazaki, A. Pulsed laser deposition of thin films and superlattices based on ZnO. *Semicond. Sci. Technol.*, 20, S1–S12 (2005).

62. Ohtomo, A., Kawasaki, M., Ohkubo, I., Kojinuma, H., Yasuda, T., and Segawa, Y.Structure and optical properties of ZnO/Mg0.2Zn0.8O superlattices. *Appl. Phys. Lett.*, **75**, 980–982 (1999)

63. Ohtomo, A., Shiroki, R., Ohkubo, I., Koinuma, H., and Kawasaki M.Thermal stability of supersaturated MgxZn1−xO alloy films and MgxZn1−xO/ZnOheterointerfaces. *Appl. Phys. Lett.*, **75**, 4088–4090 (1999).

64. Ozgur, U., Alivov, Y. I., Teke, A., Liu, C., Reshchikov, M. A., Dogan, S., Avrutin, V., Cho S. J., and Morkoc, H. A comprehensive review of ZnO materials and devices. *J. Appl. Phys.*, **98**, 041301(1)–041301(103) (20.05)

65. Padmavathy, N. and Vijayaraghavan, R. Enhanced bioactivity of ZnO nanoparticles—an antimicrobial study. *Sci. Technol. Adv. Mater.*, **9**, 035004(1)–035004(7) (2008).

66. Pearson, W. B. *A Handbook of Lattice Spacings and Structures of Metals and Alloys.Pergamon Press*, New York (1967).

67. Peng, W. Q., Qu, S.C., Cong. G.W., and Wang, Z.G. Structure and visible luminescence of ZnO nanoparticles. *Materials Science in Semiconductor Processing*, **9**, 156–159 (2006).

68. Ryu, Y. R., Lee, T. S., and White, H. W. Properties of arsenic-doped p-type ZnO grown by hybrid beam deposition. *Appl. Phys. Lett.*, **83**, 87–89 (2003).

69. Saito. N., Haneda. H., Sekiguchi. T., Ohashi. N., Sakaguchi. I., and Koumoto K. Low-Temperature Fabrication of Light-Emitting Zinc Oxide Micropatterns Using Self-Assembled Monolayers. *Adv. Mater.*, **14**, 418–421 (2002).

70. Sakka, T., Iwanaga, S., Ogata, Y. H., Matsunawa, A., and Takemoto, T. Laser ablation at solid–liquid interfaces: An approach from optical emission spectra. *J. Chem. Phys.* **112**, 8645–8653 (2000).

71. Senthil, K. E., Venkatesh, S., and Rao, M. S. R. Oxygen vacancy controlled tunable magnetic and electrical transport properties of (Li, Ni)-codopedZnO thin films. *Appl. Phys. Lett*, **96**, 232504(1)–232504(3) (2010).

72. Shafeev, G. A., Freysz, E., and Bozon-Verduraz, F. Self-influence of a femtosecond laser beam upon ablation of Ag in liquids. *Appl. Phys. A*, **78**, 307–309 (2004).

73. Shan, W., Walukiewicz, W., Ager, J. W., Yu, K. M., Yuan, H. B., Xin, H. P., Cantwell, G., and Song, J. J. Nature of room-temperature photoluminescence in ZnO. *Appl. Phys. Lett.*, **86**, 191911(1)–191911(3) (2005).

74. Simakin, A. V., Voronov, V. V., Kirichenko, N. A., and Shafeev, G. A. Nanoparticles produced by laser ablation of solids in liquid environment. *Appl. Phys. A*, **79**, 1127–1132 (2004).

75. Singh, S. C. and Gopal, R. Zinc nanoparticles in solution by laser ablation technique. *Bull. Mater. Sci.*, **30**, 291–293 (2007).

76. Snure, M. and Tiwari, A.Synthesis, Characterization, and Green Luminescence in ZnONanocages. *J. Nanosci. Nanotechnol.*, **7**, 481–485 (2007).

77. Steven, C. E., Lijun, Z., Michael, I. H., Alexander, L. E., Thomas, A. K., and David, J. N.Doping semiconductor nanocrystals. *Nature*, **436**, 91–94 (2005).

78. Sugiyama, M., Okazaki, H., and Koda, S. Size and Shape Transformation of TiO2 Nanoparticles by Irradiation of 308-nm Laser Beam.Jpn. *J. Appl. Phys.*, **41**, 4666–4674 (2002).

79. Sun, C. K., Sun, S. Z., Lin, K. H., Zhang, K. Y. J., Liu, H. L., Liu, S. C., and Wu J. J. Ultrafast carrier dynamics in ZnOnanorods. *Appl. Phys. Lett.*, **87**, 023106(1)–023106(3) (2005)

80. Sun, H. D., Makino, T., Segawa, Y., Kawasaki, M., Ohtomo, A., Tamura, K., and Koinuma, H. Enhancement of exciton binding energies in ZnO/ZnMgOmultiquantum wells. *J. Appl. Phys.*, **91**, 1993–1997 (2002).

81. Sun, H. D., Makino, T., Tuan, N. T., Segawa, Y., Tang, Z. K., Wong, G. K. L., Kawasaki, M., Ohtomo, A., Tamura, K., and Koinuma H. Stimulated emission induced by exciton–

exciton scattering in ZnO/ZnMgO multiquantum wells up to room temperature. *Appl. Phys. Lett.*, **77**, 4250–4252 (2000).

82. Sun, T., Qiu, J., and Liang, C. Controllable Fabrication and Photocatalytic Activity of ZnONanobelt Arrays. *J. Phys. Chem. C*, **112**, 715–721 (2008).

83. Sylvestre, J. P., Poulin, S., Kabashin, A. V., Sacher, E., Meunier, M., and Luong, J. H. T. Surface chemistry of gold nanoparticles produced by laser ablation in aqueous media. *J. Phys. Chem. B*, **108**, 16864–16869 (2004).

84. Takada, N., Sasaki, T., and Sasaki, K. Synthesis of crystalline TiN and Si particles by laser ablation in liquid nitrogen. *Appl. Phys. A*, **93**, 833–836 (2008).

85. Tok, A. I. Y, Boey, F. Y. C, Du. S. W., and Wong, B. K.Flame spray synthesis of ZrO_2 nanoparticles using liquid precursors. *Materials Science and Engineering: B* **130**, 114–119 (2006).

86. Tsai, M. K., Huang, C. C., Lee, Y. C., Yang, C. S., Yu, H. C., Lee, J. W., Hu, S. Y., and Chen, C. H. A study on morphology control and optical properties of ZnOnanorods synthesized by microwave heating. *J. Lumin.*, **132**, 226–230 (2012).

87. Tsuji, T., Hamagami, T., Kawamura, T., Yamaki, J., and Tsuji, M. Laser ablation of cobalt and cobalt oxides in liquids: influence of solvent on composition of prepared nanoparticles. *Appl. Surf. Sci.*, **243**, 214–219 (2005).

88. Vanheusden, K., Warren, W. L., Seager, C. H., Tallant, D. R., Voigt, J. A., and Gnade, B. E. Mechanisms behind green photoluminescence in ZnO phosphor powders. *J ApplPhys*, **79**, 7983–7990 (1996).

89. Wang, J. B., Zhang, C. Y., Zhong, X. L., and Yang, G.W. Cubic and hexagonal structures of diamond nanocrystals formed upon pulsed laser induced liquid–solid interfacial reaction. *Chem. Phys. Lett.* **361**, 86–90 (2002).

90. Wang, Z. L. Zinc oxide nanostructures: growth, properties and applications. *J. Phys.: Condens. Matter*, **16**, R829–R858 (2004).

91. Wong, E. and Searson, P. C. ZnO quantum particle thin films fabricated by electrophoretic deposition. *Appl. Phys. Lett.*, **74**, 2939–2941 (1999).

92. Xiangyang, M., Hui, Z., Yujie J., Jin, X., and Deren, Y. Sequential occurrence of ZnOnanopaticles, nanorods, and nanotips during hydrothermal process in a dilute aqueous solution. *Materials Letters*, **59**, 3393–3397 (2005).

93. Yang, G. W. Laser ablation in liquids: Applications in the synthesis of nanocrystals. *Prog. Mater. Sci.*, **52**, 648–698 (2007).

94. Yeh, M. S., Yang, Y. S., Lee, Y. P., Lee, H. F., Yeh, Y. H., and Yeh, C. S. Formation and Characteristics of Cu Colloids from CuO Powder by Laser Irradiation in 2-Propanol. *J. Phys. Chem. B*, **103**, 6851–6857 (1999).

95. Zeng, H. B., Cai, W. P., Hu, J. L., Duan, G. T., Liu, P. S., and Li, Y. Violet photoluminescence from shell layer of Zn/ZnO core-shell nanoparticles induced by laser ablation. *Appl. Phys. Lett.*, **88**, 171910(1)–171910(3) (2006).

96. Zeng, H., Cai, W., Li, Y., Hu, J., and Liu, P. Composition/Structural Evolution and Optical Properties of ZnO/Zn Nanoparticles by Laser Ablation in Liquid Media.*J.Phys. Chem. B.*, **109**, 18260–18266 (2005).

97. Zeng, H., Li, Z., Cai, W., Cao, B., Liu, P., and Yang, S. Microstructure Control of Zn/ZnO Core/Shell Nanoparticles and Their Temperature-Dependent Blue Emissions. *J. Phys. Chem. B*, **111**, 14311–14317 (2007).

98. Zhang, B. P., Binh, N. T., Wakatsuki, K., Liu, C. Y., Segawa, Y., and Usami, N. Growth of ZnO/MgZnO quantum wells on sapphire substrates and observation of the two-dimensional confinement effect. *Appl. Phys. Lett.*, **86**, 032105(1)–032105(3) (2005)

99. Zhang, J., Sun, L. D., Yin, J. L., Su, H. L., Liao, C. S., and Yan, C. H. Control of ZnO Morphology via a Simple Solution Route. *Chem. Mater*, **14**, 4172–4177 (2002).

100. Zhang, X., Liu, Y., Chen, S. A novel method for measuring distribution of orientation of one-dimensional ZnO using resonance Raman spectroscopy. *J. Raman Spectrosc.* **36**, 1101–1105 (2005).

101. Zhong, Q. P. and Matijevic, E. Preparation of uniform zinc oxide colloids by controlled double-jet precipitation. *J. Mater. Chem*, **6**, 443–447 (1996).

102. Zhou, H., Alves, H., Hofmann, D. M., Kriegseis, W., Meyer, B. K., Kaczmarczyk, G., and Hofmann, A. Behind the weak excitonic emission of ZnO quantum dots: $ZnO/Zn(OH)_2$ core-shell structure. *Appl. Phys. Lett.*, **80**, 210–212 (2002).

103. Zhu, J., Kuznetsov, A. Y., Han, M. S., Park Y. S., Ahn, H. K., Ju J. W., and Lee I. H. Structural and optical properties of $ZnO/Mg0.1Zn0.9O$ multiple quantum wells grown on ZnO substrates. *Appl. Phys. Lett.* **90**, 211909(1)–211909(3) (2007)

104. Zhu, S., Lu, Y.F., and Hong, M. H. Laser ablation of solid substrates in a water-confined environment. *Appl. Phys. Lett.* **79**, 1396–1398 (2001).

CHAPTER 7

THIN FILM AND NANOSTRUCTURED MULTIFERROIC MATERIALS

B. RANEESH and NANDAKUMAR KALARIKKAL

CONTENTS

7.1   INTRODUCTION

The novel and exciting field of multiferroics has recently attracted a great deal of interest and promising results have already been reported. The term multiferroic was first used by H. Schmid in 1994. Crystals can be defined as **multiferroic** (MF) when two or more of the primary ferroic properties are united in the same phase (Schmid et. al., 1994). These days' people have extended the meaning to incorporate some other long range orders, for example antiferromagnetic. Thus, multiferroic materials can been associated under the title magnetoelectric (ME) which was invented in 1960's. The ME effect, meaning magnetic (electric) induction of polarization P (magnetization M), was first theoretically confirmed in 1959 (Dzyaloshinsky et. al., 1959) and Asrov et. al., confirmed this prediction experimentally in 1960 (Astrov et. al., 1960). During last 10 years, a broad class associated with magnet insulator came into existence recognized as the multiferroic. This new class of materials offers coexistence of long range order associated with magnetism as well as ferroelectricity, using the feasible effects of ferroelectric (FE) behavior through the adjusting associated with magnet field. Multiferroics with combined ferroic properties can apply to specific device applications such as multiple state memory elements, spintronics, and sensors (Martin et. al., 2008). The actual ME coupling among magnetism as well as electric orders in these multiferroics, provides an opportunity to magnetic polarization by making use of electrical field as well as the other way round. This has paved way for the manufacturing of new kind of memory devices, such as electric-field controlled magnetic random access memory (MERAM), in which data can be written electrically and read magnetically.

An overview of the current state of multiferroic thin films and analysis of the key physics issues in this area plays a major role in determining the device performances. As a starting point for the discussion, it is helpful to have a clear picture of multiferroic thin films and nanoscale systems. In this chapter, we describe the fundamentals and basic characteristics of thin film multiferroics, and coupling among order parameters within single phase as well as composite multiferroics. Lastly, few functioning problems as well as suggestions within nanoscale multiferroics, along with specific focus on brand new geometries with regard to multiferroic at nanoscale such as FE nanotubes and self-patterned arrays of multiferroic nanocrystals will be detailed. The fundamental physics behind the scarcity of ferromagnetic (FM)/FE materials has been explored previously (Nicola A. Hill et. al., 2000). It was found that, in general, the d electrons in transition metals, which are essential for magnetism, reduce the tendency for off center FE distortion (Prellier et. al., 2005). As a result, an extra electric or structural driving force should exist with regard to ferromagnetism as well as ferroelectricity to occur. Figure 1 illustrates the relationship between FM, FE, MF and ME materials. The lack of ME multiferroics can be explained mainly by symmetry, electronic properties, and chemistry of the materials (Silvia Picozzi et. al., 2009). Usually, ferroelectrics through definition tend to be insulators whereas ferromagnets need conduction electrons. For example dual exchange ferromagnets like the manganites, magnetism is mediated through incompletely packed 3d shells. One can also find that just 13 point groups may produce multiferroic properties. Also there is a theoretical limit to the ME coupling, larger coupling might be expected for systems that are simultaneously FM and FE (Fiebig et. al., 2005).

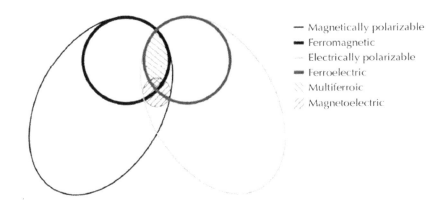

FIGURE 1   Schematic of the relationship between FM, FE, MF, and ME materials (Eerenstein et. al., 2006).

7.2 CLASSIFICATION

All of the multiferroics materials which have been analyzed up to now could be categorized in to two groups: Single phase multiferroics and multiferroic composites. In the following segments, these two classes of magentoelectric materials are discussed.

7.2.1 SINGLE PHASE MULTIFERROICS

All previous works in search for materials that are simultaneously FE and FM have been done on single phase materials. The ME effect of single phase materials comes from the asymmetry between the magnetic polar sub lattices of the crystal structure (Spaldin et. al., 2005). The ME coupling of single phase material is very weak. The ME effect in most single phase ME materials exists only at very low temperatures. The largest ME coefficient previously reported at either low temperature or at high magnetic field (H) is in the order of 1mV/cm.Oe. However, the particular value in the seen ME effect have also been promising but small to apply to any functional products. On the basis of FE and FM single phase, multiferroics can be further divided into two types:

Type-I Multiferroics: In these types of multiferroics magnetism and FE exist independently, and the coupling between the magnetism and ferroelectricity (different sources) leads the formation of some of the most studied and most promising single phase ME multiferroic materials such as

1. *Boracites*: Compounds with chemical formula $M_3B_7O_{13}X$, where M is a bivalent metal ion (M = Cr, Mn, Fe, Co, Cu, Ni) and X = Cl, Br, I, which are as a rule, ferroelectrics and antiferromagnetic/weak FM (E. Ascher et. al., 1966).
2. Lone Pair Ferroelectrics: (Bismuth-based compounds)
    Properties of the Bismuth compounds are largely determined by the Bi 6s lone pair (Sheshadri et. al., 2001). $BiFeO_3$ is an incommensurate antiferromagnet

and a commensurate FE at room temperature. The displacement of Bi, Fe, and O ions from their ideal positions results in spontaneous polarization.

3. *Improper Geometric Ferroelectrics:*

   Hexagonal rare-earth manganites with general formula $RMnO_3$ (R = Ho, Er, Tm, Yb, Lu, Y, and Sc), which show that the FE and antiferromagnetic (AFM) order parameters are coupled. $RMnO_3$ structure consists of trigonal bipyramids associated by their vertices to form layers vertical to the sixfold axis. Most studied member of this group is $YMnO_3$ (Z. J. Huang et. al., 1997).

4. *Proper Geometric Ferroelectrics:*

   Compounds with a general formula $BaMF_4$, (M = Mn, Fe, Co, Ni, Mg, and Zn) have an orthorhombic crystal structure at high-temperature. These compounds exhibit pyro- or FE properties and are antiferromagnetic/weak FM at rather low temperatures. The FE instability in the multiferroic barium fluorides arises specially due to the geometrical constraints of the underlying crystal structure (Claude Ederer et. al., 2006).

Type-II Multiferroics: In these type multiferroics, ferroelectricity is induced by magnetic ordering or charge ordering.

These includes:

1. *FE due to Magnetic Ordering*

   Ferroelectricity appears as result of magnetic ordering of the ions. A change in the ferroelectricity is observed in response to an applied magnetic field, probably mediated through a change in the magnetic structure or domain occupation (Hoyoung Jang et. al., 2011). Few such systems are $TbMnO_3$, $TbMn_2O_5$, $Ni_3V_2O_8$, $CuFe_2O_4$, and $CoCr_2O_4$.

2. *FE due to Charge Order*

   In this class of multiferroic materials, FE appears as a result of charge ordering in the system. Charge-order driven magnetic ferroelectricity is interesting, as it is expected to occur in a large number of rare earth manganites and ferrites (Fen Wang et. al., 2010). Few examples are $LuFe_2O_4$, $Pr_{1-x}Ca_xMnO_3$. RFe_2O_4 (R = a rare earth from Dy to Lu, or Y)

7.2.2   MULTIFERROIC COMPOSITES

Recently, there has been a revival of multiferroic composites (Nan et. al., 2008), in which coupling takes place indirectly *via* strain. One of the two phases should be piezomagnetic or magnetostrictive while the other should be piezoelectric or electrostrictive. Commonly, ferroelectrics in combination with ferromagnets are used for this purpose. These systems can be put into contact in the form of laminates, composites, or epitaxial layers of thin films (Martin et. al., 2010). The coupling coefficients are achieved in the order of magnitude larger than that in single-phase multiferroics. An alternative approach is to construct composite materials, for example, laminate multilayers and other three-dimensional composites, possess ME coupling coefficients three to five orders larger than those of intrinsic multiferroics and thus approach the threshold of technical applications. Among, these studied materials special attention was paid to the discovery of self-organized columnar composite such as $BaTiO_3$

- $CoFe_2O_4$, $CoFe_2O_4$ - $BiFeO_3$, or $CoFe_2O_4$ - $PbTiO_3$, and so on (Rongzheng Liu et. al., 2010).

7.3   THIN FILM MULTIFERROICS

Present electronics technology and industry are largely based on the thin film technology. For thin layers of materials on suitable substrates, materials properties can be reproduced at small scales and in the form which is amenable to making devices. Understanding of the physics of multiferroelectric thin film, mainly the physics relevant to technology, exploits the characteristic properties of multiferroelectrics. An outline of the current state of multiferroelectric devices is tailed by identification and discussion of the key physics issues that determine device performance. This chapter was set up to explore the growth of a variety of known candidates of multiferroic materials in thin film form, with potential for device applications. Such materials might either simplify the operation of current device structures or offer completely novel architectures. The chemistry and structure of these thin film systems therefore have direct impact on their electrical and magnetic properties. Also, growth defects in the films may either have a favorable or an unfavorable effect for the intended applications. Hence, understanding the structure and defects of thin films and of their interfaces becomes useful for the subsequent performance and reliability of devices based on these materials (Ramesh et. al., 2007). Multiferroic thin films as well as nanostructures happen to be created utilizing a wide selection of development methods such as sputtering, spin coating, pulsed laser deposition, solgel processes, metal–organic chemical vapor deposition, molecular beam epitaxy (Martin et. al., 2008), and so on.

Although, less advanced than the bulk multilayers mentioned above, large coupling effects have been observed in thin film heterostructures; for example, in nanopillar heterostructures, switching of the FE polarization by an applied electric field leads to a reversal of the magnetization direction (Zheng et. al., 2004) and (Zavaliche et. al., 2007). We focus specifically on approaches by which magnetic responses can be controlled through the application of an electric field in thin film heterostructures and nanostructures. Although, the number of studies has been small, we expect this to change in the immediate future and this has motivated the researchers to pay further attention to this topic. In single-phase thin film multiferroics, mainly two classes of multiferroics i.e hexagonal manganites and the Bi-based perovskites are the mostly studied.

7.3.1 $RMnO_3$

The $RMnO_3$ compounds are well-known examples of single phase multiferroics whose peculiar properties appear linked to a combination of magnetic frustration and magnetoelastic coupling. At room temperature, these materials crystallize into two types of structures (Filippetti et. al., 2002). For instance, $RMnO_3$, with R = Bi, La, Pr, Nd, Sm, Eu, Gd, and Dy, exhibit orthorhombic structure with the Pbnm space group, while $RMnO_3$ with R = Sc, Y, In, Yb, Lu and Er, crystallize in a hexagonal lattice, described by the $P6_3cm$ space group (Munoz et. al., 2000, Fiebig et. al., 2002). These conditions has allowed systematic investigation of the effects of rare earth ion size on the physical properties of the hexagonal $RMnO_3$ system. Epitaxial thin films often grow with structural characteristics different from their bulk counter parts because of the epitaxial requirement. The first multiferroic to be investigated in thin-film form (Fujimura et al. (1996) was the hexagonal manganite $YMnO_3$. Hexagonal $YMnO_3$ consists of layers of Y^{3+} ions separating layers of corner-shared MO_5 trigonal bipyramids where each manganese ion is surrounded by three in plane and two apical oxygen ion. A net electric polarization in $YMnO_3$ is due to buckling of the layered MnO_5 polyhedra, accompanied by displacements of the Y ions (Gorbenko et. al., 2002). Figure 2 shows the crystal structure of $YMnO_3$ in the paraelectric and FE phases. Below the Neel temperature (77K) the magnetic structure of $YMnO_3$ can be described as frustrated antiferromagnetism on a triangular lattice in the ab-plane (Chen et. al., 2005). The lower T_N is the main limitation for the multiferroic applications of $YMnO_3$, and it is an important issue to improve room temperature magnetic properties of the present material. Since, the FE polarization develops along the (001) directions, and the Mn magnetic moments order in the (a, b) plane in a 120° triangular AFM configuration, the films should be grown with the c-axis of the hexagonal cell perpendicular to the substrate plane. Due to this arrangement of electrical and magnetic orders, epitaxial growth of (001) textured hexagonal thin films is fundamental towards the development of functional structures (Balasubramanian et. al., 2006). Moreover, towards the integration of FE $YMnO_3$ into devices, epitaxial films onto suitable bottom electrodes must be developed. Radiofrequency magnetron sputtering was used to obtain epitaxial (001) films on (111) MgO and (001) ZnO/(001) sapphire, and polycrystalline films on (111)Pt/(111)MgO. Since, this time $YMnO_3$ has been grown on a number of other substrates including Si(001) (Yoo et. al., 2002), Pt/TiOx/SiO$_2$/Si(001) (Suzuki et. al., 2003), Y stabilized ZrO_2(111) (Dho et. al., 2004), and GaN/sapphire (0 0 1) (Posadas et. al., 2005). The multiferroic thin-films show normal decrease in FE polarizations as well as dielectric response in contrast to the corresponding single crystal value. The films also exhibit an enhanced thickness-dependent magnetism compared to the bulk. The polarization–electric field (P–E) hysteresis loops for $YMnO_3$ films have shown that saturation polarization in $YMnO_3$ is rather small and that films can have a retention time of 104 s at ±15V applied fields. Such effects suggest that $YMnO_3$ films could be a suitable material for FE transistors, but the high growth temperatures (800°C (Fujimura et. al., 2003)–850°C (Choi et. al., 2004)) make it impractical for integration into current applications. The trend in variation of magnetic as well FE properties could be mainly ascribed to stress induced during film growth process.

The $ErMnO_3$ is also a good candidate among multiferroic materials. However, there have been only a few studies about physical properties of $ErMnO_3$. An early study on $ErMnO_3$ polycrystalline films by R.F sputtering was attributed to electrical properties of the sample (Seung Yup Jang et. al., 2009). A survey of literature reveals that not much experimental information exists on the magnetic and electric properties of $ErMnO_3$ thin films. $ErMnO_3$ shows a FE transition at 588K and an AFM transition at 77K (J. R. Sahu et. al., 2009). Jang, et. al., have shown that $ErMnO_3$ fabricated on Pt (111) /Al_2O_3 (0001) and yttria stabilized zirconia (111) substrates exhibit a spin glass behavior (Jang et. al., 2008). It is succeeded in fabricating hexagonal $GdMnO_3$, $TbMnO_3$, and $DyMnO_3$ thin films using epitaxial stabilization techniques

(Jang et. al., 2009). The multiferroic transition within the orthorhombic phase is currently comprehended by using two crucial aspects: (1) magnetic frustration due to competing exchange integrals between successive neighbors stabilizes a spiral magnetic phase below the Néel temperature T_N; (2) electric polarization at T_N due to the Dzyaloshinskii–Moriya interaction, where oxygen atoms are pushed off the Mn–Mn bond (Andreev et. al., 2010). As a result, this particular spin–lattice coupling mixes polar phonon and spin waves involving deviations out of the spiral magnetic plane. Orthorhombic YMO_3 phase (space group Pnma) can be obtained by either annealing under high pressure or in epitaxial thin films grown on proper substrates (Salvador et. al., 1998). It was reported that the orthorhombic YMO_3 is AFM (T_N ~40 K) dielectric material (ME) without FE order previously (Van aken et. al., 2004). But some recent work confirms that it is also FE both in ceramic and thin films forms and can be increased up to 60% below the AFM Neel temperature under the external magnetic field (Lorenz et. al., 2007).

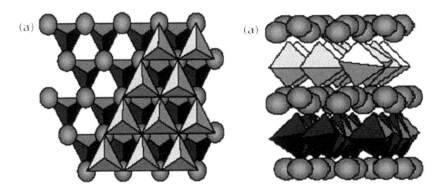

FIGURE 2   The crystal structure of $YMnO_3$ in the paraelectric and FE phases (Van Aken et. al., 2004).

7.3.2   BiFeO₃

A lot of studies have been performed on multiferroic $BiFeO_3$, which shows intrinsic multifunctionality that would apparently make it a strong candidate for nanoscale electronics applications. $BiFeO_3$ is a room temperature, single-phase, multiferroic material with a high FE Curie temperature (1103K) and AFM Néel temperature (643K) (Wang et. al., 2003). Both its spontaneous polarization and saturation magnetization are similar when compared to other single phase multiferroics. But there are still obstacles to be overcome before its applications in multiferroic devices, such as high leakage current, small spontaneous polarization Ps and remnant polarization Pr, high coercive field, FE reliability, and inhomogeneous magnetic spin structure. Bulk crystals possess a rhombohedral symmetry (Figure 3); however, in thin films, strain induced by heteroepitaxy reduces the symmetry to monoclinic (Yang et. al., 2009).

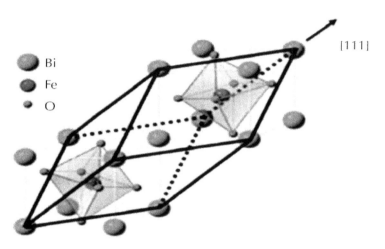

FIGURE 3   The crystal structure of perovskite $BiFeO_3$ (Hiroshi Naganuma (2011)).

Epitaxial and polycrystalline $BiFeO_3$ films have been successfully fabricated *via* various techniques such as pulsed laser deposition (PLD) (Wang et. al., 2003), chemical solution deposition (Lakovlev et. al., 2005), and magnetron sputtering (Lee et. al., (2005). A giant polarization has been demonstrated with properly tailored $BiFeO_3$ films, which can be in the order of magnitude higher than those of their bulk ceramic and single crystal counterparts (Ying Hao Chu et. al., 2007). The remanent polarization in $BiFeO_3$ single crystal has been reported to be 3.5 C/cm^2 along the (1 0 0) direction and 6.1 C/cm^2 along the (1 1 1) direction at 77k temperature and $BiFeO_3$ thin film show enhanced polarization of 50–100 C/cm^2 (Wang et. al., 2003). The magnetic properties of $BiFeO_3$ thin films show higher values of magnetization (Ms = 80 emu/cc) can also be markedly different from those of the bulk (Wang et. al., 2005). The electrical and magnetic properties reported for multiferroic $BiFeO_3$ thin films are largely dependent on synthesis techniques and processing parameters involved. Undoubtedly these electrical behaviors are determined by the film texture, orientation, and interfacial characteristics with the substrates. The unwanted high leakage current due to poor film texture is damaging to the performance of multiferroic $BiFeO_3$ thin films. The large leakage current of the films, which is likely due to defects such as oxygen vacancies, has prevented films from achieving complete saturation. Several research groups have tried Bi-site and Fe-site substitution by using lanthanide to enhance FE reliability and modify its spatially spin-modulated incommensurate structure. Recently, divalent-ion doping effects on $BiFeO_3$ have been studied, and a consensus on the associated phase diagram has been reached. This gives information to implement the electric modulation of conduction and FE states by choosing an optimized doping ratio.

The influence of the substrates on the properties of MF thin films has not been so intensively studied theoretically. The strain effects and thickness dependence of FE and magnetic properties in epitaxial $BiFeO_3$ thin films using the Landau–Devonshire theory are investigated (Hua Ma et. al., 2008). There are many studies of $BiFeO_3$ thin films using different substrates in order to enhance the polarization. Jang et al have grown epitaxial (001) $BiFeO_3$ films on (001) STO substrates, which are subjected to a compressive strain due to the lattice mismatch of -1.4%. In contrast, epitaxial (001) $BiFeO_3$ films grown on (001) Si substrates (Jang et. al., 2008) are under biaxial tensile strain due to the difference in thermal expansion between the film and the substrate. Strain-induced effects on phase transitions in $BiFeO_3$ thin films deposited on (001) STO and (001) Yttria-stabilized zirconia oxide $ZrO_2(Y_2O_3)$ (YSZ) substrates were studied (Kartavtseva et. al., 2007). A saturation magnetization much higher than that reported for the bulk is obtained in $BiFeO_3$ thin films on STO (Thery et. al., 2007).

7.3.3 $BiMnO_3$

A short while ago, there has been brand new involvement in the easy perovskite $BiMnO_3$ as the multiferroic product. Moreira Santos et al have studied the functionality within $BiMnO_3$ thin films, indicating the exact coexistence with FM plus FE properties (Moreira dos Santos et. al., 2002). The first growth of $BiMnO_3$ thin films was on $SrTiO_3$ (0 0 1) single crystal substrates using PLD (Dos Santos et. al., 2004). Thin Film of $BiMnO_3$ has reported to become FE under 450K as well as go through a unique orbital ordering resulting in ferromagnetism at 105K (Atou et. al., 1999). Temperature reliant magnet measurements have demonstrated that this ferromagnetic transition temperatures differs based on the substrate (Son et. al., 2004). This particular depressive disorder within Curie temperatures continues to be related to ideas as different because of stoichiometry problems, stress, as well as dimension results. Structural and compositional depiction in the film completed in the growth windows firmly lowered with regard to volume in $BiMnO_3$ which can be related to the occurrence of a few percent of Bi vacancies (Gajek at el., 2007).

7.3.4 $BiCrO_3$

Bismuth chromite was initially produced by Sugawara et al. and reported to be AFM below 123K having a weakened parasitic FM moment (Sugawara et. al., 1968). From first-principles density functional calculations, $BiCrO_3$ was predicted to have a G-type AFM ordering. But due to the difficult high pressure 40 kbar synthesis, its ferromagnetic and ferroelectric properties have not been unambiguously established. The magnetism and ferroelectricity of $BiCrO_3$ are still debated due to the lack of high quality single crystals of $BiCrO_3$ preventing reliable investigations for understanding its physical properties (Hill et. al., 2002). Thin films of $BiCrO_3$ were grown on $LaAlO_3$ (0 0 1), $SrTiO_3$ (0 0 1), and $NdGaO_3$ (1 1 0) substrates and were shown to be antiferromagnetic, displaying weak ferromagnetism, with an ordering temperature of 120–140K (Murakami et. al., 2006). Piezoelectric response and tenability of the dielectric constant were detected in the films at room temperature. It is found that G-type antiferromagnetism with a spin-orbit induced weak FM component; however, the weak FM component could be terminated when the AFM spins are oriented along a particular one of the three twofold rotation axes (Jun Ding et. al., 2011).

7.3.5 RFe_2O_4

The RFe_2O_4 is really a combined layered oxide compound that includes the alternating stacking of rare earth and iron oxide triangular lattice. The RFe_2O_4 (R = Y, Dy to Lu) compounds possess complex magnetic structures attributed to their spin frustrated triangular lattices. In this family, Fe ions have mixed valence states with an equal number of Fe^{2+} and Fe^{3+} ions in the hexagonal lattice, such spin and charge frustrations in the system lead to their rich magnetic phases (Figure 4). Subramanian et al. have reported the low-frequency magneto dielectric effects of $LuFe_2O_4$ at room temperature in a low magnetic field (Subramanian et. al., 2006). All of the rare earth ferrites show ferrimagnetic ordering around 250K because of the powerful magnetic interactions between the localized Fe moments. Many of the ferrites exhibit a Verwey-type phase transition due to the ordering of Fe^{2+} and Fe^{3+} ions. In this case the origin of ferroelectricity is basically related to the charge-ordering transition arising from strong electron correlation (Ikeda et. al., 1995). This suggests a strong coupling between spin moment and electric dipole, and hence, potential applications of RFe_2O_4 in which the charge and spin degrees of freedom of electrons can be controlled. The research on the RFe_2O_4 thin film is desirable not only to understand its rich magnetic phases and charge-ordered (CO) states but also to explore its applicability. R. C. Rai, et al. have reported that the optical and electronic properties of the $LuFe_2O_4$ thin films show strong temperature dependence with discontinuities around a ferrimagnetic and CO transitions (Rai et. al., 2012). The film shows electronic transitions, energy band gap, as well as resistivity, their own sensitivity to some ferrimagnetic transition in 240K and also the CO transition at 350K, confirming a strong coupling of spin, charge, and orbital degrees of freedom in the $LuFe_2O_4$ thin film.

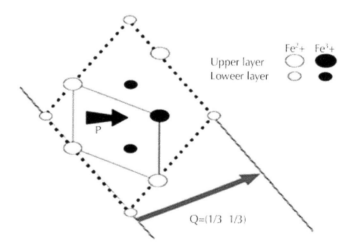

FIGURE 4   . The charge-ordering model of the double iron layer in RFe_2O_4. The polarization P is represented with a short, black arrow. A wave vector Q showing a charge wave is drawn with a long, gray arrow (Naoshi Ikeda et. al., 2005).

7.3.6 RMn₂O₅

The family RMn_2O_5 (R = rare earth, Y and Bi) was described first in 1964 but the multiferroism properties were discovered recently. These compounds possess strong ME coupling, showing remarkable new physical effects, such as the Colossal magneto dielectric (Hur et. al., 2004) and magneto-polarization-flop effects (Kimura et. al., 2003). The crystal structure of RMn_2O_5 is orthorhombic at room temperature with the Pbam space group (Munoz et. al., 2002). It contains edge-shared $Mn^{4+}O_6$ octahedra connecting along the c-axis, and pairs of $Mn^{3+}O_5$ pyramids linking with two $Mn^{4+}O_6$ chains. The Mn^{4+} Mn^{3+} and rare earth R^{3+} ions have magnetic moments, which are responsible for the complex magnetism (Noda et. al., 2008). Hur et al. report a striking interplay between ferroelectricity and magnetism in the multiferroic $TbMn_2O_5$ (Hur et. al., 2004). The dielectric constant of $TbMn_2O_5$ exhibits a maximum change of 13% at 3K and 20% at 28K (Figure 5). The magnetic-field dependence of the total electric polarization in $TbMn_2O_5$ was obtained by measuring the ME current as a function of magnetic field, which was varied uniform rate of 100 Oe s⁻¹ (Hur et. al., 2004). $BiMn_2O_5$ from this family has been widely studied. It is found that the thin film of $BiMn_2O_5$ synthesized on $LaAlO_3$ is magnetically frustrated due to substrate induced strain and demonstrates spin-glass like behavior (Shukla et. al., 2010). Recently the Colossal magnetodielectric (CMD) effect was discovered in $TbMn_2O_5$, $DyMn_2O_5$, and $HoMn_2O_5$ single crystals (Alonso et. al., 1997, Kimura et. al., 2007). The most complex phase diagram was observed for $DyMn_2O_5$ with five subsequent transitions upon decreasing temperature in zero magnetic fields.

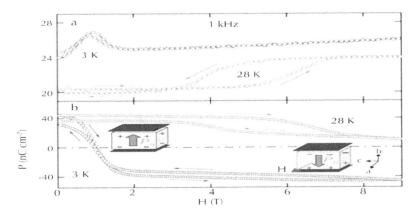

Figure 5. (a) Dielectric constant versus applied magnetic field at 3K and 28K and (b), Change of total electric polarization by applied magnetic fields at 3K and 28K (Hur N et. al., 2004).

7.4 OTHER SINGLE PHASE MULTIFERROIC THIN FILMS

Another sort of multiferroic materials is Bi_2NiMnO_6 (BNMO) which has a double-perovskite composition. Contrary to the particular fragile ferromagnetism regarding antiferromagnetically ordered spin inside $BiFeO_3$, BNMO exhibits huge moment regarding ferromagnetically together Ni and also Mn spin. The film obviously revealed multiferroic attributes, each ferromagnetic behavior having a Curie temperature of approximately 100K as well as ferroelectric conduct having an over loaded polarization of approximately 5 C/cm. Lately, utilizing first-principles thickness practical concept, Baettig and Spaldin analyzed Bi_2FeCrO_6 (BFCO) and forecasted to be ferrimagnetic as well as FE which attributes presumably much bigger than materials from the recognized multiferroic components. The thin film BFCO grown on 100-oriented $SrTiO_3$ substrates display each ferroelectricity as well as magnetism in room temperatures having a optimum dielectric polarization of 28 C/cm^2 in E_{max} = 82 kV/cm. (Riad Nechache et. al., 2006).

The $BiCoO_3$, as a perovskite structure oxide with an excellent FE property, has been studied by many groups for its application in magnetoelectronic devices (Belik et. al., 2006, Urantani et. al., 2005). The magnetic interaction of $BiCoO_3$ is dominated by the short-range AFM coupling through the super exchange interaction (Ren-Yu Tian et. al., 2011). An additional phase much like $BiCoO_3$ which has been created like a thin film is realized in $PbVO_3$ (Martin et. al., 2007). The $PbVO_3$ thin film had been developed upon $LaAlO_3$, $SrTiO_3$, NdGaO $LaAlO_3$/Si substrates and discovered to become a extremely tetragonal perovskite thin film. A number of other single phase multiferroic materials with A-sites doping and magnetic transition metal B-sites doping have been produced in the last few years. $La_{0.1}Bi_{0.9}MnO_3$ thin-films happen to be developed by Gajek et al. Such as in bulk form, the existence of La had been discovered in order to help the expansion associated with single-phase perovskite thin film extending from the depositing pressure-temperature parameter to acquire top quality thin film (Gajek et. al., 2006). The $La_{0.1}Bi_{0.9}MnO_3$ film has also been reported to be FM along with Curie temperature of 95K (Gajek et. al., 2007).

7.5   COMPOSITES THIN FILMS

Multiferroic composite films have been studied for their potential applications in ME systems. In these thin films, not only the compositions become selected but also additionally the morphological features of how these two phases tend to be put together regarding one another within the nanometer scale could be designed. An external electrical field makes a piezoelectric stress within the ferroelectric, that creates a related stress within the materials along with a following piezomagnetic modify the magnetization as well as magnet anisotropy (Martin et. al., 2010). In this particular nature, experts experimentally examined numerous components within a layered thick-film geometry, including ferroelectrics such as $Pb(Zr_xTi_{1-x})O_3$, (Ryu et. al., 2001), (Ryu et. al., 2007) $Pb(Mg_{0.33}Nb_{0.67})O_3$– $PbTiO_3$ (PMN–PT), (Ryu et. al., 2002) and ferromagnets such as $TbDyFe_2$ (Terfenol-D) (Ryu et. al., 2001), $CoFe_2O_4$ (Srinivasan et. al., 2004), $Ni_{0.8}Zn_{0.2}-Fe_2O_4$ $La_{0.7}Sr_{0.3}MnO_3$ (Srinivasan et. al., 2002), and so on. Consequently, the actual ME impact within the composites might be a lot more than several orders associated with magnitude greater than which within single-phase components. Based on the structure as well as preparing circumstances, ME coefficients from the composites differ in order to 10–100 mV/cm Oe in low-frequencies, achieving upward in order to V/cm Oe orders of magnitude associated with degree in resonance (Ce-Wen Nan et. al., 2008). An in-depth study on composites thin films is actually beyond the actual range associated with this particular chapter.

7.6   MULTIFERROIC NAOPARTICLES

The combination of a high surface area, flexibility, and superior directionality makes nano structures suitable for many applications. However, the dimension results within the multiferroism still not clear and its understanding requires more experimental evidences. Difference in properties at bulk and nano-level has provided new insight and direction to the researchers to meet the ever increasing demand of smaller and more efficient devices. Among the few single-phase multiferroics that have been discovered to date, bismuth ferrite ($BiFeO_3$) is the only material that exhibits ferroelectricity and anti-ferromagnetism at room temperature, which makes it suitable for room temperature ME applications. However, the bulk form of synthesized $BiFeO_3$ has weak magnetization and inhomogeneity, giving rise to leakage current, so that it is difficult to observe the FE loops. The low saturation magnetization of bulk $BiFeO_3$ is attributed to the residual moment arising from a canted spin structure. Multiferroic $BiFeO_3$ nanoparticles display size-dependent magnetic behavior where the actual particles tend to be similar due to the incomplete magnetic supercells that retain a net magnetic moment (Tae-Jin Park et. al., 2007). In this system, increased magnetization values at the nanoscale is due to size-confinement effects at room temperature and the presence of insignificant amounts of Fe^{2+}, indicating the absence or great suppression of defects associated with oxygen deficiency. Thus, samples derived from nanoscale systems should possess high resistivity and enhanced multiferroic properties with promising potential. In addition to the potential ME applications, $BiFeO_3$ might find applications as photocatalytic materials due to its small bandgap. In fact, regarding the photocatalytic property of $BiFeO_3$, it was demonstrated that $SrTiO_3$ coated $BiFeO_3$ nanoparticles can produce H_2 under the irradiation of visible light, whereas pure $SrTiO_3$ only responded to UV irradiation (Lou et. al., 2006). Furthermore, $BiFeO_3$ magnetic nanoparticles are employed as being a catalyst to produce a great ultrasensitive way of the particular dedication with regard to H_2O_2.

As an example, low dimensional $BaTiO_3$ nanoparticles, the particular FE Curie temperature diminishes steadily together with particles sizing. On the other hand, recent work has shown that ferromagnetism occurs strongly in nanoparticles of the non-magnetic oxides (Sundaresan et al. 2006) but decreases with increasing particle size. A recent finding of ferroelectricity in much smaller (12 nm) nanoparticles of $BaTiO_3$ (Nuraje et. al., 2006) inspired to investigate the simultaneous coexistence of ferromagnetism and ferroelectricity in a $BaTiO_3$ nanocrystalline sample where the former is expected to arise from the surface and the latter from the core. The particular multiferroic characteristics will be made achievable from the surface magnetism in the nanocrystalline $BaTiO_3$ (Mangalam et. al., 2009) The nanocrystalline $PbTiO_3$ ceramics sintered at various temperatures exhibits coexistence of ferroelectricity and weak ferromagnetism at room temperature. Improvement within the magnet moment as well as ferroelectricity having a decrease in the actual grain scale $PbTiO_3$ ceramics had been noticed. This particular outcome allows for the options of new perovskite electromagnetic devices in the nanoscale region (Min Wang et. al., 2010).

Recent studies in lab, have revealed some interesting aspects related to hexagonal $ErMnO_3$ nanomultiferroic system which indicates that AFM ordering of the Mn sublattice occurs at $T_N = 65K$ (Raneesh et. al., 2012). Indications of weak ferromagnetism in the form of a narrow hysteresis loop at 5K and a very weak ME coupling in $ErMnO_3$ nanoparticles are attributed to intrinsic effects. The results show that phonon transport through $ErMnO_3$ nanocrystalline is effectively suppressed by the enhanced phonon boundary scattering due to size effects. In addition, the strain factor on nanosystem provides a mechanism to tune the thermal conductivity of materials. Quantification of this phenomenon should afford data and criteria for the design and production of multiferroic $ErMnO_3$ nanocrystalline materials based devices. The crucial step for the incorporation within polymers is actually surface area functionalization, because particles much less compared to 100 nm often agglomerate within polymer matrices and results in the formation of hybrid nanocomposites. Sebastian Wohlrab et. al., have reported the integration of single phase multiferroic $BiFeO_3$ nanoparticles into PMMA films without agglomeration (Sebastian Wohlrab et. al., 2008). The nanocomposites have a high transmittance due to the small particle diameter. The mixing associated with $BiFeO_3$ within clear $BiFeO_3$/PMMA films had been demonstrated effectively with regard to spin as well as dip coating techniques.

7.7　MULTIFERROIC NANOCOMPOSITES

The present ME composites can not only be used as on chip use but also in the microelectronic product industry. Numerous studies in history within nanosize involving multiferroics delivers specialized in thin film. A relatively simple stain field distribution might be beneficial with regard to building a great ME coupling and predesigned preparation techniques are needed for designing new materials along with excellent properties.

It really is fascinating to understand, exactly how little the feed dimensions are as well up-and-coming small to conserve the ME effect and it is still hard to get ready thick nanoceramics. The majority of the investigations up to now in the nanosize is associated with multiferroics dedicated to thin films. Nanostructured multiferroic $[xCoFe_2O_4- (1-x)PbTiO_3]$ films were deposited on (001), (110), and (111) $SrTiO_3$ substrates using pulsed-laser deposition from the composite ceramic targets (Igor Levin et. al., 2006). The interphase boundaries in all cases were approximately vertical and extended from the top surface to the film/substrate interface. The actual morphological features from the nanostructures could be managed through changing the strain condition within the film, which may be achieved by utilizing various base orientations as well as stage phase fractions (Zheng et. al., 2004). For composite comprising $CoFe_2O_4$ nanostructures covered by a Pb(Zr,Ti)O$_3$ film, an unexpected out-of-plane magnet easy axis caused through the top Pb(Zr, Ti)O$_3$ layer along with a uniform microdomain framework could be noticed (Xingsen Gao et. al., 2010). Through very careful control of the actual depositing variables, epitaxially developed composites can be acquired.

The actual nanocomposites can have each electric as well as magnet attributes of obvious improvement within out-of-plane anisotropy caused through the top PZT layer as well as display of the ME coupling exposed through magnetocapacitance measurements. Especially for the self-assembled nanocomposite films, the thermal dynamics may have some difference with the physical vapor deposition techniques such as PLD. A better understanding on the growth mechanism will be important to control the structure and properties of these nanocomposite films. L. H. Yan et. al., have reported the preparation of the epitaxial multiferroic composite $BaTiO_3$-$CoFe_2O_4$ thin Films by using polymer-assisted deposition technique (YAN et. al., 2011). The actual self-assembled nanocomposite is actually exposed to become CFO nano-particles inlayed into the BTO matrix. The surface energy and nucleation temperatures associated with the materials are crucial parameters which determine the actual composite growth behavior and microstructures.

Among the attractive composite constructions, self-assembled nanostructures associated with ferro (i) magnetic nanopillars inlayed within a FE matrix as well as the other way round are found to be promising. From the year 2004 onwards works on $BaTiO_3$– $CoFe_2O_4$ nanostructures have started (Zheng et al.). A number of other composites along with immiscible ABO_3 perovskites as well as AB_2O_4 spinels happen to be documented (Zhan et. al., 2006). Currently, transverse nanostructures happen to be produced utilizing epitaxial self-assembly associated with ferrimagnetic $CoFe_2O_4$ as well as FE $BaTiO_3$ phases on single crystal $SrTiO_3$ substrates. These films, which consisted of $CoFe_2O_4$ nanorods in a $BaTiO_3$ matrix, were shown to exhibit substantial ME coupling (Zavaliche et. al., 2007). Conversely, the mechanical constraint arising from the substrate and the good bonding between the FM and the FE phases in the nanostructured films may well much affect the coupling interactions.

The massive difference on lattice parameters between the two phases favors the formation towards pillars by having forces around many nanometers, which determines the increased interface-to-volume relative amount, a real parameter any time working to partnership choices of materials used. Ryan Comes et. al., have demonstrated the growth of self-assembled BFO–CFO nanocomposites with the pillars patterned into a square array through the use of a substrate with CFO islands on the surface patterned using EBL (Ryan Comes et. al., 2012). The top-down lithrographic procedure is utilized in order to design CFO island destinations within the base surface area, accompanied by the bottom-up self-assembly procedure to create the actual ordered CFO-BFO nanocomposite. This technique is more flexible than the techniques used to create other ordered nanocomposites in an arbitrary pattern. Because the flexible coupling within strain-mediated multiferroics is actually mediated through the user interface, great effects of the actual interfacial features in the atomic range enables a big return associated with ME coupling. Core/shell-type nanostructures made up of FE as well as ferri-/FM components offer perfect versions to analyze ME coupling and may be promising candidates for possible applications of multiferroics at the low dimension. Rongzheng et. al., 2010 has reported a general approach for the fabrication of Fe_3O_4/ $PbTiO_3$ core/shell nanostructures with uniform size and shape by a combined hydrothermal and annealing process (Rongzheng et. al., 2010). These types of core/shell-type foundations along with quite strong interactions might work as perfect strain transfer media in order to build high-order multiferroic polymer–ceramic composites.

7.8   ONE-DIMENSIONAL MULTIFERROIC SYSTEMS

One-dimensional nanostructure materials, such as nano-rods, nanowires, and nanotubes, because of their surface effect, size effect, and quantum confinement effect, exhibit special excellent physical properties compared with the materials with conventional structures. It could not only reveal interesting one dimensional multiferroicity at nanoscale but also enable the development of novel multiferroic devices that was impossible before, such as a one-dimensional ME probe with nanometer resolution. For most of the process dedicated to low-dimensional thin films associated with multiferroic composites, the mechanised restriction associated with film substrate offers significantly restricted ME reaction. Multiferroic nanofibers happen to be suggested alternatively in order to ME thin films and also have been recently produced through sol-gel dependent electrospinning. The $BiFeO_3$ (BFO) nanotubes have been studied by several groups. Park et al. have first demonstrated the synthesis of the BFO nanotubes (Park et. al., 2004), and then Zhang et al. have studied their piezoelectric and dielectric characteristics (Zhang et. al., 2005). Wei et al. have studied the photo-absorption and magnetic properties of the BFO nanotubes (Wei et. al., 2008). Sol-gel based electro spinning offers an attractive alternative to synthesize long nanostructured $BiFeO_3$ in the form of fibers that are easier to collect and characterize. The recent studies on such multiferroic composite fibers also suggest that such electrospun ultrafine fibers are nanocrystalline with grain size of $22 \pm 5nm$ having fiber diameter of 400 ± 70 nm (Figure 6). Finer fibers are composed of finer $BiFeO_3$ particles. This, in turn, affects their domain structure and enhances their ferromagnetism and ferroelectricity. Such nanocrystalline $BiFeO_3$ ultrafine fibers remain unexplored in literature. In fiber form, $BiFeO_3$ is suitable for many applications, such as electromagnetic devices and nanosystems, since the fibers can be arranged to form ordered macro-structural arrays, which serve as promising building blocks. It is strongly recommended that extra surface area defects found within BFO nanowires are accountable for powerful noticeable visible green emission. Passivation associated with BFO nanowires surface area along with H_2 considerably enhances the actual near band emission release while controlling surface area recombination. This kind of improved surface area release guarantees numerous possible applications of BFO nanowires with photonic gadgets for example LEDs but additionally within fluorescence-based chemical substance sensing (Prashanthi et. al., 2012). Yangxue et. al., have effectively created hexagonal $YMnO_3$ nanofibers through sol–gel preparation depending on electrospinning (Yangxue et. al., 2010). Right after becoming being heated at 1100°C for 6 hr, the pure hexagonal $YMnO_3$ nanofibers were acquired with a decreased size which ranges from 200 to 800 nm and also the materials had been homogenous within chemical substance metabolism more than the size.

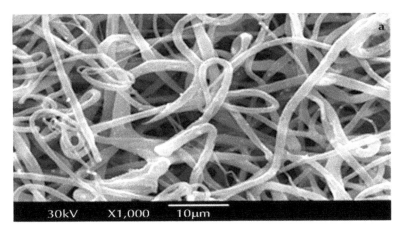

FIGURE 6 The SEM microstructures of PCL/BiFeO$_3$ fibers with 3 wt%, BiFe$_{O3}$ nanoparticles.

To be able to create multiferroic components along with powerful ferroelectricity as well as ferromagnetism at room temperature, hybrid multiferroic composites consisting each FE as well as FM phase needs to be discovered (Nan et. al., 2008). Multiferroic NiFe$_2$O$_4$– Pb(Zr$_{0.52}$Ti$_{0.48}$)O$_3$ composite nanofibers can be produced with a sol-gel based electrospinning method, along with fiber diameters which range from 100 to 400 nm. Great multiferroic properties associated with nanofibers happen to be confirmed, along with ferroelectricity from the nanofibers verified through field-displacement butterfly loop from PFM measurements as well as ferromagnetism verified through magnet hysteresis (Xie et. al., 2008). It is demonstrated that the nanofibers exhibit ME response in the order of magnitude higher than multiferroic composite thin films of similar compositions (Zhang et. al., 2009). Multiferroic nanostructures associated with cobalt ferrite-barium titanate had been produced with a two-step wet chemical substance process, combining the co-precipitation and sol-gel methods. Both level of tetragonality associated with barium titanate and the lattice parameter of cobalt ferrite considerably improved the content material associated with ferrite within the nanostructures.

7.9 CONCLUSION

The amazing advancement within thin film multiferroic as well as nanostructures is a crucial enabler encouraging refreshing results. The specific surface-to-volume percentage proportion weighing scales utilizing the inverse dimensions open up lots of functions which stick to the identical operating law. One of the biggest issues coping with region related to multiferroics nowadays is the requirement of room temperature overall performance. Still, picking out single-phase materials showing cohabitation associated with powerful ferro/ferrimagnetism as well as ferroelectricity is restricted. The physical properties of MF thin films can be adjusted by using different heteroepitaxial constraints and by using different substrates due to lattice mismatch. Composites of piezoelectric and magnetostrictive phases can be electromagnetically coupled *via* stress mediation. Domains, domain walls, and defects play a critical role in unscrambling the ME coupling phenomena. It will be worth to review the works on nanostructured multiferroics (such as nanoparticles, nanocomposites, nanofibers, and so on.) which has also undergone substantial developments in the past few years. The nanostructured multiferroics possess both electrical and magnetic properties with an apparent enhancement in out-of-plane anisotropy and large ME coupling.

KEYWORDS

- **Multiferroic composites**
- **Multiferroic naoparticles**
- **Single phase multiferroics**
- **Thin film multiferroics**

REFERENCES

1. Alonso, J. A., Casais, M. T., Martínez-Lope, M. J., and Rasines, I. High Oxygen Pressure Preparation, Structural Refinement, and Thermal Behavior of RMn_2O_5 (R=La, Pr, Nd, Sm, Eu). *J. Solid State Chem.* **129**, 105 (1997).

2. Andreev, N., Abramov, N., Chichkov, V., Pestun, A., Sviridova, T., and Mukovskii, Ya. Fabrication and Study of $GdMnO_3$ Multiferroic Thin Films. *Acta physica polonica A*, **17**, 217 (2010).

3. Ascher, E., Rieder, H., Schmid, H., and Stössel, H. Some properties of ferromagnetic nickel-iodine boracite, $Ni_3B_7O_{13}$". *J. Appl. Phys.*, **37**, 1404 (1966).

4. Astrov, D. N. The magnetoelectric effect in antiferromagnetics. *Sov. Phys. JETP*, **11**, 708 (1960).

5. Atou, T., Chiba, H., Ohoyama, K., Yamaguchi, Y., and Syono, Y. Structure Determination of Ferromagnetic Perovskite $BiMnO_3$ *J. Solid State Chem.*, **145**, 639 (1999).

6. Balasubramanian, K. R., Kai-Chieh, C, and Feroz, A. Mohammad, Lis, M. Porter, Paul, A. Salvador, Jeffrey DiMaio, and Robert, F. Davis. Growth and structural investigations of epitaxial hexagonal $YMnO_3$ thin films deposited on wurtzite GaN (001) substrates. *Thin Solid Films*, **515**, 1807–1813 (2006).

7. Belik, A. A., Iikubo, S., Kodama, K., Igawa, N., Shamota, S. S., Niitaka, Azuma M., Shimakawa, Y., Takano, M., Izumi, F., and Takayama-Muromachi. Neutron Powder Diffraction Study on the Crystal and Magnetic Structures of $BiCoO_3$. *E. Chem. Mater.*, **18**, 798 (2006).

8. Ce-Wen, Nan, Bichurin, M. I., Shuxiang, Dong, Viehland, D., and Srinivasan, G. Multiferroic magnetoelectric composites: Historical perspective, status, and future directions. *J. Appl. Phys.*, **103**, 031101 (2008).

9. Chen, W. R., Zhang, F. C., Miao, J., Xu, B., Dong, X. L., Cao, L. X., Qiu, X. G., Zhao, B. R., and Dai, P. C. Re-entrant spin glass behavior in Mn-rich $YMnO_3$. *Appl. Phys. Lett.*, **87**, 042508 (2005).

10. Choi, T. and Lee, J. Bi modification for low-temperature processing of $YMnO_3$ thin films. *Appl. Phys. Lett.*, **84**, 5043 (2004).

11. Claude, Ederer and Nicola, A. Spaldin Origin of ferroelectricity in the multiferroic barium fluorides $BaMF_4$: A first principles study. *Phys. Rev. B*, **74**, 024102 (2006).

12. Dho, J., Leung, C. W., MacManus-Driscoll, J. L., and Blamire, M. G. Epitaxial and oriented $YMnO_3$ film growth by pulsed laser deposition. *J. Cryst. Growth*, **267**, 548 (2004).

13. Dos Santos, A. M., Cheetham, A. K., Tian, W., Pan, X., Jia, Y., Murphy, N. J., Lettieri, J., and Schlom, D. G. Adsorption-controlled growth of BiMnO3 films by molecular-beam epitaxy. *Appl. Phys. Lett.*, **84**, 91 (2004).

14. Dzyaloshinskii, I. E. On the magneto-electrical effects in antiferromagnets. *Sov. Phys. JETP*, **10**, 628 (1959).

15. Eerenstein, W., Mathur, N. D., and Scott, J. F. Multiferroic and magnetoelectric materials. *Nature*, **442**, 759 (2006).

16. Fen, W., Chang-Hui, L., Tao, Z., Yi, L. and Young, S. Electrically driven magnetic relaxation in multiferroic $LuFe_2O_4$ *J. Phys.: Condens. Matter.*, **22**, 496001 (2010).

17. Fiebig, M. Revival of the magnetoelectric effect. *J. Phys. D*, **38**, R123 (2005).

18. Fiebig, M., Lottermoser, Th., Frohlich, D., Goltsev, A. V., and Pisarev, R. V. Observation of coupled magnetic and electric domains, *Nature*, **419**, 818 (2002).

19. Filippetti, A. and Hill, N. A. Coexistence of magnetism and ferroelectricity in perovskites. *Phys. Rev. B*, **65**, 195120 (2002).

20. Fujimura, N., Ishida, T., Yoshimura, T., and Ito. Epitaxially grown $YMnO_3$ film: New candidate for nonvolatile memory devices. *Appl. Phys. Lett.*, **69**, 1011 (1996).

21. Fujimura, N., Sakata H., Ito D., Yoshimura T., Yokota T. and Ito T. Ferromagnetic and ferroelectric behaviors of A-site substituted $YMnO_3$-based epitaxial thin films. *J. Appl. Phys.*, **93**, 6990 (2003).

22. Gajek, M., Bibes, M., Barthelemy, Varela A., and Fontcuberta, J. Perovskite-based heterostructures integrating ferromagnetic-insulating La0.1Bi0.9MnO3. *J. Appl. Phys.*, **97**, 103909 (2005).

23. Gajek, M., Bibes, M., Varela, M., Fontcuberta, J., Herranz, G., Fusil, S., Bouzehouane, K.., Barthelemy, A., and Fert, A. $La_{2/3}Sr_{1/3}MnO_3$–$La_{0.1}Bi_{0.9}MnO_3$ heterostructures for spin filterin. *J. Appl. Phys.*, **99**, 08E504 (2006).

24. Gajek, M., Bibes, M., Wyczisk, F., Varela, M., Fontcuberta, J., and Barthélémy, A. Growth and magnetic properties of multiferroic $La_x Bi_{1-x} MnO_3$ thin films. *Phys. Rev. B*, **75**, 174417 (2007).

25. Gajek, M., Bibes, M., Fusil, S., Bouzehouane, K., Fontcuberta, J., Barthélémy, A., and Fert, A. Tunnel junctions with multiferroic barriers. *Nature Mater.*, **6**, 296 (2007).

26. Gorbenko, O. Y., Yu, O., Samoilenkov, S.V., Graboy, I. E., and Kaul, A. R. Epitaxial Stabilization of Oxides in Thin Films. *Chem. Mater.*, **14**, 4026 (2002).

27. Hill, N. A., Battig, P., and Daul, C. First Principles Search for Multiferroism in $BiCrO_3$. *J. Phys. Chem. B*, **106**, 3383 (2002).

28. Hiroshi, N. Multifunctional Characteristics of B-site Substituted BiFeO3 Films, Ferroelectrics - Physical Effects, Dr. MickaÃ«l Lallart (Ed.), ISBN: 978-953-307-453 (2011).

29. Hoyoung, Jang, Lee, J. S., Ko, K. T., Noh, W. S., Koo, T. Y., Kim, J. Y., Lee, K. B., Park, J. H., Zhang, C. L., Sung, Baek, Kim, and Cheong, S.W. Coupled Magnetic Cycloids in Multiferroic $TbMnO_3$ and $Eu_{3/4}Y_{1/4}MnO_3$ *Phys. Rev. Lett.*, **106**, 047203 (2011).

30. Huang, Z. J., Cao, Y., Sun, Y. Y., Xue, Y. Y., and Chu, C. W. Coupling between the ferroelectric and antiferromagnetic orders in $YMnO_3$ *Phys. Rev. B*, **56**, 2623 (1997).

31. Hua, Ma, Lang, Chen, Junling, Wang, J. Ma, and Boey, F. Strain effects and thickness dependence of ferroelectric properties in epitaxial $BiFeO_3$ thin films. *Appl. Phys. Lett.*, **92**, 182902 (2008).

32. Hur, N., Park, S., Sharma, P. A, Guha, S., and Cheong, S W. Colossal Magnetodielectric Effects in $DyMn_2O_5$. *Phys. Rev. Lett.*, **93**, 107207 (2004).

33. Hur, N., Park, S., Sharma, P. A., Ahn, J. S., Guha, S., and Cheong, S. W. Electric polarization reversal and memory in a multiferroic material induced by magnetic fields. *Nature*, **429**, 392 (2004).

34. Igor, Levin, Jianhua, Li, Julia, Slutsker, and Alexander, L. R. Design of Self-Assembled Multiferroic Nanostructures in Epitaxial Films. *Adv. Mater.*, **18**, 2044 (2006).

35. Ikeda, N., Kohn, K., Kito, H., Akimitsu, J., and Siratori, K. Anisotropy of Dielectric Dispersion in $ErFe_2O_4$ Single Crystal J. Phys. Soc. Jpn. 64, 1371 (1995).

36. Izyumskaya N., Alivov Ya., and Morkoç H. (2010) Oxides, oxides, and more oxides: high-κ oxides, ferroelectrics, ferromagnetics, and multiferroics. Critical Reviews in Solid State and Materials Sciences.

37. Jang, S. Y., Lee, D., Lee, J. H., Noh, T. W., Jo, Y., Jung, M. H., and Chung, J. S. Oxygen vacancy induced re-entrant spin glass behavior in multiferroic $ErMnO_3$ thin films. *Appl. Phys. Lett.*, **93**, 162507 (2008).

38. Jang, H. W., Baek, S. H, Ortiz, D., Folkman, C. M., Das, R. R., Chu, Y. H., Shafer, P., Zhang, J. X., Choudhury, S., Vaithyanathan, V., Chen, Y. B., Felker, D. A., and Biegalski, M. D. Strain-Induced Polarization Rotation in Epitaxial (001) $BiFeO_3$ Thin Films. *Phy. Rev.lett.*, **101**, 107602 (2008).

39. Jun, D., Yugui, Y., and Leonard, K. Density functional study of weak ferromagnetism in a thick $BiCrO_3$ film. *J. Appl. Phys.*, **109**, 103905.

40. Kartavtseva, M. S., Gorbenko, O. Yu., Kaul, A. R., Murzina, T. V., Savinov, S. A., and Barthelemy, A. $BiFeO_3$ thin films prepared using metalorganic chemical vapor deposition. *Thin Solid Films*, **515**, 6416 (2007).

41. Kimura, T., Goto, T., Shintani, H., Ishizaka, K., Arima, T., and Tokura, Y. *Magnetic control of ferroelectric polarization Nature (London)*. **426**, 55 (2003).

42. Kimura, T., Kobayashi, S., Wakimoto, S., Noda, Y., and Kohn, K. Magnetically induced ferroelectricity in multiferroic compounds of RMn_2O_5. *Ferroelectrics*, **354**, 77 (2007).

43. Lakovlev, S., Solterbeck, C. H., Kuhnke, M., and Es-Souni, M. Multiferroic $BiFeO_3$ thin films processed via chemical solution deposition: Structural and electrical characterization. *J. Appl. Phys.*, **97**, 094901 (2005).

44. Lee, Y. H., Wu, J. M., Chueh, Y. L., and Chou, L. J. Low-temperature growth and interface characterization of $BiFeO_3$ thin films with reduced leakage current. *Appl. Phys. Lett.*, **87**, 172901 (2005).

45. Lou, J. and Maggard, P. A. Hydrothermal synthesis and photocat- alytic activities of Sr-TiO_3-coated Fe_2O_3 and $BiFeO_3$. *Adv. Mater.*, **18**, 514 (2006).

46. Lorenz, B., Wang, Y. Q., and Chu, C. W. Ferroelectricity in perovskite $HoMnO_3$ and $YMnO_3$. *Phys. Rev. B*, **76** 104405 (2007).

47. Martin, L. W., Zhan, Q., Suzuki, Y., Ramesh, R., Chi, M., Browning, N., Mizoguchi, T., and Kreisel, J. Growth and structure of $PbVO_3$ thin films. *Appl. Phys. Lett.*, **90**, 062903 (2007).

48. Mangalam, R. V. K, Nirat Ray, Umesh, V., Waghmare Sundaresan, A., and Rao, C. N. R. Multiferroic properties of nanocrystalline $BaTiO_3$. *Solid State Commun.*, **149**, 1 (2009).

49. Martin, L. W., Crane, S. P., Chu, Y. H., Holcomb. M. B., Gajek, M., Huijben, M., Yang, C. H., Balke. N., and Ramesh, R. Multiferroics and magnetoelectrics: thin films and nanostructures. *J. Phys. Condens. Matter*, **20**, 434220 (2008).

50. Martin, L. W., Chu, Y. H., and Ramesh, R. Advances in the Growth and Characterization of Magnetic, Ferroelectric, and Multiferroic Oxide Thin Films. *Mater. Sci. Eng.*, **68**, 89 (2010).

51. Wang, Min, Tanw, Guo-Long, and Zhang, Q. Multiferroic Properties of Nanocrystalline PbTiO$_3$ Ceramics. *J. Am. Ceram. Soc.*, **93**, 2151 (2010).

52. Moreira dos Santos, A., Parashar, S., Raju, A. R, Zhao, Y. S, Cheetham, A. K., and Rao, C. N. R. Evidence for the likely occurrence of magnetoferroelectricity in the simple perovskite, BiMnO$_3$ *Sol. StateComm.*, **122**, 49 (2002).

53. Munoz, J. A., Alonso, M. J., Casais, M. T., Martınez-Lope, M. J., Martınez, J. L., and Fernandez-Dıaz, M. T. Magnetic structure and properties of BiMn$_2$O$_5$ oxide: A neutron diffraction study. *Phys. Rev. B*, **65**, 144423 (2002).

54. Murakami, M., Fujino, S., Lim, S. H., Long, C. J., Salamanca-Riba, L. G., Wuttig, M., and Takeuch, I. Fabrication of multiferroic epitaxial BiCrO$_3$ thin films. *Appl. Phys. Lett.*, **88**, 152902 (2006).

55. Munoz, J. A. Alonso, M. J., Martinez-Lope, M. T., Casais, J. L., Martinez and M. T. Fernandez-Diaz. Magnetic structure of hexagonal RMnO$_3$ (R=Y, Sc): Thermal evolution from neutron powder diffraction data. *Phys. Rev. B*, **62** 9498 (2000).

56. Nan, C. W., Bichurin, M. I., Dong, S. X, Viehland, D., and Srinivasan, G. Multiferroic magnetoelectric composites: Historical perspective, status, and future directions. *J. Appl. Phys.*, **103**, 031101 (2008).

57. Naoshi, I., Hiroyuki, O., Kenji, O., Kenji I., Toshiya, I., Kazuhisa, K., Youichi, M., Kenji, Y., Shigeo, M., Yoichi, H., and Hijiri, K. Ferroelectricity from iron valence ordering in the charge-frustrated system LuFe$_2$O$_4$. *Nature*, **436**, 1136 (2005).

58. Nicola, A. Hill. Why Are There so Few Magnetic Ferroelectrics. *J. Phys. Chem. B.*, **104**, 6694 (2000).

59. Noda, Y., Kimura, H., Fukunaga, M., Kobayashi, S., Kagomiya, K., and Kohn, I. Magnetic and ferroelectric properties of multiferroic RMn$_2$O$_5$. *J. Phys.: Condens. Matter*, **20**, 434206 (2008).

60. Park, T. J., Mao, Y., and Wong, S. S. Synthesis and characterization of multiferroic BiFeO$_3$ nanotubes. *Chem Commun*, **23**, 2708 (2004).

61. Posadas, A., Yau, J. B., Ahn, C. H., Han, J., Gariglio, S., Johnston, K., Rabe, K. M., and Neaton, J. B. Epitaxial growth of multiferroic YMnO$_3$ on GaN. *Appl. Phys. Lett.*, **87**, 171915 (2005).

62. Prashanthi, K., Thakur, G., and Thundat, T. Surface enhanced strong visible photoluminescence from one-dimensional multiferroic BiFeO$_3$ nanostructures. *Surface Science*, **606**, 83 (2012).

63. Prellier, W and Singh, M. P. Murugavel P, *J. Phys. Condens. Matter*, **17**, R803 (2005).

64. Rai, R. C., Delmont, A., Sprow, A., Cai, B., and Nakarmi, M. L. Spin-charge-orbital coupling in multiferroic LuFe$_2$O$_4$ thin films. *Appl. Phys. Lett.*, **100**, 212904 (2012).

65. Ramesh, R. and Nicola, A. Spaldin. *Multiferroics: progress and prospects in thin films nature material.* **6**, 21 (2007).

66. Riad Nechache, Catalin Harnage, and Alain Pignolet. Growth, structure, and properties of epitaxial thin films of first-principles predicted multiferroic Bi$_2$FeCrO$_6$. *Appl. Phys. Lett.*, **89**, 102902 (2006).

67. Rongzheng, L., Yuzhen, Z., Rongxia, H., Yongjie, Z., and Heping, Z. Multiferroic ferrite/perovskite oxide core/shell nanostructures. *J. Mater. Chem.*, **20** 10665 (2010).

68. Ryu, J., Priya, S., Uchino, K., and Kim, H. E. Magnetoelectric effect in composites of magnetostrictive and piezoelectric materials. *J. Electroceram.*, **8**, 107 (2002).

69. Ryu, J., Priya, S., Carazo, A. V., Uchino, K., and Kim, H. Magnetoelectric Properties in Piezoelectric and Magnetostrictive Laminate Composites. *J. Am. Ceram. Soc.*, **84**, 2905 (2001).

70. Ryu, S., Park, J. H., and Jang, H. M. Magnetoelectric coupling of [001]-oriented Pb (Zr $_{0.4}$ Ti $_{0.6}$) O $_3$–Ni $_{0.8}$ Zn $_{0.2}$ Fe $_2$ O $_4$ multilayered thin films. *App. Phys. Lett.*, **91**, 142910 (2007).
71. Ryan, C., Hongxue, L., Mikhail, K., Richard, K., Jiwei, Lu, and Stuart, A. Wolf. Directed Self-Assembly of Epitaxial CoFe$_2$O$_4$–BiFeO$_3$ Multiferroic Nanocomposites. *Nano Lett.*, **12**, 2367 (2012).
72. Sahu, J. R., Ghosh, A., Sundaresan, A., and Rao, C. N. R. Multiferroic properties of ErMnO$_3$. *Mater. Res. Bull.*, **44**, 2123 (2009).
73. Salvador, P. A., Doan, T. D., Mercey, B., and Raveau, B. Stabilization of YMnO3 in a Perovskite Structure as a Thin Film. *Chem. Mater.*, **10**(10), 2592 (1998).
74. Seung, Yup Jang, Daesu, Lee, Jung-Hyuk, Lee, and Pattukkannu, Murugavel. Ferroelectric properties of multiferroic hexagonal ErMnO$_3$ thin films. *J. Korean Phy. Soc.*, **55**, 841 (2009).
75. Schmid, H. Multi-ferroic magnetoelectrics. *Ferroelectrics*, **162**, 665 (1994).
76. Sheshadri, R. and Hill, N. A. Visualizing the role of Bi 6s "lone pairs" in the off-center distortion in ferromagnetic BiMnO$_3$. *Chem. Mater.*, **13**, 2892 (2001).
77. Shukla, D. K., Kumar, R., Mollah, S., Choudhary, R. J., Thakur, P., Sharma, K., Brookes, N. B., and Knobel, M. Modifications in magnetic properties of BiMn$_2$O$_5$ multiferroic using swift heavy ion irradiation. *J. Appl. Phys.*, **107**, 09D903 (2010).
78. Silvia, Picozzi and Claude, Ederer. First principles studies of multiferroic materials. *J. Phys. Condens. Matter*, **21**, 303201 (2009).
79. Son, J. Y., Kim, B. G., Kim, J. H., and Cho, C. H. Writing polarization bits on the multiferroic BiMnO$_3$ thin film using Kelvin probe force microscope. *Appl. Phys. Lett.*, **84**, 497 (2004).
80. Spaldin, N. A. and Fiebig, M. The renaissance of magnetoelectric multiferroics. *Science*, **15**, 5733 (2005).
81. Sebastian, W., Hongchu, D., Margarita, W., and Stefan, K. Foam-derived multiferroic BiFeO$_3$ nanoparticles and integration into transparent polymer nanocomposites. *J. Exp. Nanosci.*, **3**, 1 (2008).
82. Subramanian, He. T., Chen, J. Z., Rogado, N. S., Calvarese, T. G., and Sleight, A. W. Giant Room–Temperature Magnetodielectric Response in the Electronic Ferroelectric LuFe$_2$O$_4$. *Adv. Mater.*, **18**, 1737 (2006).
83. Srinivasan, G., Rasmussen, E. T., Levin, B. J., and Hayes, R. Magnetoelectric effects in bilayers and multilayers of magnetostrictive and piezoelectric perovskite oxides. *Phys. Rev. B*, **65**, 134402 (2002).
84. Srinivasan, G., Rasmussen, E. T., Bush, A. A., and Kamentsev, K. E. Structural and magnetoelectric properties of MFe$_2$ O$_4$–PZT (M=Ni, Co) and La x (Ca, Sr) $_{1-x}$ MnO $_3$–PZT multilayer composites. *App. Phys. A*, **78**, 721 (2004).
85. Sundaresan, A., Bhargavi, R., Rangarajan, N., Siddesh, U., and Rao, C. N. R. Ferromagnetism as a universal feature of nanoparticles of the otherwise nonmagnetic oxides. *Phys. Rev. B*, **74**, 161306 (2006).
86. Suzuki, K., Fu, D., Nishizawa, K., Miki, T., and Kato, K. Ferroelectric Property of Alkoxy-Derived YMnO$_3$ Films Crystallized in Argon Jap. *J. Appl. Phys.*, **42**, 5692 (2003).
87. Tae-Jin Park, Georgia, C. Papaefthymiou, Arthur, J. Viescas, Arnold R. Moodenbaugh, and Stanislaus, S. Wong Size-Dependent Magnetic Properties of Single-Crystalline Multiferroic BiFeO$_3$ Nanoparticles. *Nano Lett.*, **7**, 776 (2007).
88. Thery, J., Dubourdieu, C., Baron, T., Ternon, C., Roussel, H., and Pierre, F. MOCVD of BiFeO$_3$ Thin Films on SrTiO$_3$. *Chem. Vapor Depos.*, **13**, 232 (2007).
89. Urantani, Y., Shishidou, T., Ishii, F., and Oguchi, T. First-Principles Predictions of Giant Electric Polarization. *Jpn. J. Appl. Phys.*, **44**, 7130 (2005).

90. Van Aken, B. B., Palstra, T. T. M., Filippetti, A., and Spaldin, N. A. The origin of ferroelectricity in magnetoelectric YMnO₃. *Nat. Mater.*, **3**, 164 (2004).

91. Wang, J., Neaton, J. B., Zheng, H., Nagarajan, V., Ogale, S. B., Liu, B., Viehland, D., Vaithyanathan, V., Schlom, D. G., Waghmare, U. V., Spaldin, N. A., Rabe, K. M., Wuttig, M., and Ramesh, R. Epitaxial BiFeO₃ multiferroic thin film heterostructures. *Science*, **299**, 171 (2003).

92. Wang, Scholl A, Zheng, H., Ogale, S. B, Viehland, D., Schlom, D. G., Spaldin, N. A., Rabe, K. M., Wuttig, M., Mohaddes, L. J. Neaton, Waghmare, U. V., Zhao, T., and Ramesh, R. Epitaxial BiFeO₃ Multiferroic Thin Film Heterostructures. *Science*, **307**, 1203 (2005).

93. Wei, J, Xue, D., and Xu, Y. Photoabsorption characterization and magnetic property of multiferroic BiFeO₃ nanotubes synthesized by a facile sol-gel template process. *Scrip Mater*, **58**, 45 (2008).

94. Xie, S. H., Li, J. Y., Liu, Y. Y., Lan, L. N., Jin, G., and Zhou, Y. C. Electrospinning and multiferroic properties of NiFe₂O₄–Pb(Zr₀.₅₂Ti₀.₄₈)O₃ composite nanofibers. *J. Appl. Phys.*, **104**, 024115 (2008).

95. Xingsong, G., Brian, J. R., Lifeng, L., Balaji, B., Daniel, P., Michael, Z., Marin, A., and Dietrich, H. Microstructure and properties of well-ordered multiferroic Pb (Zr, Ti) O₃/CoFe₂O₄ nanocomposites. *Nano*, 1099 (2010).

96. Xing-Yuan Chen, Ren-Yu Tian, Jian-Ming Wu, Yu-Jun Zhao, Hang-Chen Ding, and Chun-Gang Duan. Fe, Mn, and Cr doped BiCoO₃ for magnetoelectric application: a first-principles study. *J. Phys.: Condens. Matter*, **23**, 326005 (2011).

97. Yangxue, Ye, Huiqing, Fan, and Jin, Li. Fabrication and texture evolution of hexagonal YMnO₃ nanofibers by electrospinning. *Materials Letters*, **64**, 419 (2010).

98. Yan, L. H., Liang, W. Z., Liu, S. H., Huang, W., and Lin, Y. Multiferroic BaTiO₃-CoFe₂O₄ Nano Composite Thin Films Grown by Polymer-Assisted Deposition. *Integr. Ferroelectr.*, **131**, 82 (2011).

99. Yang, C. H., Seidel, J., Kim, S. Y., Rossen, B., Yu, P., Gajek, M., Chu, Y. H., Martin, L. W., Holcomb, M. B., He, Q., Maksymovych, P., Balk, S. V. Kalinin., Baddorf, A. P., Basu, S. R., Scullin, M. L., and Ramesh, R. Electric modulation of conduction in multiferroic Ca-doped BiFeO₃ films. *Nature Materials*, **8**, 485 (2009).

100. Ying-Hao, C., Lane, W. Martin, Mikel, B. Holcomb, and Ramamoorthy, R. Controlling magnetism with multiferroics. *Mat.today*, **10**, 10 (2007).

101. Yoo, D. C., Lee, J. Y., Kim, I. S., and Kim, Y. T. Microstructure control of YMnO₃ thin films on Si (100) substrates. *Thin Solid Films*, **416**, 62 (2002).

102. Zavaliche, F., Zhao, T., Zheng, H., Straub, F., Cruz, M. P., Yang. P. L., Hao, D., and Ramesh, R. Electrically assisted magnetic recording in multiferroic nanostructures. *Nano Lett.*, **7**, 1586 (2007).

103. Zhan, Q., Yu, R., Crane, S. P., Zheng, G., Kisielowski, C., and Ramesh, R. Structure and interface chemistry of perovskite-spinel nanocomposite thin films. *Appl. Phys. Lett.*, **89**, 172902 (2006).

104. Zhang, X. Y., Lai, C. W., and Zhao, X. Synthesis and ferroelectric properties of multiferroic BiFeO₃ nanotube arrays. *Appl Phys Lett*, **8**, 143102 (2005).

105. Zhang, C. L., Chen, W. Q., Xie, S. H., Yang, J. S., and Li, J. Y. The magnetoelectric effects in multiferroic composite nanobers. *Appl. Phys. Lett.*, **94**, 102907 (2009).

106. Zhao, T., Zheng, H., Straub, F., Cruz, M. P., Yang, P. L., Hao, D., and Ramesh, R. Electrically assisted magnetic recording in multiferroic nanostructures. *Nano Lett.*, **7**, 1586 (2007).

107. Zheng, H., Wang, J., Lofland, S. E., Ma, Z., Mohaddes-Ardabili, L., Zhao, T., Salamanca-Riba, L., Shinde, S. R., Ogale, S. B., Jia, Y., Schlom, D. G., Wuttig, M., Roytburd, A., and Ramesh, R. Multiferroic BaTiO$_3$-CoFe$_2$O$_4$ Nanostructures. *Science*, **303**, 661 (2004).

CHAPTER 8

CURRENT ADVANCES IN NANOMEDICINE: APPLICATIONS IN CLINICAL MEDICINE AND SURGERY

INDU RAJ, VINOD KUMAR P, and NANDAKUMAR KALARIKKAL

CONTENTS

8.1   INTRODUCTION

Materials with components less than 100 nm in at least one dimension are called nanomaterials. Nanoparticles have significant effects like surface effects, size effects, quantum effects, and so on. Nanocomposites usually exhibit improved properties like stiffness, enhanced transparency, toughness, increased scratch, abrasion, solvent, and heat resistance compared to traditional materials. Nanotechnology involves the use of nanomaterials and biotechnology in various fields (Preeti et. al., 2011).

Nanostructured materials, biotechnology, genetic engineering, complex molecular machine systems, nanorobots, and so on have wide applications in clinical medicine for preserving and improving human health. This application has led to the emergence of a novel field known as **nanomedicine.** In this field, nanomaterials and technology are utilized for preventing, diagnosing, and treating various diseases. With these tools, diagnosis and treatment are possible in the shortest possible time (Robert et. al., 1999, Preeti et. al., 2011).

In this chapter, an attempt has been made to compile currently available literature related to applications of nanomaterials and technology in the field of modern medicine. Our aim is to give an overall idea about nanomaterials and technologies utilized in medical and surgical specialties.

8.2   NANODIAGNOSTICS

The Medical diagnostic procedures can be divided into in vitro diagnostic techniques, implantable sensors and imaging techniques. Whenever nanomaterials and technology are utilized in this field, this comes under **nanodiagnostics**. Nanotechniques are faster and more sensitive and only small amount of sample is needed. There are potential diagnostic applications like biomarker discovery, cancer diagnosis, detection of microbial infectious agents, and so on (Jain et. al., 2005 (a), Hassan et. al., 2006). A specific example for the use of nanomaterials is bio-barcode assay. Bio-barcode assay is proposed as an alternative to the polymerase chain reaction (PCR) (Hassan et. al., 2006). Nanotechnology is reported to improve PCR as well as to provide non-PCR methods for rapid diagnostics. This assures improved sensitivity and reduced cost for many diagnostic procedures (Hassan et. al., 2006). With this technology, diagnosis at the single-cell and molecular levels are possible; e.g.: biochips and nano biosensors could integrate diagnostics with therapeutics, and help in the development of personalized medicine. Examples of nanodiagnostic tools are quantum dots (QDs), gold nanoparticles, and so on (Jain et. al., 2005 (a), Morrow et. al., 2007).

8.3   NANOMATERIALS

Various nanoparticles have been utilized in modern medical and surgical specialties. In this chapter, we summarize the most important nanomaterials used in nanomedicine.

8.3.1   QUANTUM DOTS (QD)

Quantum dots (QDs) are fluorescent semiconductor inorganic nanocrystals ranging in size from 1–10 nm. They contain groups II–IV (e.g., Cd-Se, Cd-Te, Cd-S, and Zn-Se) or III–V (e.g., In-P and In-As) elements of the periodic table. The QDs consist of an inorganic shell/core and an organic coating, which renders them biocompatible. The inorganic core size determines the wavelength (color) of light emitted following excitation. Inorganic, core of group III–V elements is preferable for their clinical performance in comparison to group II–IV elements. This is mainly due to lower toxicity and the higher stability of the group III–V elements (Madani et. al., 2013). The QDs are one of the most promising nanostructures for diagnostic applications. These inorganic fluorophores offer broad excitation spectra, high sensitivity, and stable fluorescence, which make them suitable for molecular diagnostics, genotyping, and multiplexed diagnostics. These are very much useful in biomedical sensors and in imaging, and are proven to be superior to traditional organic dyes (Medintz et. al., 2005).

Dual-mode imaging can be done with nanoparticles which has tremendous application in oncology. Nanoparticle probes (QDs and magnetic iron oxide nanoparticles combined) are useful for imaging tumors (Choi et. al., 2006). The QD bioconjugates are used for viewing cancer cells in living animals. The bioconjugated QDs are used for correlation with disease progression and response to therapy (Xing et. al., 2007). With all these advances, **theranostics** is now developing very fast. This field is very much important because in this branch, the effect of new drug therapy on the individual is assessed and a treatment may be tailored, based on the test results. This is the key to personalized medicine. This enables the clinician to predict the outcome of drug therapy. For example, QDs covalently linked to antibodies coated with polyacrylate caps are used for immunofluorescent labeling of breast cancer marker HER-2 (Hassan et. al., 2006).

The main disadvantage with the use of QDs in nanomedicine is its toxicity. This requires a biologically compatible surface coating which may shield the toxic core from the surroundings. This coating can increase QD size that may lead to problems of excretion and systemic sequestration. Current research is focused on mechanisms to reduce QD toxicicty. One of the promising mechanisms is the use of polyhedral oligomeric silsesquioxane (POSS)-coating. Researchers expect that POSS coating can improve photostability, biological compatibility, and colloidal stability. This coating does not alter the size and photophysical properties of QDs. The amphiphilic nature of the coating makes the particle soluble in aqueous solutions and permits rapid transfer across cell membranes, permits the use of QDs in lower concentrations for imaging applications (Rizvi et. al., 2012; Xing et. al., 2007).

8.3.2   GOLD NANOPARTICLES

Gold has a very long track record in various medical treatments with much less side effects. Gold contains electrons that are free to move throughout the metal, not tied to a particular atom. These electrons act as conductors of current when a voltage is applied.

In metallic gold, the free electrons absorb the energy from a particular wavelength of light (wavelength range from 700–800 nm) and convert it into heat. This property is utilized in treating large tumors in human body. This is known as **hyperthermia therapy.** The ease of bioconjugating gold nanoparticles with various targeting ligands suggests their use as selective photothermal agents in molecular cancer cell targeted therapy. It is found that gold nanoparticles can noninvasively detect and destruct cancer. Gold nanoparticle probes were designed to increase fluorescence intensity in fluorescence-based assays and detection techniques (Swierczewska et. al., 2011; Brown et. al., 2010 El-Sayed et. al., 2005). Light scattering ability is also exhibited by these tiny particles, which make them suitable for dark field microscopy and transmission electron microscopy. Thus gold nanoparticles have opened new avenues in various fields especially in diagnostics and therapeutics (Stuchinskaya et. al., 2011; Ali et. al., 2012; Peng G et. al., 2009; Chitrani, 2010).

8.3.3   SILVER NANOPARTICLES

Silver nanoparticles contain silver atoms and are generally smaller than 100 nm. Nanosilver exhibits unusual physical, chemical, and biological properties. It is used and marketed as a water disinfectant and room spray (Chen et. al., 2008). Majority of the biocidal silver products contain silver in nanoform (Nowack et. al., 2011). Usage of nanosilver is becoming more and more widespread in medicine (Chen et. al., 2008). Silver (Ag) is an accepted antibacterial material for treating wounds and chronic diseases. It is a strong cytotoxic agent against a broad range of microorganisms. Nanosilver also has strong anti-inflammatory effects. However, silver salt and silver metal release silver ion too rapidly or too inefficiently, which limit its usage in the biomedical field (Wei Chook et. al., 2012). Nano silver (NS) toxicity is also to be critically discussed before starting widespread usage in the medical specialties (Chaloupka et. al., 2010). The interaction of nanosilver particle with the human body after entering via different portals, their distribution, and accumulations at different organs, degradation, adverse effects, and toxicity need further discussion and evaluation (Chen et. al., 2008).

8.3.4   TITANIUM DIOXIDE

Titanium dioxide can absorb ultraviolet light very well. Titanium dioxide alone or in combination with Zinc oxide is used in sun screens. When used in micro sized forms, their opaqueness causes cosmetic problems. However, this issue was addressed when nanoparticles were used (Smijs et. al., 2011). Photocatalytic property of this nanoparticle ,ie, the capability to use light energy to catalyze reactions at reduced temperatures, is utilized in cancer therapy (Ju-Young Park et. al.,2009 Akira et. al.,2000,Wiyong et. al.,2009).

8.3.5   IRON OXIDE

Both iron and iron oxide have magnetic properties. There are two unpaired electrons for iron oxide and iron has four unpaired electrons. Since unpaired electrons make a material magnetic, iron oxide shows less magnetic properties than iron. Iron oxide is therefore called a paramagnetic material (Yi-Xiang et. al., 2001). The paramagnetic properties are shown by iron oxide both in nano form and in bulk. Advantage of these tiny particles is that they can go, where larger particles cannot. Over the past decades, superparamagnetic iron oxide nanoparticles (SPIONs) have a great role in biomedical research and clinical applications. In the presence of an applied magnetic field, SPIONs exhibit magnetic properties. They are used in both *in vivo* applications such as hyperthermia (HT), magnetic drug targeting (MDT), magnetic resonance imaging (MRI), gene delivery (GD), and *in vitro* magnetic separation. Successful applications of SPIONs rely on the particle's size, shape, and size distribution (Lin et. al., 2008).

In MRI, if paramagnetic nanoparticles are attached to the object then a better image can be obtained. Making a core made of iron oxide nanocrystals surrounded by nanoporous silica can also improve the MRI images of tumors. The SPIONs can be excellent MR contrast agents when coated with biocompatible polymers. Commonly used polymers include proteins, polysaccharides, lipids, and hydrophilic synthetic polymers. These coating agents improve the stability and biocompatibility of SPIONs and reduce their aggregation. Galactose, mannose, folic acid, and antibodies when applied to SPION surfaces, tissue specificity to hepatocytes, macrophages, and to tumor regions can be improved, which in turn improve biocompatibility and reduce non-specific uptake (Muthiah etal, 2013). The SPIONs conjugated with targeting ligands/proteins are very much used for drug delivery applications (Mahmoud et. al., 2011). It is also used as controlling agents for the release of therapeutic drugs. Researchers are developing drug delivery methods where therapeutic molecules stored inside the pores are slowly released in a diseased region of the body e.g. near a tumor.

8.3.6 CARBON NANOTUBES

The CNTs are composed only of carbon and are well-ordered, hollow graphitic nanomaterials. They have high aspect ratio, high surface area, and are ultra-light weight. They possess unique physical and chemical properties. The CNTs absorb near-infrared (NIR) light and generate heat. These unique properties facilitate the use of CNTs in drug delivery and thermal treatment of cancer (Madani et. al., 2013). The CNT act as porous vehicles such as mesh or bundle within which active drug molecules are entrapped and allows slow release of the drug. The CNTs will thereby act as channels for drug molecules (Foldvari et. al., 2008). The CNT are very useful in tissue regeneration and they are good biosensors. Researchers have developed elastic materials embedded with needle like carbon nanofibers. The material is used as balloons. Balloons are inserted next to diseased tissues, and then inflated, allowing the carbon nanofibres to penetrate into the cell and deliver therapeutic drugs. Recent reports indicated that exaggerated inflammation and mesothelioma-like lesions are produced in mice by carbon nanotubes (Yoshioka et. al., 2011, 2010), which calls for its toxicity issues as well.

8.3.7   HYDROXYAPATITE

Hydroxyapatite (HA) is a highly biocompatible, natural biomineral. It is considered as an ideal material for bone substitutions. It has got similar crystallographic structure and chemical composition to that of bone mineral. Hydroxy apatite is an osteoconductive material, which is usually coated on metallic implant surfaces to enhance osseointegration. Usually the coating is done by plasma spraying, but failures can occur due to the mismatch of the coefficient of thermal expansion of titanium alloy and HA coating. Now with the development of nano HA, efforts are going on to overcome this, with sintering technique at temperature lower than 800°C. Nano HA is useful in tissue regeneration also.

The HA's poor mechanical properties like, brittleness and low strength, hinder its use in load bearing implants and high-load applications. But nanocomposites of HA potentially improve both mechanical properties and biocompatibility of bone grafting materials. According to the kind of reinforcement required, bioactive ceramics, bioactive glass or glass-ceramic, and metals HA composites are available. Both non-biodegradable and biodegradable polymers are used. The biodegradable class includes collagen, chitosan, polylactides, gelatin, as well as polyanhydrides. Eg: Biotic bones (nanocomposites composed of nanohydroxyapatite (n-HA) and collagen). Still a major problem which exists is that the biological properties of both nanoHA and vital tissue do not match with each other (Wang et. al., 2008, Ning et. al., 2003). Attempts are going on to make nanoHA as an ideal bone repairing and replacing material. Hydroxyapatite coatings on polymeric scaffolds for tissue engineering are also widely explored. A nano-hydroxyapatite-coated chitosan scaffold was developed and effects of this scaffold on the viability and differentiation of periodontal ligament stem cells (PDLSCs) and bone repair were assessed by researchers. Based on their observations, behavior of PDLSCs on a new nanohydroxyapatite-coated genipin-chitosan conjunction scaffold (HGCCS) *in vitro* was compared with an uncoated genipin-chitosan framework, and later evaluated for the effect of PDLSC-seeded HGCCS on bone repair *in vivo*. When seeded on HGCCS, PDLSCs exhibited greater viability, better alkaline phosphatase activity. This is expected to be used as a promising tool for calvarial bone repair (Ge et. al., 2012).

Magnetic nano-hydroxyapatite-coated γ-Fe(2)O(3) (m-nHAP) particles exhibited remarkable influence on the porous structures and compressive strength of the nanocomposite hydrogels. The average pore diameter exhibited a minimum of 1.6 ± 0.3 μm and the compressive strength reached a maximum of about 29.6 ± 6.5 MPa with the m-nHAP content of around 10 wt% in the nanocomposite hydrogels. When seeded and cultured with osteoblasts it appeared most favorable to the osteoblasts. When the m-nHAP content is increased, the adhesion density and proliferation of the osteoblasts were found significantly increased (Hou et. al., 2013). The porous tri-components scaffold composed of chitosan (CS), silk fibroin (SF), and nanohydroxyapatite particles (nHA) - named CS/SF/nHA. The biodegradation characteristics of this scaffold satisfy the requirements of good biomedical materials. The study of the mechanical properties showed that the tri-components scaffold has better properties than the bi-component scaffolds (Qi et. al., 2013).

8.3.8 SILICA NANOPARTICLES

The biological activity of Silica nanoparticle (SNP) can be related to the particle shape and surface characteristics, not so dependent to particle size. The SNP can be adsorbed to cellular surfaces and it can affect membrane structures and integrity. Toxicity is related to mechanisms of interactions with outer and inner cell membranes, vesicle trafficking, and pathways signaling responses. Interaction with membranes may induce the release of endosomal substances, cytokines reactive oxygen species, and chemokines and thus induce inflammatory responses (Fruijtier-Pölloth, 2012). Silica nanoparticles show greater potential for a variety of diagnostic and therapeutic applications. Although properties and drawbacks of crystalline micron-sized silica are well documented, little information collected about the toxicity of amorphous, and nano-size forms. Because nano size possesses novel properties like kinetics and bioactivity, their biological effects may differ greatly from those of micron-size bulk materials. *In vivo* studies revealed reversible inflammatory changes to lung, forming granuloma, and emphysema, but no progressive lung fibrosis (Napierska et. al., 2010).

8.4 NANOTECHNIQUES

8.4.1 NANOFLUIDIC TECHNOLOGY

Nanotechnology on a chip is a new paradigm for total chemical analysis systems (Jain et. al., 2005 (b). Nano fluidic technology has broad applications in systems biology, pathogen detection, drug development, and in clinical research and in personalized medicine. Only small amount of sample is needed. Some examples of devices that incorporate nanotechnology-based biochips and microarrays are nano fluidic arrays and protein nano biochips. These devices can be adapted for point-of-care use, e.g. isolation and analysis of DNA. This helps in developing new detection schemes for cancer. They are used in toxicological studies, gene therapy, and also in DNA separation (Stavis et. al., 2009).

8.4.2 PHOTODYNAMIC THERAPY

Photodynamic therapy (PDT) is a form of phototherapy. The PDT kills mammalian cells as well as microbial cells like bacteria, fungi, and viruses. This is used clinically to treat a wide range of medical conditions, including malignant cancers (Wang et. al., 2002). In the PDT, a photosensitizer reaches cancer cells with the help of carrier and then it will be taken up by cancer tissue. Then this will destroy cancer tissue by photo radiation. Nanomaterials satisfy all the requirements for an ideal PDT agent (Denise et. al., 2008). The particle absorbs energy from the light, which will heat the particle and surrounding tissue. Light also produces high energy oxygen molecules which destroy most organic molecules that are next to them (tumor).This therapy is appealing for many reasons e.g. it does not leave a "toxic trail" of reactive molecules throughout the body (as in chemotherapy). Photodynamic therapy has tremendous potential as it is a noninvasive procedure (Minchin et. al., 2007; Chen et. al., 2006; Nyman et. al., 2004; Bechet et. al., 2008).

8.4.3 CYTOGENETICS

Chromosome structure analysis and identification of abnormalities are done by cytogenetics. The localization of specific gene probes by fluorescent *in situ* hybridization (FISH) combined with conventional fluorescence microscopy has reached its limit. Molecular cytogenetics is now enhanced by atomic force microscopy and quantum dot (QD) FISH (Jain et. al., 2007). This method is focusing on nanodissection and nanoextraction of chromosomal DNA. The extraction of very small amounts of chromosomal DNA by the scanning probe is possible. The photo stability and narrow emission spectra of nonorganic QD fluorophores make them desirable candidates for the use of FISH to study the expression of specific mRNA transcripts (Chan et. al., 2005). The technique also gives excellent histological results for FISH combined with immunohistochemistry. Multiple subnuclear genetic sequences use novel QD-based FISH for direct multicolor imaging (Bentolila et. al., 2006).

8.4.4 SENSOR TEST CHIPS AND SENSORS

Sensor test chips can detect proteins and other biomarkers left behind by cancer cells. This detection helps in diagnosis of cancer in the early stages, with small sample size (Zheng et. al., 2005). The nanoshell-conjugated antibodies can be used to target cancerous cells. An infrared laser can pass through flesh without heating it, and the nanoshell is heated sufficiently to destroy the cancer cells (Loo et. al., 2004). Nanoparticles of cadmium selenide (quantum dots) are used to glow cancer tumors and used as a guide for more accurate tumor removal (Vasudevanpillai et. al.,2008).

Nanosensors used for detection of chemical or biological materials are called nanobiosensors. These sensors have enormous variety of applications. At the same time these are inexpensive. These are useful in detection of nucleic acids, proteins, and ions (Jain et. al., 2003). Future concern is to develop implantable detecting and monitoring devices on the basis of these detectors. Cantilever technology provides an alternative to polymerase chain reaction (PCR). There is no need to label or copy the target molecules. Nano cantilevers can be used to design a new class of ultra-small sensors for detecting viruses, bacteria, and other pathogens (Gupta et. al., 2006). A real-time cantilever biosensor can provide continuous monitoring of clinical parameters in personalized medicine.

Optical-detectable tags can be formed by surface enhanced Raman scattering (SERS) of active molecules at the glass-metal interface (e.g. Surface Plasmon resonance (SPR). Each type of tag exploits the Raman spectrum. The SERS bands are 1/50 the width of fluorescent bands. It enables a greater degree of multiplexing and multiplexed analytic quantification. The SERS-based tags are coated with glass. So, attachment to biomolecules is straightforward, and the SERS technology can directly detect chemical agents and biological species (e.g., spores, biomarkers of pathogenic agents) (Hassan et. al., 2006).

Virus particles are biological nanoparticles. The viral particles used as sensors, are called as viral nanobiosensors. Clinically relevant viruses can be detected by herpes simplex virus and adenovirus. It is possible to detect as few as five viral particles in a 10 mL serum sample. This system is more sensitive compared to ELISA-based methods. It is an improvement over PCR-based detection. It is cheaper and faster and has less number of artifacts (Perez et. al., 2003).

8.4.5   MOLECULAR NANOTECHNOLOGY

Molecular nanotechnology is a speculative sub-field of nanotechnology. Molecular machines are developed which could re-order matter at an atomic or molecular scale. These proposed elements such as molecular assemblers and nanorobots are far beyond current capabilities. Machines or robots in the scale of nanometers are utilized in nanorobotics (Cerofolini et. al., 2010). Nanomedicine make use of nanorobots introduced into the body (for example: Computational Genes), to detect or repair infections and damages (Freitas et. al., 2005). Nanodevices working inside the body can be observed by MRI. Injected medical nanodevice first go to work in a specific organ or tissue mass. Then it is ensured that the nanodevices have gone to the correct target. During scanning the nanodevices congregated neatly around their target (e.g., a tumor mass) are clearly seen (Robert et. al., 1999, 2003). Usually drugs and surgery only encourage tissues to repair themselves. Molecular machines promote more direct repairs (Drexler et. al., 1986). This utilizes the same tasks that living systems utilize for self-repair. Access to cells is possible by inserting needles into cells without killing them. Molecular systems build or rebuild every molecule in a cell, and also can disassemble damaged molecules. Nanocomputers direct machines to perform the task. Repairing is by working structure by structure, and then by cell by cell and tissue by tissue, whole organs can be repaired. By organ by organ, health is restored finally. Molecular machines can build cells, thus cells damaged to the point of inactivity can be repaired. Thus, cell repair machines will make medical field to rely on direct repairs (Drexler et. al., 1986).

Polymer supramolecular assemblies represent a very promising strategy for enzymatic reactions. Capsules, dendrimers, Micelles, and vesicles mimic natural systems and serve as avenues for new medical approaches (Palivan et. al., 2012)

8.5   APPLICATIONS

8.5.1   DRUG DELIVERY

Drug delivery is based upon facts like:
  a)   Efficient encapsulation of the drugs,
  b)   Successful delivery to the targeted region and
  c)   Successful release there (Loo et. al., 2004).

Targeted and/or controlled delivery of protein and peptides using nanomaterials like nanoparticles and dendrimers show great promise for treatment of various diseases and disorders. These products are called nanobiopharmaceuticals and this emerging field is called nanobiopharmaceutics (Peiris et. al., 2012). The field of local or targeted, drug delivery are very much improved by the development of nanosensors, nanoswitches, nanopharmaceuticals, and nanodelivery systems Kong. A benefit of using nanoscale for medical technologies is that smaller devices are less invasive and can possibly be implanted inside the body (Boisseau et. al., 2011). Drug detoxification is also another application for nanomedicine (Bertrand et. al., 2010). Many diseases depend upon processes within the cell and can only be impeded by drugs that make their way into the cell. Nanomedical approaches can improve bioavailability of drugs. Bioavailability refers to the presence of drug molecules where they are needed in the body and where they will do the most good. Drug delivery focuses on improving bioavailability (LaVan et. al., 2003,Cavalcanti et. al., 2008). The strength of drug delivery systems is their ability to alter the pharmacokinetics and biodistribution of the drug. Lipid or polymer-based nanoparticles can be designed to improve the pharmacological and therapeutic properties of drugs (Allen et. al., 2004). Triggered response is another way to improve the efficiency of drug molecule. Drugs are placed in the body and they get activated only on encountering a particular signal.

Potential nanodrugs work by very specific and well-understood mechanisms. The major impact of nanotechnology and nanoscience is the development of completely new drugs and regulated drug release. This helps to overcome the side effects of drugs (Nagy et. al., 2011). Effects of vascular endothelial conditions, shear stress rates, and physical and chemical properties of nanoparticle must be taken into consideration for the successful design of drug-nanoparticle conjugates intended for parenteral delivery (Stephen Paul Samuel et. al., 2012). Polymer nanocarriers are developed to protect the active compounds till they reach the pathological sites. They are also helping in transportation of the drug. Nanocarriers ensure stability during transport and thus improve localization capacity, efficacy, stability, and sustainability of the drug and reduces amount requirement of the drug. These are behaving like artificial organelles. They not only protect the particular drug but allow various active molecules to combine in a single carrier, and even facilitate combination of therapeutic and detecting agents as in theragnostic approach (Onaca-Fischer O et. al., 2012).

The use of viral nanoparticles (VNPs) has evolved rapidly since their introduction 20 years ago, encompassing numerous chemistries, and modification strategies that allow the functionalization of VNPs with imaging reagents, targeting ligands, and therapeutic molecules (Steinmetz, 2010). Emptied virus cells (VNP) can carry drugs directly to cancer cells to kill them. Viral nanoparticles are developed from plant viruses, insect viruses, and animal viruses (Singh et. al., 2002). They avoid using human viruses in order to minimize the chance of the virus interacting with human proteins and causing toxic side effects, infection, and immune response. Plant viruses are used because they are easiest to produce in large quantities (Singh et. al., 2002). Plant viruses can self-assemble around a nanoparticle *in vitro* and they can hold approximately 10 cubic nanometers of particles (Franzen et. al., 2009).

Low bioavailability of proteins makes it difficult to treat the protein deficiency diseases with natural protein supplements. Delivering proteins to the specific sites by using carriers can offer promising solution to overcome decreased bioavailability. Carriers used commonly are based on lipids, polymers, and conjugates ranging from liposomes to nanoreactors with varying morphologies. The carriers should be highly flexible and biocompatible. Nanocarriers and nanoreactors based on biodegradable, biocompatible, and non-toxic polymer systems show better surface properties and multifunctionality, which favor them to cope up with the *in-vivo* biological conditions (Balasubramanian et. al., 2010). Studies are going on in targeting gene- or drug-loaded nanoparticles (NPs) to tumors and ensuring their intratumoral retention after systemic administration. Researchers exploit changes in lipid metabolism and cell membrane biophysics that occur during malignancy. Modifications to the surface of NPs increase their biophysical interaction with the membrane lipids of cancer cells. This can improve intratumoral retention and *in vivo* efficacy upon delivery of NPs loaded with a therapeutic gene. Different surfactants, added on to the NPs' surface, affect the interactions of NPs with the lipids of cancer cells which in turn decrease the efficacy of NP. But surfaces modified with materials like didodecyl-dimethyl-ammonium-bromide (DMAB) show greater interaction with cancer cell lipids (Airen et. al., 2012).

Liposomes are delivery systems that are used to formulate a vast variety of therapeutic and imaging agents. They show better pharmacokinetics, sensitivity for cancer diagnosis, and therapeutic efficacy compared to their free forms. The complex physiology of the tumor micro environment and multifactorial nature of cancer demand the development of multifunctional nanocarriers. This requirement is usually satisfied by multifunctional liposomal nanocarriers (Perche et. al., 2013).

8.5.2 TRACKING STEM CELLS

In all multicellular organisms, some biological cells that are capable of dividing into diverse specialized cell types are present. Those are called stem cells. In adult organisms, stem cells act as a repairing system of the body (Mark et. al., 1999). Nanoparticles can be used to track stem cells. This enables measurement of very low amounts of the labeled cells. Different types of cells can be labeled with different perfluorocarbon compounds (Stoll et. al., 2012). Perfluorocarbon compounds are a subset of fluorochemicals, in which the hydrogen atoms are replaced by fluorine. If all the H atoms substituted by fluorine, it is termed as perfluorocarbon molecule. Cells can retain their usual surface markers and they are functional even after the labeling process (Stoll et. al., 2012). A super paramagnetic iron oxide nanoparticle (SPIO) is emerging as ideal probe for noninvasive cell tracking. The use of 200 nm per fluorocarbon nanoparticles to label endothelial progenitor cells taken from human umbilical cord blood enables *in vivo* progenitor cell detection by MRI. The MRI scanner can be tuned to the specific frequency of the fluorine compound in the nanoparticles, and the nanoparticle-containing cells are visible in the scan. This method is ideal for medical imaging (Parttow et. al., 2007).

8.5.3 VIRAL DIAGNOSIS

Viral diagnosis is difficult with conventional techniques. Rapid and sensitive diagnosis of viral diseases is important for infection control and development of antiviral drugs. A QD system can detect the respiratory syncytial virus particles in a matter of hours. It is also more sensitive. It allows detection of the virus earlier in the course of an infection (Bentzen et. al., 2005). For example, in respiratory syncytial virus (RSV) infections, Antibody-conjugated nanoparticles are used for rapid sensitive detection of RSV (Agrawal et. al., 2005). Diagnosis of flu virus is quickly possible by antibodies attached gold nanoparticles. When light source is directed on a sample containing virus particles and the nanoparticles, the amount of light reflected back increases, and thus a much faster testing is possible than those currently used. This is made possible by clustering the nanoparticles around virus particles.

8.5.4 ONCOLOGY

An important application of nanomedicine is in cancer diagnostics and therapy. Molecular diagnosis of cancer including genetic profiling is getting widely used nowadays. Nanomedicine is well established in oncology (Liu et. al., 2007). The application of nanotechnology in medicine mainly alters pharmacokinetics. It increases the percentage of injected dose to reach the tumor, accomplishes target-specific delivery and uptake, which in turn decreases dose requirement (Allen et. al., 2006). Many nanoparticulate formulations (e.g., cytostatic agents) have shown to exhibit increased therapeutic efficacy and diminished adverse effects (Torchilin et. al., 2005). Liposome is an artificially prepared vesicle composed of a lipid bilayer. It can act as a vehicle for drug and nutrients. It is highly useful in cancer drug delivery system e.g.; Doxil, a liposomal formulation of doxorubicin has shown good result in breast carcinoma treatment (Perez et. al., 2002). More recently, a synergistic approach such as cutting off the blood supply to the tumor and thus killing tumor cells, is getting more acceptance (Sengupta et. al., 2005, Hood et. al., 2002, LaVan, et. al., 2003; Cavalcanti et. al., 2008).Carbon nanotubes (CNT) and quantum dots (QDs) are useful in diagnosing cancerous cells and in destroying them by various methods like drug delivery and thermal treatment. Fluorescent QDs are useful as imaging molecules. Localization of cancer cells is easier with their nano size and their penetrating ability and high-resolution imaging derived from their narrow emission bands. Only limitation is that in most of the cases QDs are usually made of quite toxic elements (Nie et. al., 2007). The CNTs can deliver drugs to a target site and their ability to convert optical energy into thermal energy. The CNTs can navigate to malignant tumors by attaching antibodies that bind specifically to tumor cells. Once at the tumor site, the CNTs enter into the cancer cells by endocytosis, allow drug release, and result in cancer cell death. The CNTs can be exposed to near-infrared light and thus thermal destruction of the cancer cells is also possible. The CNT's amphiphilic nature makes them penetrate the cell membrane and their large surface area allows drug releases inside the cancer cell (Madani et. al., 2013).

Using gold nanoparticles to deliver platinum to cancer tumors may reduce the side effects of platinum cancer therapy. Using nanoparticles, nitric oxide is delivered directly to cancer cells which reduce the required amount of chemotherapy drugs. Effectiveness of the chemotherapy drug was increased five times when nitric oxide was delivered by nanoparticles. Gold nanoparticles to which RNA molecules attached are used to treat skin cancer. The nanoparticles help to penetrate the skin and the RNA then attaches to a cancer related gene, stopping it from generating proteins that are used in the growth of skin tumors. But these techniques require further clinical trials (Elsabahy et. al.,2013). Nanorobotics may be applied for early detection as well as treatment of cancer. A nanodevice for combined diagnosis and therapeutics can be implanted as a prophylactic measure in individuals who do not exhibit any obvious manifestations of cancer and cancer surveillance can be conducted by external remote monitoring. This enables early detection and appropriate therapeutic intervention at the earliest. These are biodegradable and this system is highly useful in preventive personalized management of cancer. Early detection definitely increases the chances of cure (Jain et. al., 2005 (c), Perche et. al., 2013).

8.5.5   INFECTIOUS DISEASES

Most of the conventional diagnostic methods lack ultra-sensitivity. There is time delay in getting results. With nanotechnology-based bioconjugated nanoparticle-based bioassay, a single bacterium can be detected within 20 min. The method based on SERS using silver nanorods is developed for rapid detection of trace levels of viruses with a high degree of sensitivity and specificity (Zhao et. al., 2004). This is as distinct as a fingerprint. The method provides rapid detection of infectious diseases and this helps to prevent wide spread of the diseases (Shanmugh et. al., 2006).

Multi-drug resistant tuberculosis is nowadays a life threat to the human population. It demands long term medication. Available drug regime is very old, date back to 1950s. Chances of skipping of drug intake is there, which may result in treatment failure. Introduction of nanoparticle made the treatment easier, it helps in more targeted drug delivery. This in turn reduced the cost and duration of drug therapy (Smith, 2011).

8.5.6   NEUROLOGICAL DISORDERS

The incidence of neurological diseases of unknown etiology is increasing. Parkinson's disease, Alzheimer's disease, or Huntington's disease, all these are neurodegenerative diseases causing death of small niches of neurons in the encephalic mass, rendering these disorders to diffuse. Encephalopathies are disorders that affect the entire encephalic mass without specific localized foci (Ramos-Cabrer et. al., 2013). The CNS is a highly complex system, and it provides effective mechanisms of defense against foreign elements. These defensive mechanisms usually complicate therapeutic interventions in the CNS. Three main barriers are there to regulate molecular exchange between the blood and brain parenchyma, the **blood–brain barrier**, which is formed by the glial cells and endothelial cells of the brain blood vessels, the choroid plexus epithelium, which is the border between the blood and ventricular cerebrospinal fluid, and the arachnoid epithelium, that separates the blood from the subarachnoid cerebrospinal fluid. The blood-brain barrier provides protection for the brain. But it hinders the diagnosis and treatment of these neurological diseases, as the drugs must cross the blood-brain barrier to reach the lesions. Though in its infancy, application of nanoneurotechnology provides promising answers to some of these issues (Jagat et. al., 2012; Ramos-Cabrer et. al., 2013) .

Neuroelectronic interfacing is a method in which nanodevices capable of permitting computer–linking to the nervous system, are constructed. The demand for such structures is high. This system helps to overcome many effects of diseases and injuries. Two types of power sources are there for this system. They are refuelable and non-refuelable strategies. In a refuelable strategy, energy is refilled continuously or periodically with external sources such as sonic, chemical, tethered, magnetic, or electrical. In a non-refuelable strategy, all power is drawn from internal energy storage which would stop when all energy is drained (Robert et. al., 2003; Elder et. al., 2008). Advanced researches in the field of nanotechnology enabled us to build complex functionalized macromolecules with optimized loading and release characteristics. So administering therapeutic agents to the CNS in a controlled manner, and assuring their increased circulation time in the bloodstream and avoiding agglutination of the agent with plasma proteins, all these are now reality to certain extent. The most exciting advance in recent years is the development of nanoplatform for performing all the tasks like diagnosis and therapeutics together, or theranostics (Ramos-Cabrer et. al., 2013) .

8.5.7   CARDIOVASCULAR DISORDERS

In the beginning of the 21st century, use of nanoparticles for molecular imaging of cardiovascular disease started (Choudhury et. al., 2004). One of the leading causes of death and disability in the developed world is the sequelae of cardiovascular disease (Biana et. al., 2011). In atherosclerosis, the rupture of inflamed and vulnerable plaques occur, so it is better to identify vulnerable lesions before the onset of symptoms. A number of strategies have been investigated (Jaffer et. al., 2009; Nahrendorf et. al., 2009; Sinusas et. al., 2008). Epichlorohydrin-crosslinked dextran-coated iron oxide (CLIO) nanoparticles are used for the targeted imaging of lesions *in vivo*.

In atherosclerosis, molecular MRI of macrophages is accomplished using ultra small particles of iron oxide or paramagnetic immunomicelles (Ruehm et. al., 2001). Paramagnetic nanoparticles (Flacke et. al., 2001; Winter et. al., 2003; Frias et. al., 2004) are used for imaging the thrombus plaque angiogenesis and lipoproteins. Superparamagnetic cross-linked iron oxide nanoparticles are used for the studies of apoptosis after myocardial infarction and the overexpression of cell adhesion molecules in atherosclerosis (Nahrendorf et. al., 2006; Sosnovik et. al.,2005; Mulder et. al., 2007). These studies revealed the significance of integrating multiple properties within one nanoparticle to allow exploitation of the strengths of the different imaging modalities (McCarthy et. al., 2007). Particles size up to 400 nm show enhanced vascular permeability. By addition of peptides, aptamers, antibodies, and small molecules allow the modulation of their biodistribution. The large surface area–to-volume ratio of nanomaterials permit the attachment of a number of different ligands and make them multifunctional. Thus, nanoparticles, with both diagnostic and therapeutic functionalities, can be synthesized, and used for determination of the localization of the nanoagent and for the treatment of disease (Biana et. al., 2011).

Reports about plaque angiogenesis inhibition using paramagnetic perfluorocarbon nanoparticles loaded with fumagillin, was published by Winter et. al., in 2006 from the Washington University School of Medicine (St Louis). Molecular imaging was used as a noninvasive modality. Nanoparticle can interfere with the dynamics of cardiovascular disease in a number of ways. The targeting ability of its payload allows it to capture and act on the specified tissue and also allow its activity to be captured by noninvasive imaging modalities such as MRI (Perez et. al., 2003).

Collagen in the heart valve tissue affects the proper functioning of the valve. The valve may become stiff if too much collagen present and may become floppy if collagen is too less. Combining gold nanoparticles to collagen improve the mechanical properties of the valve, repairing defects in the heart valves without surgery (Alavi et. al., 2012).

8.5.8   OPHTHALMOLOGY

Ocular diseases are usually treated by topical application. But efficacy of drug therapy is usually reduced by many barriers like ocular epithelium, tear film, blood-aqueous blood-retina, and so on. Nanoparticles help to overcome this difficulty. They act as drug carriers and thus more target specific treatments are possible. Thus nano has revolutionized the ocular treatment (Diebold et. al., 2010). Aspirin loaded albumin nanoparticles are used for treating diabetic retinopathy (Das et. al., 2012). Nanosized carriers are developed for this purpose. This novel system offers manifold advantages over conventional systems as they increase the efficiency of drug delivery by improving the release profile and also reduce drug toxicity (Wadhwa et. al., 2009).

8.5.9   ORTHOPEDICS

Orthopedic surgeons treat sport injuries, musculoskeletal trauma, infections, tumors, degenerative disorders, and congenital disorders by surgical and non-surgical methods. Implants, allografts, and autografts are used in many cases so that patient can feel better esthetics and functional and psychological well-being. Skin-like structures are also used to protect the fractured site and for esthetics. But many draw backs are there like chances of rejection of implant, infections and tearing, and discoloration of artificial skin. Nano orthopedics is developing now as a solution to all these problems. Nanoparticles are used as part of implants and make them more osteo-inductive and conductive. Nano skin is also getting popularized in mimicking human skin in texture and feel (Saxl et. al., 2011,Si-Feng et. al., 2013). Growth of replacing bone is speeded up when a nanotube-polymer nanocomposite is placed as a kind of scaffold which guides growth of replacement bone.

RENAL DISEASES

Nanonephrology is a branch of nanomedicine and nanotechnology that seeks to use nano-materials and nanodevices for the diagnosis, therapy, and management of renal diseases.

Its goals include:
   a)   To study protein structures of kidney at the atomic level,
   b)   To study the kidney cellular processes, and
   c)   Utilize nanoparticles to treat various kidney diseases.

Advances in Nanonephrology are based on nanoscale information on the cellular, molecular machinery in the normal kidney processes and in pathological states. By understanding the physico-chemical properties of proteins and other macromolecules at the atomic level in renal cells, newer therapeutic approaches can be designed to treat major renal diseases. The nanoscale artificial kidney may be possible in future. Nano-scale engineering advances can create nanoscale robots with curative and reconstruc-tive capabilities. Renal patients may suffer from severe pain also. The ability to direct events in a controlled and programmed fashion at the cellular nanolevel will definitely improve the lives of patients with kidney diseases (Santoro et. al.,2013).

NANODENTISTRY

Nanodentistry makes use of materials, biotechnology, tissue engineering, and nanoro-botics based on nanotechnology for improving oral health. Hard and soft tissues re-generation around a solid implant, or implanted biodegradable materials replaced by new tissues, all these are possible with this advanced technology (Kong et. al., 2006).

SURGICAL FIELD

In order to induce local anesthesia, a colloidal suspension containing millions of active analgesic nanorobot particles are installed on the patient's gingival. Nanorobots reach the dentin by migrating through the gingival sulcus and then pass painlessly through the lamina propria. Further the nanorobots enter dentinal tubule holes that are 1–4 micrometers in diameter and proceed toward the pulp. This movement is guided by chemical gradients, temperature differentials, and so on. All under the control of the nanocomputer directed by the dentist (Kong et. al.,2006; Preeti et. al., 2011). Nanon-eedles: Nano-sized stainless steel crystals incorporated suture needles have been de-veloped. Development of surgical instruments at nano scale level will make cell-sur-gery possible (Kumar et. al., 2006).

ORTHODONTICS

Tooth straightening procedures can be directly performed by Orthodontic nanorobots by manipulating the periodontal tissues. This makes painless tooth straightening, ro-tating, and vertical repositioning in a very short period of time like minutes to hours (Kong et. al., 2006)

CONSERVATIVE DENTISTRY

Replacing the upper layers of enamel with covalently bonded artificial materials such as sapphire or diamond, which has 20–100 times the hardness and strength of natu-ral enamel, can improve the appearance and durability of tooth (Kong et. al., 2006). Hypersensitivity is an acute pain condition caused by hydrodynamic pressure of the pulp, that occurs when the surface of the root dentin get exposed. In this condition, the dentinal tubules are open. Mechanical, chemical, or thermal stimuli can result in fluid flow along the open tubules resulting in an uncomfortable pain. Surface density and diameter of dentinal tubules are much higher in such teeth. Primary approach of treatment is occluding the dentinal tubules. Usual therapeutic agents provide temporary relief for this painful condition. This is achieved by sealing and isolating the dentinal tubules from external stimuli. This prevents fluid movement from triggering a pain response (Preeti et. al., 2011, Cummins et. al., 2009; Hassan et. al., 2006). But with reconstructive dental nanorobots, using native biological materials, it is possible to occlude specific tubules within minutes, thus patients get quick and permanent cure (Kong et. al., 2006) .

The Bonding agents help in the binding of composite restorations to the teeth. Dispersible nanoparticles are used in bonding agents. These nanoparticles ensure homogeneity of the mix and ensure that the adhesive is perfectly mixed every time (Rybachuk et. al., 2009). The latest generation of bonding agents provides a stable, dispersed, and filled adhesive that prevents particle settling and thus eliminates the need to be shaken prior to use. They are self-etching, one step materials (Preeti et. al., 2011; Kumar et. al., 2006; Rybachuk et. al., 2009).

PERIODONTICS

Nanorobots are useful for maintaining gingival health. A subocclusal dwelling nanorobotic dentifrice is effective in calculus debridement. This can be incorporated in mouthwash or toothpaste. Using this for cleaning all supragingival and sub gingival surfaces at least once a day, can prevent conventional tooth decay and gingival disease. Main advantage is the preservation of 500 or so species of harmless oral micro flora which are necessary for a healthy ecosystem. (Preeti et. al., 2011).

PROSTHODONTICS

Nano robots are useful in oral rehabilitations like crown and bridge restorations. Impression materials are needed for precise reproduction of oral tissues for fabrication of restorations. By adding nanofillers in vinylpolysiloxanes, a unique siloxane impression material is developed which shows better flow, improved hydrophilic properties, and enhanced detail precision (Rybachuk et. al., 2009; Kumar et. al., 2006). Development of new tissues, to replace implanted biodegradable materials and regeneration of hard and soft tissues around a solid implant are opening new vistas in the field of tissue regeneration. Nanoparticles, govern the macroscopic behavior of these novel materials (Rybachuk et. al., 2009; Kumar et. al., 2006).

DENTAL IMPLANTS

Dental implantology has a long, well documented history reaching back over thousands of years from ancient times. The accidental discovery of **osseointegration** by Brånemark in 1952 brought a new awake to this field. Osseointegration is the process in which, a direct interphase between bone and implant occurs without soft tissue intervening. Implants using nanotechnology can definitely show increased predictability. This is achieved by increased surface contact area and by physiochemical bond formation. Titanium implants are used most commonly. The addition of nanoscale deposits of hydroxyapatite and calcium phosphate creates a more complex implant surface which will favor bonding (Albrektsson et. al., 2000; Goene et. al., 2007; Preeti et. al., 2011).

ORAL PATHOLOGY

The atomic force microscopy is considered as a qualitative and quantitative analysis technique. AFM is useful in studying the collagen network and also useful for analyzing dentine surface changes caused by different chemical agents. Advantages of AFM is its ability to "zoom" over the magnification range of both optical and electron microscopes. Thus AFM based structural analysis of dentine and its collagen components are critical in understanding the structure of fully mineralized, skeletal substrates. It enables early discrimination of various pathophysiological states and disease progression such as osteoporosis (Sharma et. al., 2010, Preeti et. al., 2011). This is effective in the treatment for periodontal disease prevention, disease progression, and development of bone, cartilage, tendons, skin, and collagen-based materials in tissue engineering. (Sharma et. al., 2010; Preeti et. al., 2011). The AFM provides unique data regarding biochemical and adhesion properties of bacterial biofilms which are not measurable by optical microscopy (Sharma et. al., 2010). It is an accepted nanotechnology technique in use, for analyses of cells and biofilm surfaces. It provides topographic imaging, microphysical and nanophysical probing and characterization of biofilm surfaces (Sharma et. al., 2010).

Saliva is an inexpensive, noninvasive, and easy diagnostic medium. It contains proteomic as well as genomic markers which help in identification of diseases at molecular level (Wong et. al., 2006). A unique type of sub 100 nm membrane bound secretory vesicles called "exosomes" are specialized class of biomarkers found in human saliva which are secreted by salivary gland epithelium and released into the salivary fluid *via* exocytosis. Malignancy and other diseases cause elevated exosome secretion. Tumor-antigen enrichment of exosomes is associated with cancer cells. Due to their small size, sensitive quantitative detection tools are needed for detection and characterization of salivary exosomes. Saliva exosomes are potential non-invasive biomarker resource for oral cancer. It can also be analyzed by AFM (Sharma et. al., 2010).

8.6 CONCLUSION

In this review chapter, we have given a nutshell description about medical as well as surgical nanomaterials and technologies. Further evaluation of the toxicological aspects, is needed before starting their wide usage at the human level.

KEYWORDS

- **Dental implantology**
- **Quantum dots**
- **Silica nanoparticle**
- **Superparamagnetic iron oxide nanoparticles**
- **Titanium dioxide**

REFERENCES

1. Agrawal, A, Tripp, R. A, Andersen, L. J, and Nie, S. Real-time detection of virus particles and viral protein expression with two-color nanoparticle probes. *J Virol*, **79**, 8625–8 (2005).
2. Airen, Xu, Mingfei, Yao, Guangkui, Xu, Jingyan, Ying, Weicheng, Ma, Bo, Li, and Yi, Jin, A physical model for the size-dependent cellular uptake of nanoparticles modified with cationic surfactants. *Int J Nanomedicine*, **7**, 3547–3554 (2012) Published online 2012 July 10.doi:10.2147/IJN.S32188.
3. Fujishima, Akira, Rao, Tata, N. A. Tryk, Donald. Titanium dioxide photocatalysis
4. Journal of Photochemistry and Photobiology C: *Photochemistry Reviews*, 1, 1–21(2000).
5. Alavi, S. H, Ruiz, V. Krasieva, T. Botvinick, E. L., and Kheradvar. A, Characterizing the collagen fiber orientation in pericardial leaflets under mechanical loading conditions, *Ann Biomed Eng.*, **41**(3),547–61 (Mar 2013). doi: 10.1007/s10439-012-0696-z. Epub 2012 Nov 21.
6. Albrektsson, T., Sennerby, L., and Wennerberg, A. State of the art of oral implants. *Periodontology*, **2008**. 47, 15–26 (2000) doi: 10.1111/j.1600-0757.2007.00247.x.
7. Ali, M. E.; Hashim, U.; Mustafa, S.; Che Man, Y. B., and Islam, Kh N. *Journal of Nanomaterials*, (2012). Article ID 103607.
8. *Allen, T. M., Cheng, W. W., Hare, J. I., and Laginha K. M. Pharmacokinetics and pharmacodynamics of lipidic nano-particles in cancer. Anticancer Agents Med Chem., 6, 513–523 (2006).*
9. Allen, T. M. and Cullis, P. R. *Drug Delivery Systems: Entering the Mainstream. Science*, **303**(5665), 1818–1822 (2004). doi:10.1126/science.1095833.
10. *Amirbekian, V., Lipinski, M. J., Briley-Saebo, K. C., Amirbekian. S. Aguinaldo, J. G., Weinreb, D. B., Vucic, E., Frias. J. C., Hyafil, F., Mani, V, Fisher, E. A., and Fayad, Z. A. Detecting and assessing macrophages in vivo to evaluate atherosclerosis noninvasively using molecular MRI. Proc Natl Acad Sci., U S A.; 104, 961–966 (2007).*
11. Balasubramanian, V., Onaca, O., Enea, R, Hughes, D. W., and Palivan, C. G. Protein delivery: from conventional drug delivery carriers to polymeric nanoreactors. *Expert*, 7(1), 63–78 (Jan , 2010) doi: 10.1517/17425240903394520.

12. Bechet, D., Couleaud, P., Frochot, C, Viriot, M. L., Guillemin, F, Barberi-Heyob, M, Nanoparticles as vehicles for delivery of photodynamic therapy agents. *Trends Biotechnol.*, **26**(11), 612–21. (2008 Nov) doi: 10.1016/j.tibtech.2008.07.007. Epub 2008 Sep 17.

13. Bentollla, L. A., and Weiss, S. Single-step multicolor fluorescence in situ hybridization using semiconductor quantum dot-DNA conjugates. *Cell Biochem Biophys*, **45**, 59–70 (2006).

14. Bentzen, E. L., House, F., Utley, T. J., Crowe, J. E., Wright, and D. W. Progression of respiratory syncytial virus infection monitored by fluorescent quantum dot probes. *Nano Lett.*, **5**, 591–5 (2005).

15. Bertrand, N., Bouvet, C., Moreau, P., and Leroux J. C. Transmembrane pH-Gradient Liposomes to Treat Cardiovascular Drug Intoxication. *ACS Nano* **4**(12), 7552–7558. doi:10.1021/nn101924a (2010).

16. Bertrand, N. and Leroux, J. C. The journey of a drug carrier in the body: an anatomo-physiological perspective. *Journal of Controlled Release*. (2011) doi:10.1016/j .jconrel.2011.09.098.

17. Bertrand, N. and Leroux, J. C. The journey of a drug carrier in the body: an anatomo-physiological perspective. *Journal of Controlled Release* (2011). doi:10.1016/j. jconrel.2011.09.098.

18. Biana, Godin, Jason, H. Sakamoto Rita, E. Serda, Alessandro Grattoni, Ali Bouamrani, and Mauro, Ferrari. *Emerging applications of nanomedicine for the diagnosis and treatment of cardiovascular diseases incomplete* (2011).

19. Boisseau, P., and Loubaton, B. Nanomedicine, nanotechnology in medicine. *Comptes Rendus Physique*, **12** (7), 620 (2011). doi:10.1016/j.crhy.2011.06.001.

20. Brown, S. D., Nativo, P., Smith, J. A., Stirling, D., Edwards, P. R., Venugopal, B.; Flint, D. J., Plumb, J. A., Graham, D., and Wheate, N. J. *J. Am. Chem. Soc.*, ,**132**, 4678–4684 (2010).

21. Cavalcanti, A., Shirinzadeh, B, Freitas, R. A., and Hogg, T. Nanorobot architecture for medical target identification. *Nanotechnology*, **19**(1), 015103(15pp). Bibcode 2008 Nanot.19a5103C. (2008). doi: 10.1088/0957-4484/19/01/015103.

22. Cerofolini, G., Amato, P., Masserini, M., and Mauri, G. A. Surveillance System for Early-Stage Diagnosis of Endogenous Diseases by Swarms of Nanobots. *Advanced Science Letters*, **3**(4), 345–352. doi:10.1166/asl.2010.1138 (2010).

23. Chaloupka, K., Malam, Y., and Seifalian, A. M. Nanosilver as a new generation of nano-product in biomedical applications. *Trends Biotechnol.*, **28**(11), 580–8. doi: 10.1016/j. tibtech.2010.07.006. Epub 2010 Aug 18 (2010 Nov).

24. Chan, P., Yuen, T., Ruf, F, Gonzalez-Maeso, J., and Sealfon, S. C. Method for multiplex cellular detection of mRNAs using quantum dot fluorescent in situ hybridization. *Nucleic Acids Res*, **33**, e161 (2005).

25. Chen, B. Pogue, B. W. Hoopes, and Hasan, P. J. TVascular and cellular targeting for photodynamic therapy. *Crit Rev Eukaryot Gene Expr.*, **16**(4), 279–305 (2006).

26. Chen, X. Schluesener, H. J.,Nanosilver: a nanoproduct in medical application. *Toxicol Lett.* , **176**(1), 1–12. Epub 2007 Oct 16 (2008 Jan 4).

27. Chithrani, D. B. Intracellular uptake, transport, and processing of gold nanostructures. *Mol Membr Biol.*, **27**(7), 299–311. Epub 2010 Oct 7 (2010 Oct).

28. Choi, J., Jun, Y, Yeon, S, Kim, H. C., Shin, J. S. , and Cheon, J. Biocompatible heterostructured nanoparticles for multimodal biological detection. *J Am Chem Soc*, **128**, 15982–3 (2006).

29. Choudhury, R. P. Fuster, V, and Fayad, Z. A. *Molecular, cellular and functional imaging of atherothrombosis.* Nat Rev Drug Discov., **3**, 913–*925 (*2004)

30. Cummins, D. Dentin hypersensitivity: from diagnosis to a breakthrough therapy for everyday sensitivity relief. *J Clin Dent.*, **20**(1), 1–9 (2009)

31. Denise, Bechet, Pierre, Couleaud, Céline Frochot, Marie-Laure, Viriot, François, Guillemin, and Muriel, Barberi-Heyob. **Nanoparticles as vehicles for delivery of photodynamic therapy agents**. *Trends in Biotechnology,* **26**(11), 612–621 (November 1, 2008)

32. Diebold, Y. and Calonge, M. Applications of nanoparticles in ophthalmology. *Prog Retin Eye Res.* **29**(6), 596–609. (Nov, 2010) doi: 10.1016/j.preteyeres.2010.08.002. Epub 2010 Sep 6.

33. Elsabahy, M. and Foldvari, M. Needle-free gene delivery through the skin: an overview of recent strategies, *Curr Pharm Des.* [Epub ahead of print] (2013, Mar 12)

34. Elder, J. B., Liu, C. Y, and Apuzzo, M. L. Neurosurgery in the realm of **10**(-9), part 1: stardust and nanotechnology in neuroscience, Neurosurgery.2008 Jan;**62**(1), 1–20(2008). doi: 10.1227/01.NEU.0000311058.80249.6B.

35. El-Sayed, I. H., Huang, X., and El-Saved, M. A. Surface plasmon resonance scattering and absorption of anti-EGFR antibody conjugated gold nanoparticles in cancer diagnostics: applications in oral cancer. *Nano Lett,* **5**, 829-34 (2005).

36. El-Sayed, I. H., Huang, X., El-Sayed, M. Selective laser photo-thermal therapy of epithelial carcinoma using anti-EGFR antibody conjugated gold nanoparticles. *Cancer Lett,* **239**, 129–35 (2006).

37. Fan, R., Kamik, R., Yue, M., Li, D., Majumdar, A, and Yang, P. DNA translocation in inorganic nanotubes. *Nano Lett,* **5**, 1633–7 (2005).

38. *Flacke, S, Fischer, S, Scott, M. J., Fuhrhop, R. J., Allen, J. S., McLean. M, Winter, P., Sicard, G. A., Gaffney, P. J., Wickline. S. A., and Lanza, G. M. Novel MRI contrast agent for molecular imaging of fibrin implications for detecting vulnerable plaques. Circulation.,* **104**, 1280*–1285 (*2001).

39. Foldvari, Marianna, and Bagonluri, Mukasa. Carbon nanotubes as functional excipients for nanomedicines: II. Drug delivery and biocompatibility issues. *Nanomedicine: Nanotechnology, Biology and Medicine,* **4**(3) **Elsevier** (Sep 1, 2008).

40. Freitas, Robert A., Jr.; Havukkala, Ilkka (2005). Current Status of Nanomedicine and Medical Nanorobotics. *Journal of Computational and Theoretical Nanoscience* **2**(4), 1–25.doi:10.1166/jctn.2005.001.

41. Freitas, R. *Nanodentistry, JADA,* **131**, 1559–1565 (2000).

42. *Frias, J. C., Williams, K. J., Fisher. E. A., and Fayad Z. A. Recombinant HDL-like nanoparticles: a specific contrast agent for MRI of atherosclerotic plaques. J Am Chem Soc.,* **126**, 16316*–16317* (2004).

43. Fruijtier-Pölloth, C, The toxicological mode of action and the safety of synthetic amorphous silica-a nanostructured material, *SourceToxicology,* **294**(2–3), 61–79(Apr 11, 2012) . doi: 10.1016/j.tox.2012.02.001. Epub 2012 Feb 13.

44. Ge, S, Zhao, N, Wang, L, Yu, M, Liu, H, Song, A, Huang, J, Wang, G, and Yang, P. Bone repair by periodontal ligament stem cellseeded nanohydroxyapatite-chitosan scaffold, *Int J Nanomedicine,* **7**, 5405–5414 (2012). doi: 10.2147/IJN.S36714. Epub 2012 Oct 10.

45. Global facts on tobacco or oral health. Global oral health programme. World Health Organization: Geneva, Switzerland, 2005. Available at: http://www.who.int/oral_health/publications/orh_factsheet_wntd.pdf (accessed 17 June 2010)

46. Goene, R. J, Testori, T, and Trisi, P. Influence of a nanometer-scale surface enhancement on de novo bone formation on titanium implants: a histomorphometric study in human maxillae. *Int J Periodontics Restor Dent.,* **27** 211–219 (2007).

47. Gupta A. K., Nair, P. R., Akin, D., Ladisch. M. R., Broyles, S, Alam, M. A., et al. Anomalous resonance in a nanomechanical biosensor. *Proc Natl Acad Sci* USA, **103**, 13362–7 (2006).

48. Hassan, M. E. Azzazy, Mai, M. H. Mansour, Kazmierczak, and Steven, C. Nanodiagnostics: A New Frontier for Clinical Laboratory Medicine. *Clin Chem.*, **52**(7), 1238–1246 (Jul, 2006).

49. *Hood, J. D., Bednarski, M., Frausto, R., Guccione, S, Reisfeld, R. A., Xiang, R, and Cheresh, D. A. Tumor regression by targeted gene delivery to the neovasculature.*Science, **296**, 2404–2407 (2002*).*

50. Hou, R, Zhang, G. Du, G, Zhan, D., Cong, Y, Cheng, Y., and Fu, J, Magnetic nanohydroxyapatite/PVA composite hydrogels for promoted osteoblast adhesion and proliferation, Colloids Surf B Biointerfaces., **103**, 318–25(Mar 1, 2013). doi: 10.1016/j.colsurfb.2012.10.067. Epub 2012 Nov 14

51. *Hyafil, F., Cornily, J. C., Feig, J. E., Gordon, R., Vucic, E., Amirbekian, V., Fisher, E. A., Fuster, V., Feldman, L. J., and Fayad, Z. A. Noninvasive detection of macrophages using a nanoparticulate contrast agent for computed tomography. Nat Med.,* **13**, 636–641 *(*2007).

52. Jaffer, F. A, Libby, P, and Weissleder, R. Optical and multimodality molecular imaging: insights into atherosclerosis. *Arterioscler Thromb Vasc Biol.*, 29, 1017–1024 (2009).

53. Kanwar, Jagat R. PhD Nanoparticles in the treatment and diagnosis of neurological disorders: untamed dragon with fire power to heal. **Nanomedicine: Nanotechnology, Biology and Medicine, 8**(4), 399–414, (May, 2012).

54. Jain, K. K. Current status of molecular biosensors. *Med Device Technol*, **14**, 10–5 (2003).

55. Jain, K. K. Nanotechnology in clinical laboratory diagnostics. *Clin Chim Acta*, **358**, 37–54 (2005).

56. Jain, K. K. Nanotechnology-based lab-on-a-chip devices. In: J, Fuchs, M. Podda, (Eds). Encyclopedia of Diagnostic Genomics and Proteomics. Marcel Dekkar Inc., New York, 891–5 (2005).

57. Jain, K. K. Role of nanobiotechnology in developing personalized medicine for cancer. *Technol Cancer Res Treat*, **4**, 407–16 (2005).

58. Jain, Kewal K. *Applications of Nanobiotechnology in Clinical Diagnostics* (November 13, 2007).

59. Park, Ju-Young, Lee, Changhoon, Jung, Kwang-Woo, and Dongwoon Jung. Structure Related Photocatalytic Properties of TiO$_2$. *Bull. Korean Chem. Soc.*, **30**, 2 (2009).

60. Eric, Drexler K., *Engines of Creation: The Coming Era of Nanotechnology*, ISBN 0-385-19973-2 (1986).

61. Morrow, K. John, (Jr, PhDa), Bawa, Raj (MS, PhD b,c,d), and Wei, Chiming (MD, PhD, FACC, FAHA, FAANe). Recent Advances in Basic and Clinical Nanomedicine, *Med Clin N Am*, **91** 805–843 (2007).

62. Kong, L. X. and Peng, Z. Nanotechnology and its role in the management of periodontal diseases. *Periodontology 2000*, **40**, 184–196 (2006).

63. Kumar, S. R. and Vijayalakshmi, R. Nanotechnology in dentistry. *Indian J Dent Res.*, **17**, 62–69. doi: 10.4103/0970-9290.29890 (2006)

64. LaVan, D. A., McGuire, T., and Langer, R. Small-scale systems for in vivo drug delivery. *Nat Biotechnol.*, **21**(10), 1184–1191 doi:10.1038/nbt876 (2003).

65. LaVan. D. A. , McGuire, T., Langer, R. Small-scale systems for in vivo drug delivery. *Nat Biotechnol.*, **21**(10), 1184–1191. doi:10.1038/nbt876 (2003).

66. Lin, M. M., Kim do, K, El Haj, A. J., and Dobson, J. Development of superparamagnetic iron oxide nanoparticles (SPIONS) for translation to clinical applications, IEEE Trans. *Nanobioscience*, **7**(4), 298–305(Dec, 2008) doi: 10.1109/TNB.2008.2011864.

67. *Liu, Y, Miyoshi, H, and Nakamura, M. Nanomedicine for drug delivery and imaging: a promising avenue for cancer therapy and diagnosis using targeted functional nanoparticles. Int J Cancer,* **120**, 2527–2537 *(*2007).

68. Loo, C, Lin, A, Hirsch, L, Lee, M. H., Barton, J, Halas, N, West, J, Drezek, R. Nanoshell-enabled photonics-based imaging and therapy of cancer. *Technol Cancer Res Treat.,* **3**(1), 33–40 (2004)

69. Loo, C, Lin, A, Hirsch, L, Lee, M. H., Barton, J, Halas, N, West, J, and Drezek, R. Nanoshell-enabled photonics-based imaging and therapy of cancer". *Technol Cancer Res Treat.,* **3**(1), 33–40 (2004).

70. Madani, S. Y., Shabani, F, Dwek, M. V., Seifalian. A. M. Conjugation of quantum dots on carbon nanotubes for medical diagnosis and treatment. *International Journal of Nanomedicine,* **2013**(8) 941–950 **(**March, 2013).

71. Mahmoudi, M., Sant, S, Wang, B, Laurent, S., and Sen, T. Superparamagnetic iron oxide nanoparticles (SPIONs): development, surface modification and applications in chemotherapy. *Adv Drug Deliv Rev.,* **63**(1–2), 24–46 (Jan- Feb, 2011). doi: 10.1016/j. addr.2010.05.006. Epub 2010 May 26.

72. Pittenger, Mark F., Mackay, Alastair M., Beck, Stephen C., Jaiswal, Rama K. et al,Multilineage potential of adult human mesenchymal stem cells,Science, **284**, 5411 (Apr 2, 1999) Research Library Core,pg. 143

73. *McCarthy, J. R., Kelly, K. A., Sun, E. Y., and Weissleder, R. Targeted delivery of multifunctional magnetic nanoparticles.* Nanomedicine, **2**, 153–*167 (*2007)

74. Medintz, I. L., Uyeda, H. T., Goldman, E. R., and Mattoussi, H. *Nature Materials,* **4**(6), 435 (2005).

75. Minchin, Rod. Sizing up targets with nanoparticles. *Nature nanotechnology,* **3**(1), 12–13 (2008). doi:10.1038/nnano.2007.433.

76. *Mulder, W. J., Griffioen, A. W., Strijkers, G. J., Cormode, P., Nicolay, K., and Fayad Z. A. Magnetic and fluorescent nanoparticles for multimodality imaging.*Nanomedicine, **2**, 307–*324 (*2007).

77. Muthiah M, Park, I. K., and Cho, C. S. Surface modification of iron oxide nanoparticles by biocompatible polymers for tissue imaging and targeting. *Biotechnol Adv.* (Mar 22, 2013). pii: S0734-9750(13)00064-5. doi: 10.1016/j.biotechadv.2013.03.005.

78. Na, H. B., Lee, J. H. , An, K, Park, Y. I., Park, M, Lee, I. S., et al. Development of a T1 contrast agent for magnetic resonance imaging using MnO nanoparticles. *Angew Chem Int Ed Engl.,* **46**, 5397–401 (2007)

79. Nagano, K. Biodistribution of nanosilica particles in pregnant mice and the potential risk on the reproductive development. *Yakugaku Zasshi.,* **131**(2), 225–8 (Feb, 2011).

80. Nagy, Z. K., Zsombor, K.; Balogh, A., Vajna, B., Farkas, A., Patyi, G., Kramarics, A, and Marosi, G. Comparison of Electrospun and Extruded Soluplus-Based Solid Dosage Forms of Improved Dissolution. *Journal of Pharmaceutical Sciences*: n/a., (2011) doi:10.1002/jps.22731.

81. *Nahrendorf, M, Jaffer, F. A., Kelly, K. A., Sosnovik, D. E., Aikawa, E, Libby, P, and Weissleder, R. Noninvasive vascular cell adhesion molecule-1 imaging identifies inflammatory activation of cells in atherosclerosis. Circulation,* **114**, 1504–*1511 (*2006*)*

82. Nahrendorf, M, Sosnovik, D. E., and French, B. A., et al. Multimodality cardiovascular molecular imaging, Part II. *Circ Cardiovasc Imaging,* **2**, 56–70 (2009).

83. Napierska, D., Thomassen, L. C. , Lison, D., Martens, J. A., and Hoet. P. H. The nanosilica hazard: another variable entity. Part Fibre Toxicol., **7**(1), 39 (Dec 3, 2010). doi: 10.1186/1743-8977-7-39.

84. Nie, Shuming, Xing, Yun Kim, Gloria J., and Simmons, Jonathan W. ."Nanotechnology Applications in Cancer". *Annual Review of Biomedical Engineering*, **9**, 257–88 (2007). doi:10.1146/annurev.bioeng.9.060906.152025.

85. Ning, C. and Dai, K. Research development of hydroxyapatite-based composites used as hard tissue replacement. *Sheng Wu Yi Xue Gong Cheng Xue Za Zhi.Sep*, **20**(3), 550–554 (2003).

86. Nowack, B, Krug, H. F., Height, M. 120 Years of Nanosilver History: Implications for Policy Makers. *Environ Sci Technol*. (Jan 10, 2011).

87. Nyman, E. S, Hynninen, P. H., Research advances in the use of tetrapyrrolic photosensitizers for photodynamic therapy. *J Photochem Photobiol B*., 73(1–2), 1–28 (Jan 23, 2004).

88. Onaca-Fischer, O, Liu, J., Inglin, M., and Palivan, C. G. Polymeric nanocarriers and nanoreactors: a survey of possible therapeutic applications, *Curr Pharm Des*., **18**(18), 2622–43 (2012).

89. Saxl, Ottilia, Nanoskin Saves, Lives, and Limbs. *Nano the magazine for small sciencess*, **11**, 32 (12 October, 2011)

90. Singh, P., G. Destito, A., and Schneemann, M. Nanobiotechnology today: focus on nanoparticles. *J. Nanobiotechnology*, Manchester 4 (2006).

91. Palivan, C. G.,Fischer-Onaca, O, Delcea, M.,Itel, F, Meier, W. Protein-polymer nanoreactors for medical applications. *Chem Soc Rev*., (Apr 7, 2012), **41**(7), 2800–23. doi: 10.1039/c1cs15240h. Epub 2011 Nov 15.

92. Paquette, D. W., Hanlon, A, Lessem, J, and Williams, R. C. Clinical relevance of adjunctive minocycline microspheres in patients with chronic periodontitis: secondary analysis of a phase 3 trial. *J Periodontol*., **75**, 531–536.("Nanodentistry" Article published in Dental Dialogue, Vol. XXXII, No. 4, Oct. – Dec. 2006, Pgs. 120–121)s (2004).

93. Parttow, K. C., Chen, J, Brant, J. A. , Neubauer, A. M., Meyerrose, T. E., Creer, M. H., et al. 19F magnetic resonance imaging for stem/progenitor cell tracking with multiple unique perfluorocarbon nanobeacons. *FASEB J*, **21**, 1647–54 (2007).

94. Ramos-Cabrer, Pedro, and Campos, Francisco. Liposomes and nanotechnology in drug development: focus on neurological targets. *Int J Nanomedicine*., **8**, 951–960 (2013). Published online 2013 March 3.doi:10.2147/IJN.S30721.

95. Peiris, Pubudu M., Lisa, Bauer, Randall, Toy, Emily, Tran, Jenna, Pansky, Elizabethm, Doolittle, Erik, Schmidt, Elliott, Hayden, et. al., 2012. Enhanced Delivery of Chemotherapy to Tumors Using a Multicomponent Nanochain with RadioFrequency-Tunable Drug Release *ACS NANO* (AmericanChemical ociety). doi:10.1021/nn300652p.

96. Peng, G., Tisch, U., Adams, O., Hakim, M., Shehada, N., Broza, Y. Y., Bilan, S., Abdah-Bortnyak, R., Kuten, A., and Haick, H. *Nature Nanotech*., **4**, 669–673 (2009).

97. Perche, F. and Torchilin, V. P. Recent trends in multifunctional liposomal nanocarriers for enhanced tumor targeting. *J Drug Deliv*., 2013, 2013, 705265. doi: 10.1155/2013/705265. Epub 2013 Mar 7.

98. *Perez, A. T., Domenech, G. H., Frankel, C, and Vogel, C. L. Pegylated liposomal doxorubicin (Doxil) for metastatic breast cancer: the Cancer Research Network Inc, experience. Cancer Invest.*, **20**, 22–29 (2002).

99. Perez, J. M., Simeone, F. J., Saeki, Y., Josephson, L., and Weissleder, R. Viral-induced self-assembly of magnetic nanoparticles allows the detection of viral particles in biological media. *J Am Chem Soc*, **125**, 10192–3 (2003).

100. Pinon-Segundo, E., Ganem-Quintanar, A, and Alonso-Perez, V. Preparation and characterization of triclosan nanoparticles for periodontal treatment. *Int J Pharm*: **294**, 217–232 (2005).

101. Kumar, Preeti Satheesh , Kumar, Satheesh, Savadi, Ravindra, C., and John, Jins. Nanodentistry: A Paradigm Shift-from Fiction to Reality. *J Indian Prosthodont Soc.*, **11**(1) (March, 2011). 1–6.Published online 2011 April 20, doi: 10.1007/s13191-011-00620.

102. Qi, X. N. ,Mou, Z. L. ,Zhang, J., Zhang, Z. Q. Preparation of chitosan/silk fibroin/hydroxyapatite porous scaffold and its characteristics in comparison to bi-component scaffolds. *J Biomed Mater Res A.* (Mar 27, 2013). doi: 10.1002/jbm.a.34710.

103. Robert, A. and Freitas. *Jr. Nanomedicine*, Volume I: Basic Capabilities, (1999). ISBN 1-57059-645-X.

104. Robert, A. and Freitas. *Jr. Nanomedicine*, Volume IIA: Biocompatibility, (2003). ISBN 1-57059-700-6.

105. *Ruehm, S. G., Corot, C., Vogt, P., Kolb, S., and Debatin. J. F. Magnetic resonance imaging of atherosclerotic plaque with ultrasmall superparamagnetic particles of iron oxide in hyperlipidemic rabbits. Circulation.,* **103**, 415–422 (2001).

106. Rybachuk, A. V., Chekman, I. S., and Nebesna, T. Y. Nanotechnology and nanoparticles in dentistry. Pharmacol Pharm., **1**, 18–20 (2009).

107. Chan, S. D., Chan, W. C. W. *Proc. Nat. Acad. Sci. USA*, **107**, 11194–11199 (2010).

108. Franzen, S., Lommel, S. *Nanomedicine*, **4**, 575–588 (2009).

109. Das, Saikat, Bellare, Jayesh R. , and Banerjee, Rinti. Protein based nanoparticles as platforms   for aspirin delivery for ophthalmologic applications. *Colloids and Surfaces B:Biointerfaces*, **93**, Pages 161–168 (May 1, 2012).

110. Santoro, D., Satta, E., Messina, S., Costantino, G., Savica, V., and Bellinghieri, G, Pain in end-stage renal disease: a frequent and neglected clinical problem. *Clin Nephrol.*, **79** Suppl 1, S2–11 (Jan, 2013)

111. Rizvi, Sarwat, B. , Yildirimer, Lara, Shirin Ghaderi, Ramesh, Bala , M Seifalian, Alexander, and Keshtgar, Mo. A novel POSS-coated quantum dot for biological application. *Int J Nanomedicine.*, **7**, 3915–3927(2012).Published online 2012 August 2.doi: 10.2147/IJN. S28577.

112. *Sengupta, S., Eavarone, D, Capila, I., Zhao, G., Watson, N., Kiziltepe, T., and Sasisekharan, R. Temporal targeting of tumour cells and neovasculature with a nanoscale delivery system. Nature.,* **436**, 568–572 (2005).

113. Shanmukh, S., Jones, L., Driskell, J., Zhao, Y, Dluhy, R, and Tripp, R. A. Rapid and sensitive detection of respiratory virus molecular signatures using a silver nanorod array SERS substrate. *Nano Lett.*, **6**, 2630–2636 (2006).

114. Sharma, S, Cross, S. E., Hsueh, C, Wali, R. P., Stieg, A. Z., and Gimzewski J. K. Nanocharacterization in dentistry. *Int J Mol Sci.*, **11**(6), 2523–2545 (2010). doi: 10.3390/ijms11062523.

115. Sinusas, A. J., Bengel, F, Nahrendorf, M., et al. Multimodality cardiovascular molecular imaging, part I. *Circ Cardiovasc Imaging.*, 1, 244–256 (2008).

116. Smijs, T. G. and Pavel, S, *Titanium dioxide and zinc oxide nanoparticles in sunscreens: focus on their safety and effectiveness Nanotechnology, Science and Applications*, **Published Volume 2011**(4), 95– 112 (October, 2011).

117. Smith, J. P. Nanoparticle delivery of anti-tuberculosis chemotherapy as a potential mediator against drug-resistant tuberculosis. *Yale J Biol Med.* , **84**(4), 361–369 (Dec, 2011).

118. Chook, Soon Wei, Chia, Chin Hua, Zakaria, Sarani, Ayob, Mohd. Khan, Chee, Kah Leong , Huang, Nay Ming, Neoh, Hui Min , Lim, Hong Ngee , Jamal, Rahman, andRahman, Raha Mohd Fadhil Raja Abdul. Antibacterial performance of Ag nanoparticles and AgGO nanocomposites prepared via rapid microwave-assisted synthesis method. *Nanoscale Res Lett.*, **7**(1), 541 (2012). Published online 2012 September 28. doi:10.1186/1556-276X-7-541.

119. *Sosnovik, D. E. , Schellenberger, E. A. , Nahrendorf, M, Novikov, M. S., Matsui, T, Dai, G, Reynolds, F, Grazette, L, Rosenzweig, A, Weissleder, R, and Josephson, L. Magnetic resonance imaging of cardiomyocyte apoptosis with a novel magneto-optical nanoparticle. Magn Reson Med.,* 54, 718–724 (2005).

120. Stavis, S. Strychalski, E. A., Gaitan, M. Nanofluidic structures with complex three-dimensional surfaces. *Nanotechnology*, 20(16) (2009).

121. Samuel, Stephen Paul , Jain, Namrata, O'Dowd, Frank, Paul, Toby, Kashanin, Dmitry, Gerard, Valerie A , Gun'ko, Yurii K , Prina-Mello, Adriele, andVolkov,Yuri. Multifactorial determinants that govern nanoparticle uptake by human endothelial cells under flow, *Int J Nanomedicine.,* 7, 2943–2956(2012).Published online 2012 June 14.doi: 10.2147/IJN. S30624

122. Steinmetz, N. F. Viral nanoparticles as platforms for next-generation therapeutics and imaging devices. *Nanomedicine.,* 6(5), 634-41 (Oct, 2010). Epub 2010 Apr 28.

123. Stoll, G., Basse-Lüsebrink, T., Weise, G. Jakob, P. Visualization of inflammation using (19) F-magnetic resonance imaging and perfluorocarbons, Wiley Interdiscip Rev Nanomed Nanobiotechnol., 4(4), 438–47. (Jul-Aug, 2012) doi: 10.1002/wnan.1168. Epub 2012 Mar 15.

124. Stuchinskaya, T., Moreno, M., Cook, M. J., Edwards, D. R., and Russell, D. *A. Photochem. Photobiol. Sci.,* **10**, 822 (2011).

125. Swierczewska, M. Lee, S, Chen, X. The design and application of fluorophore-gold nanoparticle activatable probes. *Phys Chem Chem Phys.,* (Jun 7, 2011), 13(21), 9929-9941. Epub 2011 Mar 7.

126. *Torchilin, V. P. Recent advances with liposomes as pharmaceutical carriers. Nat Rev Drug Discov.,* **4***, 145–160 (*2005).

127. Ulrich, K. E., Cannizzaro, S. M., Langer, R. S., and Shakeshelf, K. M. Polymeric systems for controlled drug release. *Chem Rev,* 99, 3181–3198 (1999). University of Waterloo, Nanotechnology in Targeted Cancer Therapy, http://www.youtube.com/watch?v=RBjWwlnq3cA 15 January 2010.

128. Vasudevanpillai, Biju, Tamitake, Itoh, Abdulaziz, Anas, Athiyanathil, Sujith, and Mitsuru, Ishikawa. Semiconductor quantum dots and metal nanoparticles: syntheses, optical properties, and biological applications. *Analytical and Bioanalytical Chemistry*, **391**(7), 2469–2495 (August, 2008).

129. Wadhwa, S., Paliwal, R. Paliwal, S. R., and Vyas, S. P. Nanocarriers in ocular drug delivery: an update review. *Curr Pharm Des.,* **15**(23), 2724–50 (2009).

130. Wang, R. Wen, D. Xie, X., and Zhong, Y. Development of nanohydroxyapatite composites as bone grafting materials. *Sheng Wu Yi Xue Gong Cheng Xue Za Zhi.,* 25(5), 1231–4 (Oct, 2008).

131. Wang, S. S., Chen, J., Keltner, L. Christophersen, J., Zheng, F. Krouse, and M, A Singhal. "New technology for deep light distribution in tissue for phototherapy". *Cancer Journal,* **8**(2), 154–63 (2002). doi:10.1097/00130404-200203000-00009.PMID 11999949.

132. *Winter, P. M., Morawski, A. M., Caruthers, S. D. , Fuhrhop, R. W., Zhang H, Williams, T. A., Allen, J. S., Lacy, E. K., Robertson, J. D., Lanza, G. M., and Wickline, S. A. Molecular imaging of angiogenesis in early-stage atherosclerosis with alpha(v) beta3-integrin-targeted nanoparticles. Circulation.,* **108***, 2270–2274 (*2003).

133. Kangwansupamonkon, Wiyong (PhD), Lauruengtana, Vichuta (MSc), Surassmo, Suvimol, (MSc), Ruktanonchai, Uracha. Antibacterial effect of apatite-coated titanium dioxide for textiles applications Nanomedicine: Nanotechnology, Biology and Medicine, 5(2), 240–249 (June, 2009).

134. Wong, D. T. Salivary diagnostics powered by nanotechnologies, proteomics and genomics. *J Am Dent Assoc.*, 137, 313–321 (2006).

135. Xing, Y, Chaudry, Q, Shen, C, Kong, K. Y. , Zhau, H. E. , Chung, L. W., et al. Bioconjugated quantum dots for multiplexed and quantitative immunohistochemistry. *Nat Protoc*, **2**, 1152–1165 (2007).

136. Wang, Yi-Xiang J., Shahid, M. Hussain, Gabriel and P., Krestin. Super paramagnetic iron oxide contrast agents: physicochemical characteristics and applications in MR imaging. *European Radiology*, **11**(I11), 2319–2331 (November 2001).

137. Yoshioka, Y. Nanosafety studies of nanomaterials about biodistribution and immunotoxicity, **131**(2), 221–4 (Feb, 2011).

138. Yoshioka, Y., Yoshikawa, T., and Tsutsumi, Y. Nano-safety science for assuring the safety of nanomaterials. *Nihon Eiseigaku Zasshi.*, **65**(4), 487–92 (Sep, 2010).

139. Zhao, X, Milliard, L. R., Mechery, S. J., Wang, Y., Bagwe. R. P., Jin, S, et al. A rapid bioassay for single bacterial cell quantitation using bioconjugated nanoparticles. *Proc Natl Acad Sci* USA,101, 15027-15032 (2004).

140. Zheng, G,, Patolsky, F, Cui, Y, Wang, W. U., and Lieber C.M. Multiplexed electrical detection of cancer markers with nanowire sensor arrays". *Nat Biotechnol.* **23**(10), 1294–1301(2005). doi:10.1038/nbt1138.

CHAPTER 9

STUDY OF PLASMA INDUCED GAS PHASE GROWTH MECHANISM OF TIO$_2$ MULTILAYER THIN FILMS AND ITS CORRELATION WITH ITS MORPHOLOGICAL AND ELECTRICAL PROPERTIES

P. LAHA, A. B. PANDA, S. K. MAHAPATRA, P. K. BARHAI, A. K. DAS, and I. BANERJEE

CONTENTS

9.1   INTRODUCTION

Advanced semiconductor technology demands continuous shrinkage of device dimensions with high performance efficiency (Plummer and Griffin, 2001, Ronen et. al., 2001, Pinacho et. al., 2005). Although, a large number of devices are used in the semiconductor technology, fabrication of MOS device has opened up a new room in miniaturization of equipment (Plummer and Griffin, 2001, Marwedel, 2003). Generally, the performance of the MOS device is known by the performance of its SiO_2 oxide layer in terms of charge storage/unit area value (Wilk et. al., 2001, Ariel et.al., 2005). It was observed that in traditional MOS device, as the thickness of the oxide layer reduces, the leakage current increases, which lowers the device performance (Lo et. al., 1997, Wilk et. al., 2001). Therefore high dielectric material is desirable for the MOS device as an alternative for SiO_2. Although single or multilayer of TiO_2 and Al_2O_3 are gaining considerable attention to increase the performance of MOS device (Campbell et. al., 1997, Alers et. al., 1998), the reports are still in scarce. The advantage of the combinations of the oxide layers is to control the dielectric constant in the form of interfacial interaction and modification in the band gap (Fermi level) so as to control the device performance. The high dielectric materials would be chosen for the combination in such a way that the required band gap should be the average of the individual band gaps (Jaros et. al., 1985). Moreover, materials' crystallinity is equally important in the device performance. Amorphous dielectric material is acceptable due to profound inter diffusion of single or multi layers at the interface to the silicon or among themselves. This enhances the pinning of Fermi level and grain boundaries affecting the leakage current (Mohammad et. al., 2006, Chandra Sekhar et. al., 2012). However, a controllable interdiffusion for a small scale MOS device of individual layer thickness <50 nm is a real challenge. Amongst several other methods like chemical vapor deposition, pulsed laser deposition, molecular beam epitaxy, and so on, plasma assisted reactive radio-frequency magnetron sputtering is widely used for fabrication of MOS devices (Chandra Sekhar et. al., 2012). The method provides large area uniformity of film thickness and strong adhesion to substrate (Chandra Sekhar et. al., 2012). The material can be sputtered to the growing surface layer in the correct proportion with sufficient energy to ensure the stoichiometry of the elements. The optical and electrical properties of the multilayer thin film are controlled by varying the sputtering parameters such as substrate temperature, partial pressures of sputtering and reactive gases, target substrate distance, sputtering power, and deposition rate. However, too many manipulative parameters involved in the process cause the deposited films not being competent for desired applications (Ellmer, 2000). In order to overcome these disadvantages and to have a control over the plasma process parameters understanding of the film growth process in the viewpoint of plasma chemistry is necessary. Optical Emission Spectroscopy (OES) is a non-intrusive, simple, and useful method to study the behavior of discharge plasma. The chemical reactions taking place among the species and their variations with plasma parameters can be easily investigated in real time scale without interfering the process.

In the present context, *in situ* OES investigation has been performed in order to understand the process of vapor-solid interaction deposition of amorphous TiO_2 and

Al_2O_3 thin films on n type silicon substrate for MOS structure by reactive magnetron sputtering. The variation of oxygen: argon gas ratio was followed during deposition. The electron temperature (T_e) and dependencies of electron (v_e)/ion velocities (v_i) on gas pressure during deposition has been correlated with the materials characteristics and transport properties of the deposited films.

Multilayered $TiO_2/Al_2O_3/TiO_2/Al_2O_3$ thin films were fabricated by reactive RF magnetron sputtering at different oxygen: argon gas ratios. The variations of film characteristics have been understood from the perception of plasma induced gas phase deposition mechanism at different deposition parameters. The plasma chemistry controls the films characteristics by controlling the interface inter diffusion of atoms during layer transition from TiO_2 to Al_2O_3. *In -situ* investigation of the plasma chemistry was performed using OES. The plasma parameters were determined and correlated with films' morphology and electrical properties. A motivating correlation is established between plasma parameters like electron temperature, electron, and ion velocities with films' morphology and transport properties. Proper understanding and optimization of plasma chemistry is found to be useful for nano fabrication of heterostructured devices.

9.2   EXPERIMENTAL DETAILS

Heterostructure of $TiO_2/Al_2O_3/TiO_2/Al_2O_3$ thin films were prepared using radio frequency (RF) magnetron sputtering deposition system (MSC-3, JE Plasma Consult GmbH). The schematic of the multilayer RF magnetron sputtering system has been shown in Figure 1.

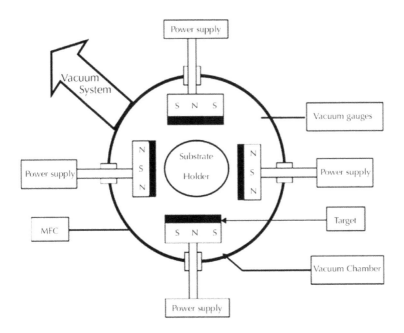

FIGURE 1   Schematic of RF magnetron sputtering system.

The oxygen argon gas flow rate was maintained at 20:80, 30:70, 50:50, and 60:40 sccm with the help of mass flow controller at constant RF power of 200 Watt for deposition. The deposition was performed at working pressure of ~6.2 × 10^{-2} mbar and deposited for 20 min for each layer without breaking the vacuum and switching on from one magnetron to other at transition. In this way, four equally thick films (~400 nm) were obtained at different oxygen, argon gas ratio but constant RF power and deposition time of 20 min for each layer.

Thickness of the oxide layers was measured by Spectroscopic Ellipsometer (Nano - View Inc., Korea, SEMG1000-VIS). The OES (get Spec UV/Vis 0172 grating UA, 200–720 nm, CCD detector with 3648 pixels, 10 μm entrance slit and spectral resolution of 0.27 nm) was used for *in situ* investigation of the plasma chemistry showing the intermixing of the two metals during interface formation of the heterostructure. The surface morphology and the particle size was determined using Atomic Force microscope (NT-MDT Moscow, Russia, Model No. Solver P47-PRO).The equivalent conductance (G), dielectric constant (ε'), and tan δ was measured from the impedance analyzer at 500 kHz frequency. Conductivity (σ) was calculated from the G value by knowing the thickness (d) and area (A) of the sample (σ = G. d/A).

9.3   FILM GROWTH PROCESS IN REACTIVE RF MAGNETRON SPUTTERING

The TiO_2/Al_2O_3 multilayer thin films were deposited by reactive RF magnetron sputtering process. In the case of RF reactive sputtering process, the major disadvantage is target poisoning (Depla, Gryse, 2004). The mechanism assumes that the metal target gets readily dissociated by electron impact, producing either metal atoms (M), or partial metal oxides (MO). These reactive species then get fully oxidized (MO_n) either in the gas-phase or on the surface. The gas-phase oxidation renders the target inactive for deposition (Safi, 2000), therefore affecting the film growth process. The sputtering yield of metal atoms from a poisoned surface is less than that from a pure metallic target, the metal flux from the target decreases with increasing reactive gas partial pressure until the entire target is poisoned. The competition between these alternatives is controlled by the atomic oxygen density in the plasma. The deposition rate is proportional to O atom density at very low concentrations because the surface oxidation step is limiting. However, as the production of O atoms increases the rate quickly reaches a maximum and declines sharply. As the O atom density further increases the rate saturates at a level that is dependent on the specific reaction time considerations. Moreover, the variation of oxygen, argon gas ratio provides information about the evolution of the target state under different sputtering conditions at the time of deposition. A uniform partial pressure P of the reactive gas will cause a uniform bombardment of neutral reactive molecules F (molecules/unit area and time) to all surfaces in the plasma chamber. From the gas kinetics, the relationship between F and

P is, $F = \dfrac{P}{\sqrt{2kT\pi m}}$ where, k is the Boltzman constant, T is the temperature, and m is

the mass of the gas molecule. Although reactive RF magnetron sputtering with metal oxide target is a complex process, the experimental results could be interpreted using the simple mechanism described later.

9.4 DISCUSSION AND RESULT

9.4.1 MORPHOLOGICAL ANALYSIS

Figure 2 shows 3D AFM images for morphology and Figure 3 shows grain size and RMS roughness of the particles in the deposited films for TiO_2/Al_2O_3 multilayer (4 layers) deposited at different oxygen, argon gas ratio (a) 20:80 (b) 30:70 (c) 50:50, and (d) 60:40 sccm with RF power of 200 Watt. Figure 3 shows grain size dependencies with oxygen, argon gas ratio for the deposited samples. The grain size increases from 33.3 nm to 66.2 nm with oxygen contents.

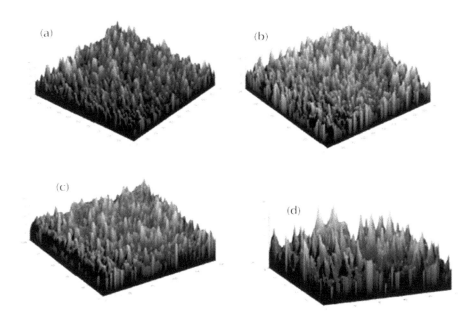

FIGURE 2 AFM images TiO_2/Al_2O_3 multilayer (4 layers) with different oxygen, argon gas ratio (a) 20:80 (b) 30:70 (c) 50:50, and (d) 60:40 sccm.

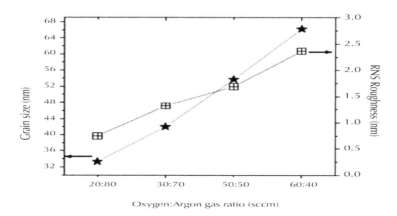

FIGURE 3   Grain size and RMS roughness dependencies with (a) 20:80 (b) 30:70 (c) 50:50, and (d) 60:40 sccm oxygen, argon gas ratio, 200 Watt RF power.

The figure indicates that the surface morphology gets more smoothened and the size of grains forming the film increases with increase in oxygen content. It could be due to decrease in deposition rate with increasing oxygen pressure and can be explained on the basis of the relation between sputtering pressure and mean free path (Chawla et. al., 2010).

The mean free path of the sputtering gas molecule is given as:

$$\lambda = 2.33 X 10^{-20} \frac{T}{P\delta_m^2}. \tag{1}$$

Where, T is the temperature, P is the pressure, and δ_m is the molecule diameter. Therefore, as the oxygen percent increases there will be increase in the collision frequency of oxygen atoms and hence more reaction is possible. Due to a large number of collisions with the oxygen atoms the sputtered atoms on the substrate content more oxygen, and there is an increase in the crystallite size and the surface roughness of the films. The results reveal that the oxygen and argon gas ratio affected the size of the deposited particles in the films. The partial pressure in terms of flow rates of oxygen and argon (total pressure remaining constant) are the factors affecting the trajectory of the sputtered atoms followed by the oxides before deposition. Since argon is the main sputtering gas, it is partial pressure imparts the kinetic energy to the sputtered atoms whereas, the partial pressure of oxygen governs collision affecting the mobility and also the diffusion of the species approaching the substrate (Petrov et. al., 2003). Numerous changes and processes at a microscopic level may be achieved via energetic particle bombardment prior to and during film formation and growth. That also helps in removal of contaminants, alteration of surface chemistry, enhancement of nucleation, and re-nucleation (due to generation of nucleation sites via defects, implanted, and recoil implanted species) process. Higher surface mobility of ad-atoms and elevated film temperatures with attendant acceleration of atomic reaction and interdiffusion rates may be the other governing phenomena (Ohring, 1992). When oxygen percentage increases the collision frequency also increases which enhance the film temperature, due to this, surface mobility of the atoms increases, which significantly enhances the atomic reaction and inter diffusion rates.

Scattering of charge carriers by high angle grain boundaries may bind the mobility. Also, better-aligned films or even epitaxial ones possess improved conductivity and light transmittance as reported by Wang et. al., (Wang et. al., 2005). According to Wang et. al., higher conductivity was obtained for films prepared using shorter target–substrate distances. The lower substrate position gave a lower growth ,rate; however, the films were of better conductivity. At high carrier concentrations above the most critical density, it is well known that the electronic states of the crystal get modified because of carrier–carrier and carrier–impurity interactions (Ruiping et. al., 1996). Many body effects such as exchange and Coulomb interactions lead to narrowing of the band gap (Lu et. al., 2006). The temperature independent conductivity also indicates that all carriers are delocalized. This band gap shrinkage represents the exchange energy due to electron–electron interaction. The onset of the gap shrinkage can be related to a semiconductor-to-metal transition accompanied by merging of donor and conduction bands. The changes in the electrical parameters due to more oxygen incorporation are attributed to decrease in grain size (Wu and Chiou, 1996). Decrease in electrical conductivity on oxygen incorporation is due to the chemisorption of oxygen at the grain boundaries, which in turn leads to the formation of extrinsic trap states localized at the grain boundaries. These states of trapped free carriers form the bulk of the grains and create potential barriers by causing depletion in the region adjacent to the grain boundaries (Basu and Basu, 2009). These potential barriers decrease the mobility of carriers. This is specially observed in wide band gap semiconductors (Basu and Basu, 2009). The increase in the oxygen flow rate decreases the mobility of the sputtered ions with increase in particle sizes. The decrease in grain size together with a high density of trap states may lead to the extension of the depletion region throughout the bulk of the grains. This has been supported by real time measurements of ion and electron velocity using OES and has been correlated with the particle size.

9.4.2 GAS PHASE STUDY THROUGH OPTICAL EMISSION SPECTROSCOPY

The OES (getSpec UV/Vis 0172 grating UA, spectral resolution of 0.27 nm) has been used for *in situ* investigation of the plasma chemistry showing the intermixing of the two metals during interface formation of the heterostructure.

Light is collected from a small well-defined zone in the plasma, where the optical fiber is being fixed. The magnification of the image is made one so that a real one to one correspondence is maintained between object and image. The fiber optics is mounted on a traversing microscope (least count, 0.05 mm). Figure 4 shows the detail of the spectrometer coupled with the reactor chamber. The system can be moved along the X- and Y- direction independently. X-Y movement is provided over a range of 7 cms in both directions in order to scan over the different points in the radial and axial directions of the plasma column.

The focused image of a narrow object plane is transported to the CCD detector through the optical fiber. The other end of the fiber optic cable is fixed to the entrance slit of the spectrometer. The getSpec UV/Vis 0172 spectrometer is fitted with a grating and blazed in the UV-Visible region (200–720 nm). The detector is a TE cooled, front illuminated CCD of 3648 pixel array with 10 micron entrance. Figure 4 illustrates the pictorial representation of the Optical Emission Spectrometer in the present case.

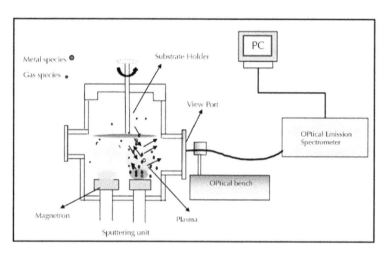

FIGURE 4   Schematic of OES measurement set up.

Optical Emission Spectra was recorded during reactive sputtering of the Ti and Al target simultaneously using different O_2, Ar gas ratio. Magnetron sputtering constitutes low pressure (in our case $\sim 10^{-2}$ mbar = 1Pa) plasmas where, the line-ratio method with corona model may be used to find out the electron temperature (Xi-Ming and Yi-Kang., 2010). The line-ratio method must be carefully applied as it also depends upon the characteristics of excited levels, excitation cross sections besides the operating conditions or process parameters. Characteristic emission lines were observed for the different species present in the plasma. The most intense ones were chosen, ArI at 603.29 nm (4p-5d), and 714.7 nm (4s-4p) [NIST database] with upper energy levels 15.13 and 13.28 eV, which represents the moderate electrons have been used for determining the electron temperature using relative intensity method given by [NIST database].

$$T_e = \frac{E_m - E_p}{k \ln\left(\dfrac{I_{pn} A_{ms} g_m V_{ms}}{I_{ms} A_{pn} g_p V_{pn}}\right)}$$

(2)

Where, I_{ms} and I_{pn} are the measured intensities from m-s and p-n transitions, respectively, A_{ms} and A_{pn} the transitions probabilities, v_{ms} and v_{pn} wave numbers, E_{m} and E_{p} are the upper level energies, g_{m} and g_{p} their statistical weights, and k the Boltzmann constant. Table 1 gives the detailed transitions of the observe demission lines. The different parameters of the Ar lines used for T_{e} calculation are taken from NIST database and presented in Table 2 [NIST database].

In doing so we are considering the plasma to be a partial local thermo dynamical equilibrium (PLTE). However, this assumption may not be correct, as the corona model is not tied to Electron Energy Distribution Function (EEDF) and OES measurements are only sensitivity to EEDF over specific electron energy of minimum threshold energy (Boffard et. al., 2004). So, the calculated T_{e} value is questionable. It may also happen that the electron temperature calculated from the OES method differs from that of the Langmuir Probes. However, Langmuir probe measurements have not been performed here.

TABLE 1

Species	Transition system	Upper level energy(eV)	Emitted wavelength(nm)
Ti I	$a^3F - y^5D^0$	3.198495	393.42
Ti I	$a^5P - y^5S^0$	4.631941	428.49
Ti I	$a^3H - x^3H^0$	4.8497604	474.27
Ti I	$z^3D^0 - e^3F$	4.6896708	564.85
Al I	$^2P^0 - ^2S$	3.1427210	396.15
Al I	$^2S - ^2P^0$	4.9938207	669.60
Ar I	$^2[5/2] - ^2[7/2]^0$	15.1305434	603.21
Ar I	$^2[3/2] - ^2[3/2]^0$	14.9715214	688.81
Ar I	$^2[3/2]^0 - ^2[1/2]$	13.3278562	696.54
Ar I	$^2[3/2]^0 - ^2[3/2]$	13.2826382	714.70
Ar I	$^2[3/2]^0 - ^2[3/2]^0$	14.859229	726.51
O I	$^5P - ^5S^0$	12.6608561	645.44

TABLE 2

λ(nm)	E(eV)	g	A(S^{-1})
603.29	15.1305434	9	2.46X10^6
714.7	13.2826382	3	6.25X10^5

As said before, the main goal of this work is the use of Ti and Al target to deposit TiO_2/Al_2O_3 multilayer films at the presence of oxygen and argon gas mixture and make a correlation between the gas phase study and transport properties. So, it was interesting to examine the influence of the reactive atmosphere on the process parameters. This result seems to indicate that deposition of TiO_2 and Al_2O_3 proceeds with not only Ti or Al atoms and ions but also with incoming oxide molecules and molecular ions such as Ti-O or Al-O. Figure 5 shows the variations of OES peak intensity with different reaction zone of the plasma at different oxygen, argon gas ratio (a) 20:80 (b) 30:70 (c) 50:50, and (d) 60:40 sccm, 200 Watt RF power. In reactive magnetron sputtering process, the plasma enhanced deposition is controlled by three processes. Initially at (a) metallic phase ($P_{O2} < 1.86$ Pa) the target is purely Ti and Al so it sputtered Ti and Al to form Ti-O and Al-O reacting with the oxygen in the ambience. The Ti-O and Al-O concentration decreases rapidly in (b) transition phase (1.86 Pa $< P_{O2} < 3.1$ Pa) of plasma, where the target has started getting oxidized. Concentration of Ti and Al sputtered atoms further decreases hence decreasing the Ti-O and Al-O formation. In the (c) compound phase ($P_{O2} > 3.1$ Pa) target is no more metallic but fully oxidized, hence the sputtered atoms are not Ti and Al but suboxides of Ti and Al.

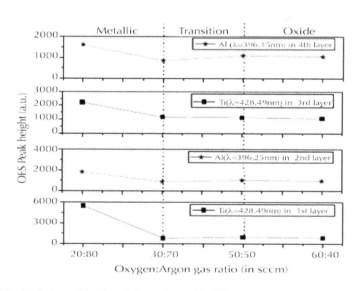

FIGURE 5   Variations of OES peak intensities with different oxygen: argon gas ratios during deposition of four different consecutive layers.

At 60:40 sccm O_2, Ar gas ratio, the Ti, and Al targets are completely poisoned, and the Ti-O 559.7 nm, Al-O 486.6 nm emission lines gradually vanish behind the detectable level. Literature reveals that, during the growth of TiO_x films, both metallic Ti atom and Ti–O molecular species are sputtered from the Ti target when it is operated under the metallic or transition mode. In metallic and transition mode, less number of oxygen molecules is present in the chamber. In the oxide mode, the emitted species are dominated by TiO and TiO_2 molecular species and the level intensity of the Ti atomic species get reduced to a minor level (for example $\leq 2\%$) that could be undetectable by the spectrometer due to complete poisoning of the Ti target. Figure 6 shows the variations of Ti-O band intensity and the variation of strongest band head intensity with different oxygen, argon gas ratio (a) 20:80 (b) 30:70 (c) 50:50, and (d) 60:40 sccm, 200 Watt RF power. Ti-O bands with strongest heads at 544.8 nm ($v' = 0$, $v'' = 1$), 549.7 nm ($v' = 1$, $v'' = 2$), $\lambda = 559.7$ nm ($v'=0$, $v'' = 0$), and $\lambda = 562.9$ nm ($v' = 1$, $v'' = 1$) are distinctly identified at 540–570 nm and the peak intensities gradually decrease from the 1st layer to 4th layer in Figure (Kakati et. al., 2009). Emission band of Ti-O was observed at this position indicating that Ti oxide formation gets initiated in the near vicinity of target much before the growing surface of the films. The decrease in Ti-O band intensities at a constant O_2, Ar gas ratio reveals the target poisoning mechanism. Figure 7 shows the variations of Al-O band intensity and the variation of strongest band head intensity with different oxygen, argon gas ratio (a) 20:80, (b) 30:70, (c) 50:50, and (d) 60:40 sccm, 200 Watt RF power. Similarly, for Al-O bands with strongest heads at 467.2 nm ($v' = 2$, $v'' = 1$), 471.5 nm ($v' = 4$, $v'' = 3$), and $\lambda = 486.6$ nm ($v' = 1$, $v'' = 1$) are distinctly identified at 460–490 nm and the peak intensities are gradually decreasing with from the 1st layer to 4th layer in Figure. Similarly, the target poisoning mechanism is for Al_2O_3 deposition.

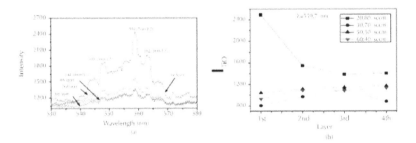

FIGURE 6 (a) Optical emission spectra of Ti-O molecular band, (b) Variations of Ti-O band intensity ($\lambda = 559.7$ nm) with different oxygen, argon gas ratios.

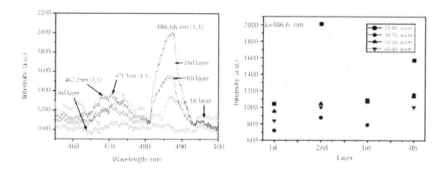

FIGURE 7 (a) Optical emission spectra of Al-O molecular band, (b) Variations of Al-O band intensity ($\lambda = 486.6$ nm) with different oxygen, argon gas ratios.

Figure 8 shows the variation of electron temperature with oxygen, argon gas ratio of 20:80, 30:70, 50:50, and 60:40 sccm during deposition of four layers. The temperature decreases with increase in oxygen gas and also for successive deposition of four layers. This indicates the target positioning phenomenon. This is due to the plasma getting more metal atoms in the ambience, which lowers down the average electron temperature of the plasma during successive layer deposition.

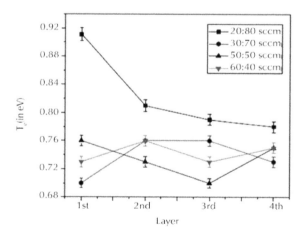

FIGURE 8 Variation of electron temperature with different oxygen, argon gas ratios of 20:80, 30:70, 50:50 and 60:40 sccm, 200 Watt RF power for four different layers.

The average electron and ion velocities were determined using the electron temperature (T_e) from the given expression (Hendron et. al., 1997).

$$v_e = \left(\frac{8kT_e}{\pi m_e}\right)^{1/2} \tag{3}$$

$$v_i = \left(\frac{8kT_e}{\pi m_i}\right)^{1/2} \tag{4}$$

The average velocity signifies the randomness of the system and thereby giving an idea of the ion mobility in the plasma zone.

9.4.3   CORRELATION OF GAS PHASE GROWTH MECHANISM WITH THE MORPHOLOGY OF THE FILMS

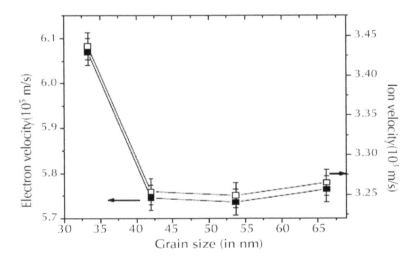

FIGURE 9   Variations of electron and ion velocities with grain size.

The OES efficiently gives evidence of the thin film growth mechanism. The metal vapor (Ti, 47.88 atomic mass, Al, 26.98 atomic mass) produced through sputtering gets reacted with reactive oxygen (16 atomic mass) to form oxides. However, Argon (39.95 atomic mass) represents as the main sputtering gas. The vapor of the compounds formed gets deposited atom by atom on the substrate through heterogeneous nucleation. The formation of the atomic layers and the nucleation sites, size of the critical nucleus are dependent on the collision and the mobility of the moving atomic species under the given field. Calculation shows that the average random velocity of the metallic ions (Ti and Al) has direct relation with the size of the particles getting deposited on the substrate. It is an interesting result explaining the phenomenon that higher oxygen flow decreases the ion velocity, thereby availing enough time for particles to grow larger (Linfeng et. al., 2010). The film growth process is driven by the solid – vapor transformation under heterogeneous nucleation through the heat released to the substrate (Nag and Haglund, 2008). There are certain factors that govern the deposition out of, which target poisoning factor is quite effective. At low reactive gas flow, the target condition remains metallic on account of its complete reaction with sputtered metal atoms. Further addition of the reactive gas results not only in the formation of a compound on the target or cathode surface. This results in sudden decrease of the deposition rate with an abrupt change in the partial pressure of the reactive gas, is called hysteresis or poisoning effect (Musil et. al., 2005). Therefore, a too low supply of the reactive gas may give rise to an under stoichiometric composition of deposited films whereas high supply of reactive gas may reduce the deposition rate due to target poisoning (Musil et. al., 2005). Hence for desired deposition of the oxide films, the controlled flow of the reactive and sputtered gas has to be maintained and monitored during deposition.

As stated above the heterogeneous nucleation process that is responsible for the thin film deposition in reactive magnetron sputtering process depends on the kinetic behavior of atoms, group of atoms or molecules through their micro structural evolution presented schematically in Figure 10. The time rate of change of cluster densities or coalescence in terms of the processes that occur on the substrate surface is basically the kinetics of chemical reactions that has been occurring inside the vacuum chamber. The surface morphology and the growth mechanism of the as deposited materials affect the electrical response of the films. Collisions between separate island like crystallites (or droplets) results coalescence, which in turn execute random motion and can be controlled through the applied electric field during deposition. These phenomena can be explained on the basis of the interaction of the field with electrically charged islands. Ionized vapor atoms as a result of sputtering and/or the potential at the substrate interface may be assumed to be the cause of island charge. Taking the islands or coalescence as a spherical particle of diameter "d" and having a surface free energy, the presence of a charge "q"contributes additional electrostatic energy (that is, q^2/d). This increase in total energy is accommodated by an increase in surface area causing distortions to the sphere into a flattened oblate spheroid. The exact shape however, being determined by the balance of various free energies. Further coalescence by ripening, sintering, or cluster mobility processes is thus the net effect of charging. It also happens that the cluster may break up in much the same way that a charged droplet of mercury does with greater charge or higher fields. The atomic force microscope, which has the capability of resolving individual atoms, has revealed the migration of clusters as evidenced in the Figure 9. Interestingly, the mobility of metal particles can be significantly altered in oxygen gas ambience (Ohring, 1992). The clusters translations, rotation as well as hopping with each other and sometimes re-separate thereafter induce a great impact on the current transport mechanism of the as deposited films.

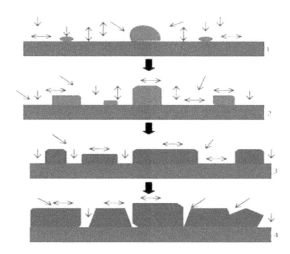

FIGURE 10 Schematic diagram illustrating fundamental growth process controlling micro structural evolution: nucleation, island growth, impingement and coalescence of islands, grain coarsening, formation of polycrystalline islands and channels, development of a continuous structure, and film growth.

9.4.4   CORRELATION OF GAS PHASE GROWTH MECHANISM WITH ELECTRICAL PROPERTIES OF THE FILMS

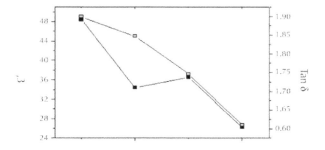

FIGURE 11 Also shows dielectric constant (ε'), tan (δ), conductivity (σ), and conductance (G/ω) of TiO_2/Al_2O_3 multilayer (4 layers) deposited at different oxygen, argon gas ratio (a) 20:80 (b) 30:70 (c) 50:50 and (d) 60:40 sccm with RF power of 200 Watt. The ε' value decreases with increasing oxygen incorporation. That is the films become less conducting with more oxygen incorporation. The magnitude of the dielectric constant depends on the ability of dipoles units in the material to orient with the oscillations of an external alternating electric field.

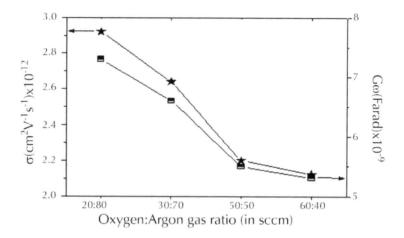

FIGURE 12   Variations of dielectric constant (ε'), tan δ, conductivity (σ), and conductance (G/ω) with different oxygen, argon gas ratios of 20:80, 30:70, 50:50, and 60:40 sccm at 200 Watt RF power.

The types of polarization may be sub divided by this following types- space charge (electronic), atomic and dipolar. Space charge polarization involves a limited transport of charge carriers (electron) until they are stopped at a potential barrier, may be due to a grain boundary or phase boundary. The polarizability (α) is generally additive, (i.e. $\alpha = \alpha_{\text{space charge}} + \alpha_{\text{dipolar}} + \alpha_{\text{atomic}}$) of each polarization mode to the dielectric constant (ε), (i.e. $\varepsilon = \varepsilon_{\text{space charge}} + \varepsilon_{\text{dipolar}} + \varepsilon_{\text{atomic}}$). Generally, dipoles are produced in the oxide material, which governs the dielectric properties. When an electric field of high frequency is applied, the probabilities of these space charges to drift and accumulate at the interface become highly remote. Therefore to get the insight of interfacial effects in heterostructured thin films the present measurements have been done in 500 kHz of frequency. Since, interfacial phenomenon is an additional polarization mechanism apart from ionic, electronic, and dipolar, a distinct variation in the trends (a steep rise) of tan δ with respect to frequency is observed. Tan δ value decreases with the increasing oxygen incorporation. It may be due to the scattering loss in the grain boundaries and the oxygen vacancies. The potential barrier interrupts the electron transport through the oxide layer Thus, our experiments unveil the analogy between the multilayer of TiO_2/Al_2O_3 by oxygen vacancies during thin film growth and the well-known interface inter-diffusion process. In other words, the mobility could be controlled by growth conditions by varying oxygen incorporation of the film. TheRFmagnetron deposition process provides many sources of defects and some of them can diffuse deeply into the substrate. During the growth period, the kinetic energy of the electrons in the sputtering process is in the range from few eV up to hundreds of eV. These mainly impart energy for collision of the reactive species, which gives impact on the substrate. Out of these impacts, variety of point defects, including oxygen vacancies gets generated in the deposited films. This highly alters the films' transport properties. Due to structural imperfections there may actually be a relatively high density of charge carriers. They tend to be localized or trapped at these centers for long times, and the insulating behavior originates from such carriers having very low mobility.

9.5 CONCLUSION

Four layers of Al_2O_3/TiO_2 films have been developed with controlled variation of O_2: Ar gas ratio. The electrical parameters of these films were found to be strongly dependent on the oxygn,: argon gas ratio. The interdiffusion of Ti and Al at the interface has been investigated usiopy OES. The plasma temperature, ion velocity, and electron velocity have been found to affect the morphology of the deposited films. We have shown that high electrical conductivity and mobility values of Al_2O_3/T_2 heterostructure are contributed by the oxygen vacancies. The variation of dielectric constant could explain the transport properties to be stress induced or an effect of polar discontinuiy.

KEYWORDS

- **Al_2O_3/TiO_2**
- **Electrical property**
- **Gas phase study**
- **Optical emission spectroscoy**

ACKNOWLEDGEMET

We express our gratitude to the Administration of Birla Institute of Technology, Mesra, Ranchi, India and BRNS, DAE, Mumbai, India for providing the necessary facilities and support for the present wok.

REFERENES

1. Alers, G. B., Werder, D. J., Chabal, Y., Lu, H. C., Gusev, E. P., Garfunkel, E., Gustafsson, T., and Urdahl, R. S. Intermixing at the tantalum oxide silicon interface in gate dielectric structures. *Appl. Phys. Lett.*, **73**, 1517 (1998).
2. Ariel, Nava, CederGerbrand, Sadoway, Donald R., and Fitzgerald, Eugene A. Electrochemically controlled transport of lithium through ultrathin SiO_2. *Journal of Apllied Physics,* **98**, 023516 (2005).
3. Basu, S. and Basu, P. K. Nanocrystalline Metal Oxides for Methane Sensors, Role of Noble Metals. *Journal of Sensors*, 861968 (2009).
4. Boffard, John B., Lin, Chun C., and Charles, A. DeJoseph Jr. Application of excitation cross sections to optical plasma diagnostics. *J. Phys. D: Appl. Phys.*, **37**, 143–161 (2004).
5. Campbell, S. A., Gilmer, D. C., Wang, X., Hsieh, M., Kim, H. S., Gladfelter, W. L., and Yan, J. Titanium dioxide (TiO_2)-based gate insulators. *IEEE Trans. Electron Devices,* **44**, 104 (1997).

6. Chandra Sekhar, M., Kondaiah, P., Jagadeesh Chandra, S. V., Mohan Rao, G., and Uthanna, S. Substrate temperature influenced physical properties of silicon MOS devices with TiO$_2$ gate dielectric. *Surf. Interface Anal,* (2012).
7. Chawla, Vipin, Jayaganthan, R., and Chandra, Ramesh. Infuence of Sputtering Pressure on the Structure and Mechanical Properties of Nanocomposite Ti-Si-N Thin Films. *J. Mater. Sci. Technol.,* **26**(8), 673–678 (2010).
8. Depla, D., Gryse, R. De. Target poisoning during reactive magnetron sputtering, Part II, the influence of chemisorption and gettering. *Surface and Coatings Technology,* **83**, 90–195 (2004).
9. Ellmer, K. Magnetron sputtering of transparent conductive zinc oxide relation between the sputtering parameters and the electronic properties. *J. Phys. D: Appl. Phys.,* **33**, 17 (2000).
10. Hendron, J. M., Mahony, C. M. O., Morrow, T., and Graham, W. G. Langmuir probe measurements of plasma parameters in the late stages of a laser ablated plume. *J. Appl. Phys.,* 81, 2131 (1997). http://physics.nist.gov/cgi-bin/AtData/lines_form.
11. Jaros, M. Electronic properties of semiconductor alloy systems. *Rep. Prog. Phys.,* **48**, 1091–1154 (1985).
12. Kakati, M., Bora, B., Deshpande, U. P., Phase, D. M., Sathe, V., Lalla, N. P., Shripathi, T., Sarma, S., Joshi, N. K., and Das, A. K. Study of a supersonic thermal plasma expansion process for synthesis of nanostructured TiO$_2$. *Thin Solid Films,* **518**, 84–90 (2009).
13. Linfeng, Lu., HonglieShen., Hui, Zhang., Feng, Jiang., Binbin, Li., and Long, Lin. Effects of Ar flow rate and substrate temperature on the properties of AZO thin films by RF magnetron sputtering. *Optoelectronics and Advanced Materials-Rapid Communications,* 4(**4**), 596–60 (2010).
14. Lo, S. H., Buchanan, D. A., Taur, Y., and Wang, W. Quantum-Mechanical Modeling of Electron Tunneling Current from the Inversion Layer of Ultra-Thin-Oxide nMOSFET's. *IEEE Electr. Device L.,* **18**, 209 (1997).
15. Lu, J. G., Ye, Z. Z., Zeng, Y. J., Zhu, L. P., Wang, L., Yuan, J., Zhao, B. H., and Liang, Q. L. Structural, optical, and electrical properties of (Zn,Al)O films over a wide range of compositions. *J. Appl. Phys.,* **100**, 073714 (2006).
16. Marwedel, P. *Embedded System Design.* Kluwer Academic Pub., Boston (2003).
17. Mohammad, Alim A., Shengtao, Li., Fuyi, L., and Pengfei, Chen. Electrical barriers in the ZnO varistor grain boundaries. *phys. stat. sol.,* **203**(2), 410–427 (2006).
18. Musil, J. et al. Reactive magnetron sputtering of thin films present status and trends. *Thin Solid Films,* **475**(1-2), 208–218 (2005).
19. Nag, J. and Haglund Jr, R. F. Synthesis of vanadium dioxide thin films and nanoparticles. *J. Phys. Condens. Matter* **20**, 264016 (2008).
20. Ohring, Milton. *Materials Science of Thin Films.* Academic Press (1992).
21. Petrov, I., Barna, P. B., Hultman, L., Greene, J. E. Microstructural evolution during film growth. *J. Vac. Sci. Technol.,* A**21**, 5 (2003).
22. Pinacho, R., Jaraiz, M., Castrillo, P., Martin-Bragado, I., Rubio, J. E., and Barbolla, J. Modeling arsenic deactivation through arsenic-vacancy clusters using an atomistic kinetic Monte Carlo approach. *Applied Physics Letters,* **86**, 252103 (2005).
23. Plummer, J., Griffin, P. Material and Process Limits in Silicon VLSI Technology, *Proceedings of the IEEE.,* **89**(3), 240–258 (2001).
24. Ronen, R., Mendelson, A., Lai, K., Pollack, F., and Shen, J. Comming Chal-lenges in Microarchitecture and Architecture, *Proc. of the IEEE,* **83**(3), 325–339 (2001).
25. Wang, Ruiping, King Laura, L. H., and Sleight, A. W. Highly conducting transparent thin films based on zinc oxide. *Journal of Materials Research,* 11, 1659–1664 (1996).

26. Safi, I. Recent aspects concerning DC reactive magnetron sputtering of thin films a review. *Surface and Coatings Technology*, **127**, 203–219 (2000).

27. Wang, W. W., Diao, X. G., Wang, Z., Yang, M., Wang, T. M., and Wu, Z. Preparation and characterization of high-performance direct current magnetron sputtered ZnO:Al films. *Thin Solid Films*, 491, 54–60 (2005).

28. Wilk, G., Wallace, R. M., and Anthony, J. M. High-kgate dielectrics Current status and materials properties considerations. *J. Appl. Phys.*, **89**, 5243 (2001).

29. Wu, Wen-Fa, and Chiou, Bi-Shiou. Effect of oxygen concentration in thesputtering ambient on the microstructure, electrical and optical properties of radio-frequencymagnetron-sputtered indium tin oxide films. *Semicond. Sci. Technol.*, 11, 196–202 (1996).

30. Xi-Ming, Zhu, and Yi-Kang, Pu. Optical emission spectroscopy in low-temperature plasmas containing argon and nitrogen, determination of the electron temperature and density by the line-ratio method. *J. Phys. D: Appl. Phys.*, 43, 403001 (2010).

CHAPTER 10

ROLE OF INTERLAYER INTERACTION ON MECHANICAL PROPERTIES OF CARBON NANOTUBES

KEKA TALUKDAR and APURBA KRISHNA MITRA

CONTENTS

10.1   INTRODUCTION

The special molecular structure of the carbon nanotubes (CNTs) has imparted in them a specialty in their chemical and physical properties. They have caught the attention of the researchers of the past decade for their exceptional properties such as high Young's modulus, high tensile and compressive strength, extraordinary flexibility and resilience, high electrical and thermal conductivity, and so on. The more understanding of the relationship between the structural order of the nanotubes and their mechanical properties will be deeper, the more will be the progress in the quality of the carbon nanotube based composites and other applications. Again, the electrical properties of the CNTs may be tuned by mechanical deformation. Such properties are of great interest for applications such as sensor or smart materials. The mechanical properties of both the single wall carbon nanotubs (SWCNTs) and multi wall carbon nanotubes (MWCNTs) have been studied which confirmed that the tensile strength of the CNTs is 1000 times greater than that of steel. The in-plane C-C bond in CNTs is strong covalent σ bond. In contrary, there exists a weak π bond out of the plane which acts in between the shells of a multi-walled carbon nanotube or in between different single-walled CNTs in a bundle. The combination of high strength, high stiffness, and high elastic modulus along the axial direction with low density and high aspect ratio of the tubes has imparted in them, excellent mechanical properties such that they may be used as reinforcing fibers in a polymer matrix to prepare low weight, high strength structural composites. As a component of a fiber-filled composite, the exact knowledge of the mechanical characteristics of the CNTs is necessary to tailor them for specific use. However, the proper strength and failure behavior of the material's can only perfectly be understood by atomistic simulation. In the nano-regime, the continuum mechanics is inadequate and an atomistic description of the system is necessary. Moreover, SWCNT bundles or MWCNTs are often taken as experimental samples and their breaking or the buckling process is very complex phenomena which can be handled by atomistic simulation technique. Here the power of molecular dynamics simulation technique is exploited in investigating the mechanical characteristics of three different types of SWCNTs, their bundles, and MWCNTs to compare their mechanical properties and failure mechanism.

10.2   CARBON NANOTUBES AND THEIR APPLICATIONS

The CNTs are tubular structure with hexagonal pattern repeating itself in space. In carbon nanotubes, each atom is bonded with other three carbon atoms by sp^2 hybridization. One s and two p orbitals combine to form three sp^2 orbitals at 120° to each other within a plane.

The covalent σ bond is a strong chemical bond. In contrary, there exists a weak π bond out of the plane which acts in between different shells of a MWCNT or in between different different shells of the SWCNTs in a SWCNT bundle.

Due to the surface curvature of the nanotubes, a rehybridization process, including a certain amount of σ character in a π type orbital, changes both their chemical and physical properties.

Since, their discovery, CNTs have received tremendous attention of the researchers due to their light weight, high strength, and high ductility. In nanotechnology, they can play important role due to their special structure and extraordinary mechanical properties. Many structures using CNTs have been proposed including everyday items like clothes and sports materials to combat jacket, bodies of aircrafts, and ships and space elevators (Edwards and Westling, 2003). For this, tensile strength of the tubes should be increased to a sufficient amount. Nanotube based field effect transistors (Chung et. al., 2002), nano tweezers (Kim and Lieber, 1999), memory circuits (Rueckes et. al., 2000), gas sensors (Suehiro et. al., 2006, Bondavalli and Feugnet, 2012), actuators (Baughman et. al., 1999), and field-emitting flat panel displays (Nakayama and Akita, 2001), are the examples of the applications of the CNTs where their electrical and thermal properties may be used. The application of CNTs in diagnosis and therapy of dreadful diseases like cancer is going to be successful. The SWCNTs can be successfully used (Liu et. al., 2007, Gannon et. al., 2007) to treat several types of cancers, with minimal, or no toxic effects to normal cells. The combination of their high strength, high stiffness, and high thermal and electrical conductivity along the axial direction and their low density with high aspect ratio has rendered them as strong candidates as reinforcing fibers for a whole new range of nanocomposites. All types of CNTs can be used for low weight structural composites.

The NASA is developing materials using nanotubes for space applications, where weight driven cost is the major concern (Odegard et. al., 2003, Frankland et. al., 2003). Sufficient improvement in the research work in this field has been achieved to enhance the adhesion of the CNTs with the host to build composite materials for various purposes (Kearns and Shambaugh, 2002, Ruan et. al., 2006, Bakshi et. al., 2010,) Wu and Luo, 2011, Kukich, 2012). Unlike the ceramic and metallic nanocomposites, carbon reinforcements do not degrade the quality of the polymer nanocomposites. CNTs represent ideal AFM tip materials (Druzhinina et. al., 2010) due to their small diameter, high aspect ratio, large Young's modulus, mechanical properties, well-defined structure, and unique chemical properties. Nanotube probes fabricated by chemical vapor deposition provide higher resolution probes. The SPM tips (Nguyen et. al., 2001) can be produced using SWCNTs and MWCNTs for their remarkable mechanical and electronic properties. The stable structure required by the sp^2 bonded carbon and the great strength of nanotubes indicates virtually indestructible tips.

10.3   THEORETICAL PREDICTIONS AND EXPERIMENTAL OBSERVATIONS

The nanoelectromechanical systems or super strong composite materials, which are of great interest in the present days, can be realized in practice by fabrication of materials using CNTs. The choice of CNTs in these fields is found to be very much beneficial for their tremendous high strength and low density. Young's modulus of a solid depends on the nature of the chemical bonding of the constituent atoms. Due to the presence of strong covalent σ bond the axial Young's modulus should be equal to the in plane elastic modulus of graphite (1.04 TPa). Measurements with HRTEM or AFM reveal that CNTs have high Young's modulus which is close to 1 TPa, that is 100 times that of steel. All experimental observations (Krishnan et. al., 1998; Salvetat et. al., 1999a, b; Treacy et. al., 1996) predict such high stiffness of the CNTs or their bundles. Very high Young's modulus of 2.8–3.6 TPa, for SWCNT and 1.7–2.4 TPa for MWCNT have also been observed in some studies (Lourie et. al., 1998). Y value of 1.28 ± 0.59 TPa was found in one investigation (Wong et. al., 1997). Scanning electron microscopy was used by Yu et. al., (Yu et. al., 1999a, 2000, a, b, c) for direct measurement of the tensile properties. Young's modulus obtained ranges from 0.32–1.47 TPa with a mean value of 1.002 TPa for single wall carbon nanotubes (SWCNTs) and 0.27–0.95 TPa for MWCNT.

Tang et. al., (Tang et. al., 2011) has investigated the structural and mechanical properties of partially unzipped CNTs. In some very recent study Talukdar and Mitra (Talukdar and Mitra, 2012) have shown the role of potential functions in the mechanical properties of CNTs. The buckling of CNTs can be well reviewed by Sima H. (Shima, 2012).

Nanoelectromechanical devices can be realized in practice with MWCNTs due to their interwall van der Waals interaction which is expected to offer very low resistance for the sliding of the inner wall of the tube through the others. Nanoscale motors (Fennimore et. al., 2003) and variable resistors (Cumings et. al., 2004) can be designed with MWCNTs. In 1995, Chopra et. al., (Chopra et. al., 1995) measured the interlayer cohesive energy by TEM study. They observed that MWCNTs sometimes collapsed like twisted ribbons. Curvature of the collapsed nanotubes at the edges was also reported by them. Before that, Yu et. al., (Yu et. al., 2000) revealed that sliding resistance depended on the interlayer distance. According to them, the surface tension, the shear elastic force and edge effect force, and so on were responsible for the effective sliding resistance. Two types of sliding, 'smooth pull-out' and a 'stick-slip' type sliding were noticed by the authors. Cumings and Zettl (Cumings et. al., 2004) attached a MWCNT inside a piezoelectrically driven manipulator in a TEM. When a CNT, attached to a contact was opened at one end by applying a voltage pulse, the protruding core was moved in the axial direction attached with the mobile probe and then released; it immediately got into the tube. That happened due to the existence of attractive van der Waals force acting between the layers. In another study (Cumings et. al., 2000) they found the interlayer static and dynamic shear strengths as 0.66 MPa and 0.43 MPa respectively. Kis et. al., (Kis et. al., 2006), following the technique of Cumings et. al., found the interlayer cohesive energy density in the range of 0.14–0.2 J m^{-2} corresponding to the interlayer cohesive energies of 23–33 MeV per atom. Scanning tunneling spectroscopy (Wildoer et. al., 1998), resonant Raman scattering (Jorio et. al., 2001), and optical absorption or emission measurements (O'Connell t al., 2002) confirmed the electronic density of states in carbon nanotubes. Ouyang et. al., (Ouyang et. al., 2001) experimentally found a pronounced dip in the electronic DOS of a (8, 8) SWCNT at Fermi level inside a bundle, compared to an isolated tube. The closing of band gap was observed for a (10, 0) tube in a bundle by Reich et. al., (Reich et. al., 2002).

The mechanical properties of SWCNTs and MWCNTs (Arryo et. al., 2003; Qian et. al., 2003; Liew et. al., 2004; Pantano et. al., 2004) were investigated by many researchers which revealed that there are some differences in the mechanical properties of the two types of tubes. In spite of the weak van Der Waals force acting in between the different layers of the MWCNTs, the walls cannot be peeled apart so easily and they possess high strength and can be efficiently be used as reinforcing fibers in composite materials.

The deformation of single and multiwalled CNTs under radial pressure were investigated by Xiao et. al. (Xiao et. al., 2006). The bending buckling behaviors of SWCNTS and MWCNTs were studied by Yao et. al. (Yao et. al., 2008). They found the dependence of the critical bending buckling curvature of CNTs on their diameter. The buckling of MWCNTs under axial compression was examined by Sears and Batra (Sears and Batra, 2006) by three different methods and they have observed a significant role of van der Waals forces on the buckling behavior of the MWCNTs. But Ruoff et. al. (Ruoff et. al., 2005) after finding the fracture strength of nanotubes got a strong evidence of the presence of defects in the structure of MWCNTs. However, Nichols et. al. (Nichols et. al., 2007) have shown how defects can be introduced into vertically aligned MWCNT ensembles by argon and hydrogen ion irradiation. That the sp^3 interwall- bridging enhances the mechanical properties of MWCNTs, was revealed in (Xia et. al., 2007). The mechanical properties of thick MWCNTs under torsional strain were investigated by Zou et. al. (Zou et. al., 2009). One more theoretical attempt to predict the mechanical behaviors of MWCNTs are examined recently in the work of Chou et. al. (Chou et. al., 2010). However, theoretical studies on SWCNT bundles are rarely reported.

It is now confirmed that CNTs possess extraordinarily high tensile strength of the order of 100 GPa or more, which is at least an order of magnitude higher than earlier measured values of their tensile strength. For this reason, there has been a tremendous motivation towards the improvements of the CNT based composite materials. Looking on the mechanical properties, both SWCNTs and MWCNTs are now used in building up of composite materials for aircraft and sports industries though high thermal and electrical conductivity in the axial direction along with their mechanical excellence have also being exploited in electronic devices and transparent and flexible conductors. Composite materials may be defined as materials made up of two or more components and consisting of two or more phases. The efficiency of the composites depends on the load transfer efficiency between the matrix and the CNTs. If the adhesion between the filler and the matrix is not improved, the benefits of the high strength of CNTs are lost. The prerequisite of any composite is good dispersion of the fibers in the hosting matrix and the stabilization of the dispersion to prevent reaggregation of the filler. For dispersion, sonication or mechanical mixing is generally used, while different functionalization processes prevents reaggregation.

Nanocomposites were first made in fiber form by Andrews et. al., (Andrews et. al., 1999). The mechanical properties of carbon nanotube based composites have been evaluated by finite element analysis (Joshi et. al., 2010). Based on finite element method, MD simulation study (Kuronuma et. al., 2010) has been done to explore the fracture behavior of cracked carbon nanotube based polymer composites. Seo et. al., (Seo et. al., 2010) concluded that the MWCNTs were the good and appropriate materials to improve elastic behavior of the composites with a large increase in their flexural strength and Young's modulus to about 60%. In spite of so many research attempts discussed above, there are still some lack of knowledge about the properties of the CNTs which can lead to better strength and stiffness of the CNT composites.

The enhancement of the performance of the mechanical properties of CNT composites is reported in several literatures (Koziol et. al., 2007; Mora et. al., 2009). Recent advancement in the CNT composite industry can be viewed in the review article of Huang and Terentjev (Huang and Terentjev, 2012).

10.4   THE AIM OF THIS STUDY

In spite of so many attempts to explain the discrepancies between the theoretical and experimental data regarding the mechanical properties of SWCNTs, some observed discrepancies between theory and experiment are still unanswered. Also, the expectation of achieving high tensile strength and stiffness of CNT polymer composites is still not fulfilled. One reason of this problem may be like that SWCNTs have a natural tendency to form bundles and thus in the most of the experiments undertaken by various investigators, SWCNT bundles or MWCNTs have been used as samples. So we must take into account the influence of interlayer interaction in bundle formation or in an MWCNT while calculating the mechanical properties of CNTs. In this work, we have compared the observed mechanical properties of three different types of SWCNTs separately and also a bundle of their mixture. Also the mechanical properties of armchair, chiral, and zigzag MWCNTs are investigated. The results are explained with the overlapping of density of states (DOS) of the CNTs in a bundle. So, this is an attempt to carry out more realistic theoretical investigations to facilitate comparison with experimental data.

10.5   ATOMISTIC MODELING AND SIMULATION

By atomistic modeling one can understand the interaction between the constituent molecules of a material and thus the behavior of the material with various external constraints, their deformation and failure process can also be understood. Starting from the simple laws of Physics to describe the position and momentum of each atom in a material and solving the equations of motion of the system of atoms or molecules, one can obtain the dynamic behavior of all the particles of the system. Thermodynamical behavior of gases and liquids (Alder and Wainwright, 1957; Alder and Wainwright, 1959; Allen and Tildesly, 1989; Rahman, 1964) was the first, where molecular dynamics simulation was performed. Later, this process has been used to find the mechanical behavior of solids. The failure process of solids is associated with so many computational complexities that modeling and simulation have now become a very exciting area of research. At first, a mathematical model is to be developed for a physical problem. Then the equations resulting from that model building (Ashby, 1996) are to be solved. The model should be such that it should be able to capture the physical features of the problem. Sidney Yip of MIT (Yip, 2005) has stated that modeling is the physicalization of a concept, simulation its computational realization. Modeling requires the knowledge about the physics of the system that is about its constituents or the behavior of the particles. Simulation requires the technique to solve enormous numerical equations related to the complicated systems. The behavior of cracks, dislocations, and grain boundary processes can be very successfully investigated by atomistic modeling (Buehler, 2008). In modern materials modeling this process has gained immense importance. Any complex problem can be solved by knowing Newton's laws and the nature of interaction of the atoms. In the present work mechanical properties of the CNTs are investigated by modeling and simulation. Nowadays, researchers have considered the computer as a tool to do experiment, similar as experimentalists do in their laboratory. Computational experiments thus need to build a suitable model of a physical problem, set up some equations to represent the problem, run simulation and to interpret the results of the simulation process.

10.5.1 MOLECULAR DYNAMICS SIMULATION

Molecular dynamics simulation is a form of computer simulation in which atoms and molecules are allowed to interact for a period of time by approximations of known physics, giving a view of the motion of the particles. Position and momentum of each particle is updated using a suitable algorithm. In materials science, this is a very effective process. We call *molecular dynamics* simulation a computer simulation technique where the time evolution of a set of interacting atoms is followed by integrating their equations of motion. Material behavior and their deformation are described in atomistic modeling. The purpose of molecular dynamics simulation is to understand the properties of assemblies of molecules in terms of their structure and the microscopic interactions between them. Something new, which cannot be found in other ways, can be learnt by simulation.

10.5.2 PRESENT METHOD OF SIMULATION

We have performed MD simulation using the software 'Brenner code' (Brenner et. al., 2002) which is an open source code and which has been customized and modified as necessitated by our scheme of work. The coordinates of a carbon nanotube generated in the program can be used for simulation. The programs are all written in FORTRAN language. All programs are to be compiled separately to run the main program. Microsoft developer studio is used as the platform of manipulating programs. Time step of 0.5 fs is chosen for the molecular dynamics simulation presented here. We have tested that slight changes in the time step affect the simulation result negligibly. Also The evolution of the system energy with respect to time is noticed and it is observed that 30000–50000 time steps are required to achieve convergence in energy minimization and hence to get equilibrium condition. However, simulation can be run using other software also, like TINKER, LAMMPS, and so on. The coordinate of the sample of MWCNTs and SWCNT bundles are generated and either tensile force or compressive force is applied at one side of the tube. While stretching or compressing a bundle or an MWCNT, all the tubes are equally stretched or compressed in the axial direction, the constraints in other directions are set equal to zero.

To visualize the MD simulation result we must have a platform where we can visualize the breaking mechanisms, buckling, and so on. For this, we have used software called 'Nanotube Modeler'. It can generate XYZ coordinates for nanotubes and nanocones. Generated geometries can be viewed using the integrated viewer or by calling the viewer program. The program is based on the JNanotube Applet by Steffen Weber.

It has extended features. Here, atomic coordinates can be generated on the basis of geometric criteria.

10.5.3   POTENTIAL FUNCTIONS USED

The different materials show different chemical complexity and hence undergo different chemical bonding. Typical nature of chemical bonding is that it shows attraction for large distances and repulsion for a distance less than the equilibrium separation.

For MD simulation, the mathematical expressions for the energy distribution representing the chemical interaction are necessary. How the interaction changes with the geometry of the atoms in the material is to be known. These interactions are given by different interacting potentials. The foundation of the computer simulation is based on the interatomic potential functions as the potentials contain the physics of the model system. The accuracy of the simulation result that is the reliability of the result to represent the behavior of the system under consideration depends significantly on the realistic choice of the potential function. The potential energy functions are simplified mathematical expressions that attempt to model interatomic forces arising from the interaction of electrons and nuclei.

2nd GENERATION REBO POTENTIAL

In 2nd generation REBO potential (Brenner et. al., 2002), the form of the potential is:

$$V(r_{ij}) = f_c(r_{ij})[V^R(r_{ij}) + b_{ij}V^A(r_{ij})] \quad (1)$$

$$\text{where, } V^R(r_{ij}) = f_c(r_{ij})(1 + \frac{Q_{ij}}{r_{ij}}).A_{ij}e^{-\alpha_{ij}r_{ij}} \quad (2)$$

$$\text{and } V_A(r_{ij}) = -f_c(r_{ij})\sum B_n e^{-\beta_{ijn}r_{ij}} \quad (3)$$

$$(n = 1, 3)$$

In equations 1 and 2, $V^R(r_{ij})$ is a pair-wise term that models the core-core and electron-electron repulsive interactions and $V^A(r_{ij})$ is a pair-wise term that models core-electron attractive interactions, where r_{ij} is the distance between nearest neighbor atoms i and j, $f_c(r_{ij})$ is the cut-off function which reduces to zero interaction beyond 2.0 Å.

here

$$b_{ij} = \left(\bar{p}_{ij}^{\sigma\pi} + p_{ji}^{\sigma\pi}\right)/2 \quad (4)$$

$$p_{ji}^{\pi} = \pi_{ij}^{rc} + \pi_{ij}^{dh} \quad (5)$$

$$p_{ji}^{\sigma\pi} = [1 + G_{ij} + P_{ij}(N_i^{(H)}, N_i^{(C)})]^{-1/2} \quad (6)$$

$$Gij = \sum_{k \neq i,j} f_c(r_{ik}) G_i(\cos(\theta_{ijk})) e^{\lambda_{ijk}(r_{ij}-r_{ik})} \quad (7)$$

$$\pi_{ij}^{rc} = F_{ij}(N_i^{(t)}, N_j^{(t)}, N_{ij}^{conj}) \quad (8)$$

The angular function

$$g_c = G_c(\cos(\theta)) + Q(N_i^t) \cdot \left[\gamma_c(\cos(\theta)) - G_c(\cos(\theta))\right] \quad (9)$$

$\gamma_c(\cos(\theta))$ is a second spline function. $Q(N_i^t)$ is defined as:

$$Q(N_i^t) = \begin{cases} 1 & N_i^t \leq 3.2 \\ \frac{1}{2} + \frac{1}{2}\cos\pi(N_i^t - 3.2)/(3.7-3.2) & 3.2 < N_i^t < 3.7 \\ 0 & N_i^t \geq 3.7 \end{cases} \quad (10)$$

$$N_{ij}^{conj} = 1 + \left[\sum f_c(r_{ik}) F(\chi_{ik})\right]^2 + \left[\sum f_c(r_{jl}) F(\chi_{jl})\right]^2 \quad (11)$$

$$\pi_{ij}^{hd} = T_{ij}\left(N_i^{(t)}, N_j^{(t)}, N_{ij}^{conj}\right)\left[\sum_{k \neq i,j}\sum_{l \neq i,j}(1-\cos^2\omega_{ijkl})f_c(r_{ik})f_c(r_{jl})\right] \quad (12)$$

$$\omega_{ijkl} = e_{ijk} \cdot e_{ijl} \quad (13)$$

The function $f_c(r)$ includes the interactions of nearest neighbor only. P represents bicubic spline. N_{ij}^{conj} is one when all of the carbon atoms that are bonded to a pair of carbon atoms i and j have four or more neighbors. As the coordination number of the neighboring atoms decreases, N_{ij}^{conj} becomes greater than 1. Here the term N_{ij}^{conj}

differentiates between different configurations. Also the term accounts for bond formation and breaking. The function $T_{ij}(N_i^{(t)}, N_j^{(t)}, N_{ij}^{conj})$ is a tricubic spline, e_{ijk} and e_{ijl} are the unit vectors in the direction of $R_{ji} \times R_{ik}$ and $R_{ij} \times R_{jl}$ respectively where the R vectors connects the subscripted atoms. The subscript n refers to the sum in Equation (2), and π_{ij}^{dh} depends on the dihedral angle for the C-C double bonds. $G_c(Cos(\theta_{ijk}))$ modulates the contribution that each nearest-neighbor makes to b_i. For diamond and graphite which contain angles 109.47° and 120° respectively, $G_c(Cos(\theta))$ is found for these values, whereas for simple cubic lattice the value of the function is found for bond angles between nearest neighbors which are 90° and 180°. The FCC lattice contains angles of 600°, 120°, 180°, and 90° for which the value can be found. Sixth-order polynomial splines in $cos(\theta)$ were used in three regions of bond angle θ, 0°<θ<109.47°, 109.47°<θ< 120° and 120°<θ< 180°. The values of second derivatives of $G_c(Cos(\theta))$ at 109.47° and 120° are fitted to the elastic constant c_{11} for diamond and the in plane elastic constant c_{11} for graphite respectively. The first derivatives are chosen to suppress the spurious oscillations in the splines. To complete the determination of the polynomials splines, it is first determined in the range 120°<θ<180° and then found by fitting the ethyne bending mode and by spline oscillations.

The modified potential includes both modified analytical functions for the intramolecular interactions and an expanded fitting database. This gives a better description of bond energies, lengths, and especially forces constants for carbon-carbon bonds. It has produced an improved fit to the elastic properties of diamond and graphite. Better prediction can be made for the energies of several surface reconstructions and interstitial defects. The rotation about dihedral angles for carbon-carbon double bonds and the angular interactions associated with hydrogen centers have been modeled in the 2nd generation REBO potential.

In the present simulation, we have used smoothing of the cut-off in the rotational potential in the 2nd generation REBO potential.

We have used a "torsional interaction" for CC bonds.

The added torsional potential is:

$$V^{tor} = V_A\left(|r_{ij}|\right) A_{ij}\left(N_i^{(t)}, N_j^{(t)}, N_{ij}^{conj}\right) \Sigma_{kl} sin^2\left(\varphi_{ijkl}\right) f_{ik} f_{jl} \qquad (14)$$

In the 2nd generation REBO potential. Where φ_{ijkl} is the angle between the ijk and the ijl planes, and the sum extends over k≠ij, l≠ij, $|sin(\theta_{ijk})| > 0.1$, and $|sin(\theta_{jil})| > 0.1$. The f-values are hardcoded to special values for H.

The conditions on the sines in V^{tor} imply that energy is not conserved when three atoms become almost along the same line. So, if there is a problem with energy conservation,

We have allowed the restrictions that the potential function is smoothed off at the cut-off region.

Thus:

$$V^{tor} = V_A \left(|\, r_{ij} \,| \right) A_{ij} \left(N_i^{(t)}, N_j^{(t)}, N_{ij}^{conj} \right) \Sigma_{kl} \sin^2 \left(\varphi_{ijkl} \right) f_{ik} f_{jl} Z_{ijk} Z_{jil} \qquad (15)$$

Adding Z_{ijk} and Z_{jil} in V^{tor} the smoothing is achieved and we get energy conserved.

C-C NON-BONDING POTENTIALS

In some simulation studies, the non-bonding interactions between carbon atoms are required. Various types of potentials are there to model the non-bonding potential. In the present study Lennard- Jones potential (Lennard-Jones, 1924) is adopted to describe the van der Waals inter-molecular interactions between different CNTs in a multiwalled carbon nanotube or in between the tubes of a single walled carbon nano-tube bundle. The form of the potential was first proposed by John Lennard-Jones and is given by:

$$V_{LJ} = 4\, \varepsilon_{ij} \left[\left(\sigma_{ij} / r_{ij} \right)^{12} - \left(\sigma_{ij} / r_{ij} \right)^6 \right] \qquad (16)$$

The ε is the depth of the potential well, σ is the distance at which the inter-particle potential is zero, and r is the distance between the particles. i, j stand for i^{th} and j^{th} atoms. The term r^{-12} describes Pauli repulsion at short ranges due to overlapping of electronic orbitals and r^{-6} describes the long range attraction. According to Pauli exclusion principle, the energy of the system increases due to the overlapping of the electronic wave functions surrounding the atoms. Generally repulsive term should be exponential, but to facilitate the computation, the exponent is taken as 12. Attractive long range term is derived from dispersion interactions.

To get the energy conserved, we have used L-J potential which is truncated at the cut off distance.

$$r_c = 2.5\sigma \qquad (17)$$

where

$$V(r_c) = V(2.5\sigma) = 4\, \varepsilon_{ij} \left[\left(\sigma_{ij} / 2.5\, \sigma_{ij} \right)^{12} - \left(\sigma_{ij} / 2.5\sigma_{ij} \right)^6 \right] = -0.0163\, \varepsilon = -1/61.3\, \varepsilon \qquad (18)$$

So, at $r_c = 2.5\,\sigma$, the potential V is about 1/60th of its minimum value ε. Interactions between two different type atoms use a geometrically average of the ε values of each atom and a straight average σ. The Lennard–Jones potential is set to zero for $r_{ij} > 2.5\,\sigma_{ij}$ and in the range $r_{ij} < R_{ij}^{(2)}$ in which the bond order potential is nonzero. In the range $R_{ij}^{(2)} < r_{ij} < 0.95\,\sigma_{ij}$, the potential is furthermore replaced by a cubic term which vanishes quadratically at $R_{ij}^{(2)}$ and which meets the Lennard–Jones expression with continuous first derivative at $0.95\,\sigma_{ij}$. So, simulation is carried out taking the following form of the potential:

$$V_{MD} = \left(\begin{array}{cc} V_{LJ}(r) - V_{LJ}(r_c)| & \text{for } r \leq r_c \\ 0 & \text{for } r > r_c \end{array} \right)$$

So, the discontinuity at $r = r_c$ is eliminated.

10.5.4   THERMOSTAT

A thermostat is required to lead a system to a desired temperature that is to enable a NVT or NPT ensemble. To modify the equations of motion to obtain a specific thermodynamical ensemble, the velocities of the atoms are rescaled by a scaling factor. We have used Berendsen thermostat (Berendsen et. al., 1984) in this study to maintain a fixed temperature. In this thermostat, velocity is rescaled by a factor $\lambda = \sqrt{(T_0/T(t)}$

In each step velocity is rescaled so that the rate of increase of temperature is proportional to the difference in temperature:

$$dT(t)/dt = 1/\tau \left(T_0 - T(t) \right) \qquad (20)$$

So the scale factor is given by:

$$\lambda = \left[1 + \Delta t / \tau \left(T / T_0 - 1 \right) \right]^{1/2} \qquad (21)$$

Where Δt is the time step and τ is a coupling constant. It is called rise time which gives the strength of the coupling of the system with the hypothetical heat bath at temperature T_0.

10.6   DISCUSSION AND RESULTS

In spite of so many attempts to explain the discrepancies between the theoretical and experimental data regarding the mechanical properties of SWCNTs, some observed discrepancies between theory and experiment are still unanswered. The SWCNTs have a natural tendency to form bundles and thus in the most of the experiments undertaken by various investigators, SWCNT bundles, or MWCNTs have been used as samples. So we must take into account the influence of interlayer interaction in our calculations of mechanical properties of CNTs. In this work, we have compared the observed mechanical properties of three different types of SWCNTs separately and also the bundle of the mixture of three different types of tubes. The results are explained with the overlapping of density of states (DOS) of the CNTs in a bundle. So this is an attempt to carry out more realistic theoretical investigations to facilitate comparison with experimental data. 2nd generation reactive empirical bond order potential is adopted for simulation along with Lennard–Jones 6–12 potential function for interlayer interaction.

10.6.1   SWCNT BUNDLE

Using the 2nd generation REBO potential, MD simulation is carried out for three different types of tubes. Lengths of the single zigzag, armchair, and chiral tubes are 85.18 Å, 98.58 Å, and 119.26 Å respectively. Room temperature is maintained by Berendsen thermostat. By stretching the tubes in small strain increments, the equilibrium potential energy is calculated by simulation. Keeping one end fixed the other end of the bundle is stretched gradually from the unstretched condition in such a way that all the tubes are stretched equally. Intertube interaction is modeled with Lennard–Jones potential. For a SWCNT bundle, volume of each tube is calculated separately and then added to get the total volume.

MECHANICAL CHARACTERISTICS OF ISOLATED TUBES

A zigzag SWCNT exhibits a high Young's modulus value of 1.47 TPa with 16% ductility. Calculated failure stress is 76.77 GPa. Stress strain curves of the three tubes are shown in Figure 1. For the (5, 0) tube, stress increases slowly up to 8% more or less linearly and then flattens giving a yield point near 9% strain. On straining the tube beyond 12% strain for a zigzag SWNT, as reported by Zhang et. al., (Zhang et. al., 1998), we have observed SW rotation that ultimately leads to rupture of the tube. After 18% strain, the tube breaks totally with the formation of a series of SW rotations.

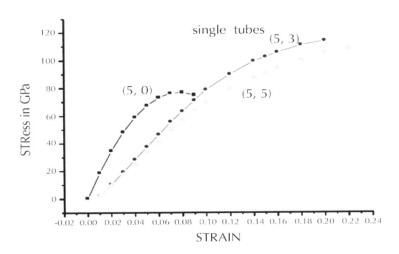

FIGURE 1   Stress-strain curves of three different types of SWCNTs under tensile loading

The stress strain curves of isolated (5, 5) and (5, 3) tubes [Figure 1] are somewhat different in nature. Also, their failure stress and failure strain values are different. The single (5, 5) tube has the highest fracture strain of 26% compared to the other two. Table 1 gives the detailed picture of the failure stress and strain values. However, the chiral SWCNT shows the maximum breaking stress of 115.65 GPa.

Young's modulus values (Table 1) agree very well with the experimental data of Krishnan et. al., (Krishnan et. al., 1998) Wong et. al., (Wong et. al., 1997), and Treacy et. al., (Treacy et. al., 1996) Failure stress or strain of the single tubes does not match with the experimental findings of Yu et. al., match with experiment when mixture of bundles of nanotubes are formed. Yu et. al., observed 10–13% maximum strain for SWCNT bundle with failure stress varying in between 13–52 GPa. For single tubes, our calculated values of failure stress matches with Mielke et. al., (Mielke et. al., 2004) and Troya et. al., (Troya et. al., 2003) with the same potential.

The failure pattern of these tubes (Figure 2) does not show complete rupture of the tubes. Rather they exhibit breaking of bonds all over the tubes at the breaking point.

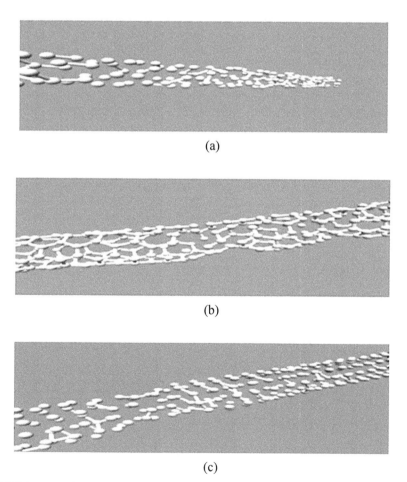

(a)

(b)

(c)

FIGURE 2   Breaking patterns of an isolated (a) zigzag (b) chiral, and (c) armchair SWCNT.

The buckling behavior of the individual SWCNTs is investigated. (5, 0) tube shows the same trend of low critical buckling stress as that of its breaking stress. Here the buckling stress of the (5, 5) tube is highest among the three. Figure 3 depicts their buckling stress *vs.* strain curves. In Figure 4 their buckling pattern are modeled by Nanomodeler. The narrowest tube, that is the (5, 0) tube is bent under compression. Other two tubes form kinks on the application of compressive force. Maximum bond breakage before collapsing is observed for the armchair tube.

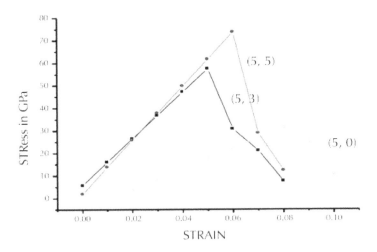

FIGURE 3   Stress-strain curves for the isolated SWCNTs under compressive loading.

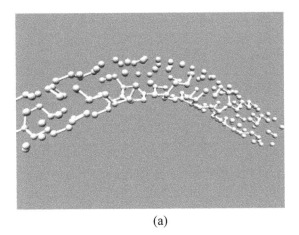

(a)

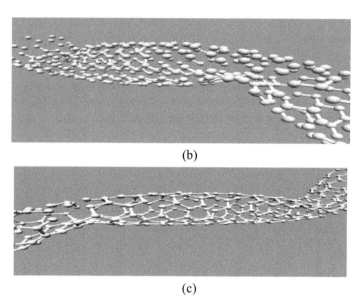

(b)

(c)

FIGURE 4   Buckling patterns of (a) a zigzag tube (b) an armchair tube, and (c) a chiral tube.

MECHANICAL CHARACTERISTICS OF THE BUNDLE OF SWCNTS

We have investigated the mechanical behavior of a bundle of nanotubes taking one tube from each group. Surprisingly, Young's modulus is noticeably increased to 1.60 TPa again and failure strain is reduced to 9% (Figure 5). Failure stress is also reduced much giving a value of 68.50 GPa. Such changes are not observed in any other case.

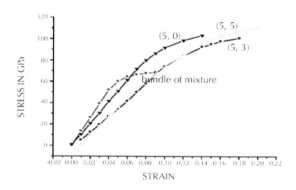

FIGURE 5   The comparison of the stress- strain curves of the isolated SWCNTs and their bundle

UNDER TENSILE LOADING

All the tubes are equally broken from the middle part. The complete fracture is reported at the breaking strain. The breaking of the bundle of their mixture is observed near its middle part and at 9% strain complete breakage is reported. Due to the presence of the van der Waals force the tubes of a bundle deform as a unit.

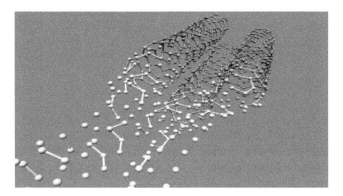

FIGURE 6   Breaking pattern of the bundle of the three different types of SWCNTs.

The changes are also observed on the application of the compressive force. In this case the stress-strain curves are shown in Figure 7. Here critical buckling stress and strain have not exceeded the maximum stress of the isolated tubes. The buckling pattern of the bundle is depicted in Figure 8. Kinks are still observed in this picture. So the main difference is observed in the tensile properties of the bundle compared to the isolated tubes.

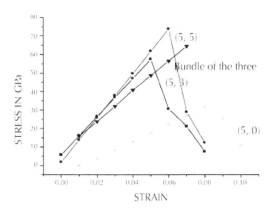

FIGURE 7   The comparison of the stress vs. strain curves of the isolated SWCNTs and their mixture under compressive loading.

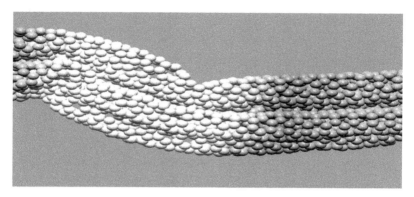

FIGURE 8   Buckling behavior of the bundle of SWCNTs.

This changed mechanical behavior of the bundle may be explained by overlapping of energy bands in nanotube bundles. Scanning tunneling spectroscopy (Wildoer et. al., 1998), resonant Raman scattering (Jorio et. al., 2001), and optical absorption or emission measurements (O'Connell et. al., 2002) confirmed the electronic DOS in carbon nanotubes. Ouyang et. al., (Ouyang et. al., 2001) experimentally found a pronounced dip in the electronic DOS of a (8, 8) SWCNT at Fermi level inside a bundle, compared to an isolated tube. This is due to symmetry breaking by other tubes in proximity. Not always a gap is produced, but gap may also be closed for such a tube. Closing of band gap is observed for a (10, 0) tube in a bundle in (Reich et. al., 2002). Due to this type of overlapping of electronic bands, attraction or repulsion arises inside the SWCNT bundle. All intermolecular/van der Waals forces are anisotropic which means that they depend on the relative orientation of the molecules. The induction and dispersion interactions are always attractive and irrespective of orientation, but the electrostatic interaction changes sign upon rotation of the molecules. That is, the electrostatic force can be attractive or repulsive, depending on the mutual orientation of the molecules giving rise to different interaction between the tubes of a bundle.

The electronic band structure of nanotube bundles differs significantly from the band structure of the isolated tube. In the low energy part of the band structure, the bundling of the nanotubes changes the electronic properties by symmetry breaking and by the intratube dispersion perpendicaular to K_z. That also holds good for larger electronic energies also. Again, when isolated tubes of different symmetries are present, energy bands are strongly split. Reich et. al., (Reich et. al., 2002) showed that in a high symmetry packing of (6, 6) nanotube the degenerate bands of isolated tube remained degenerate by symmetry in the crystal. In their study, it was also revealed that the dispersion of the electronic bands perpendicular to k_z is less in a zigzag (10, 0) tubes than a (6, 6) tubes. The first two valance states at the Γ point of the brillouin zone results in a strong dispersion in the corresponding states perpendicular to k_z for armchair tubes. Chiral tubes are likely to be more complicatedly influenced by symmetry breaking and band splitting. The change of behavior of the curves for an isolated tube and their bundles can thus be explained by the change of DOS which must be taken into account while comparing a theoretical result with the experimental findings.

10.6.2   MECHANICAL PROPERTIES OF THE MWCNTS

Fracture strength and strain of the defect- free MWCNTs are remarkably high and show completely brittle strain curves (Figure 9). For a zigzag tube, the linear portion of the curve extends up to a high strain of 9% though it shows a yield point around 14% strain and its rupture begins at 18%. But linear portions are not extended so much for the other two tubes. After the sharp rise of the stress, the stiffness is reduced for the armchair and chiral MWCNTs. Highest Young's modulus is observed for a chiral tube as 0.93 TPa which is close to 1.0 TPa and it is in agreement with the experimental results of Krishnan et. al., (Krishnan et. al., 1998) and Treacy et. al., (Treacy et. al., 1996). Failure stresses are much higher than the results of Yu et. al., but match well with the results of Demczyk et. al., (Demczyk et. al., 2002). Here all the MWCNTs give larger values of failure stress than that of the single tubes. Interlayer interaction plays a different role in binding the tubes.

Under axial compressive stress, the three tubes follow the stress-strain curves as shown in Figure 10. The three curves show more or less the same pattern with maximum value of critical buckling strain of 12% for a zigzag tube. But their critical buckling stress values are close to each other. The strain of 12% for the zigzag MWCNT can be compared with the result of Srivastava et. al.,

Srivastava et. al., 1999 who got the same value with generalized tight binding molecular dynamics simulation. The critical buckling stress in our calculation (~ 100 GPa) is in the range of the experimental observation of Lourie et. al., (Lourie et. al., 1998). Here, the critical buckling strain of armchair and chiral tubes are 11% and 10% respectively, after which the tubes collapse completely.

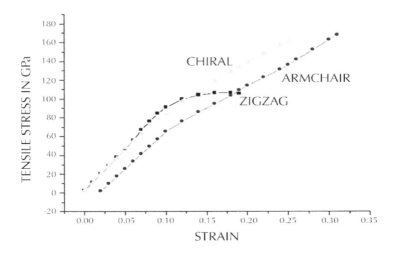

FIGURE 9   Stress-strain curves of the MWCNTs under tensile loading.

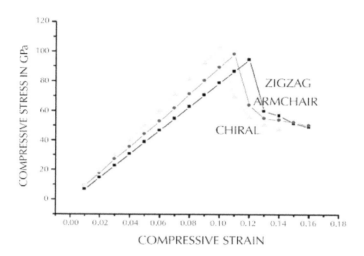

FIGURE 10   Stress-strain curves of MWCNTs under compressive loading.

The pristine MWCNTs break in the manner as shown in Figure 11. The outer shell breaks first and then the breakage proceeds to the other inner layers. Our observation is in agreement with Yu's experimental findings. Failure occurs at the edge of the chiral and armchair tubes in brittle (Figure 11(b), (c)) manner, while the zigzag tube shows necking before failure (Figure 11(a)).

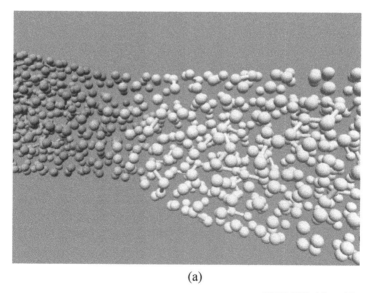

(a)

FIGURE 11   *(Continued)*

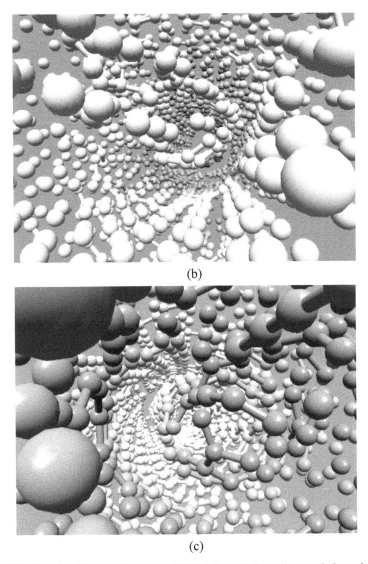

(b)

(c)

FIGURE 11   The breaking mechanisms of defect free a) zigzag b) armchair, and c) chiral MWCNTs under axial tension.

The maximum bonds of the outer shells of the three tubes are broken before complete rupture and the innermost tubes are less affected. The inner tube breaks on complete failure as depicted in Figure 11 (b) and 11(c).

Due to the presence of van der Waals force in between the layers of a MWCNT, different shells of an MWCNT are buckled as a single unit. Otherwise, they would buckle in different manner. The buckling of a DWCNT in the absence of Van der Wals force is shown by Sears and Batra (Sears and Batra, 2006). In the present study, a zigzag MWCNT buckles with the formation of a kink at a place nearer to the loading edge (Figure 12(a)). In absence of any defect, the kink formed in the zigzag tube becomes more and more sharp and ultimately some atoms from that part of the tube gets separated from the tube due to pressure at about 12% strain. An armchair tube is depressed at the middle (shell wall buckling) such that it produces wavy structure (Figure 12(b)) and the buckling of a chiral tube is accompanied with twist. At 10% compression, the chiral tube takes the shape as depicted in Figure 12(c). Beyond that strain it collapses by breaking of bonds. However, Sears and Batra (Sears and Batra, 2006) observed a critical strain value for the buckling of a zigzag MWCNT as 8% by a continuum model following Euler buckling theory.

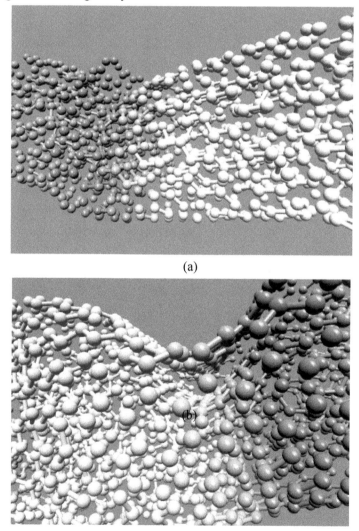

(a)

(b)

FIGURE 12   *(Continued)*

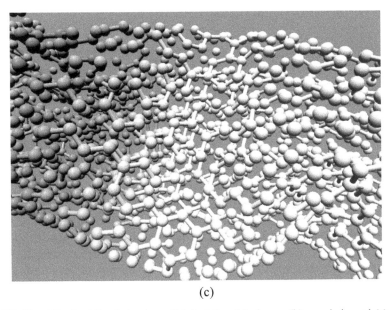
(c)

FIGURE 12   The buckling behaviors of defect free (a) zigzag, (b) armchair, and **(c)** chiral MWCNTs under axial compression.

The results are given below in tabulated form. Comparing Table 1 with Table 2, it can be said that the nature of interlayer interaction plays a different role in the SWCNT bundle and in MWCNTs.

TABLE 1   Tensile and compressive properties of the isolated SWCNTs and their bundle (Y.M.-Young's modulus, F.S. –Failure stress, F.Sr.-Failure strain, C.M.-Compressive modulus, C.S.- Critical buckling stress, C.Sr.- Critical buckling strain)

SWCNT	Y.M. (TPa)	F.S. (GPa)	F.Sr.	C.M. (TPa)	C.S. (GPa)	C.Sr.
Zigzag (5,0) Single:	1.47	76.77	14%	0.46	31.55	8%
Armchair (5,5) Single:	0.79	107.91	26%	1.20	73.71	6%
Chiral (5,3) Single:	0.83	115.65	22%	1.03	57.45	5%
Mixture of three Different tubes	1.60	68.50	9%	0.839	64.84	6%

TABLE 2   Tensile and compressive properties of the three-walled MWCNTs

MWCNT	Y.M. (TPa)	F.S. (GPa)	F.Sr.	C.M. (TPa)	C.S. (GPa)	C.Sr.
Zigzag (5,0)	0 .92	106.59	18%	0.87	102.58	12%
Armchair (5,5)	0.79	168.24	31%	0 .81	98.26	11%
Chiral (5,3)	0.93	161.56	25%	0. 96	102.95	10%

10.7   CONCLUSION

Comparing the MD simulation results of different SWCNTs and their bundles, it can be concluded that a major reason of non-compatibility of the theoretical results with the experimentally obtained low values of failure strain and failure stress is the existence of the SWCNTs in the form of bundles where different types of tubes may be present. The reduction is maximum for a bundle of the mixture of three types and we get 43.75% decrease of the strain from the lowest strain of 16% for a single zigzag tube. Overlapping of density of states is responsible for the changed behavior of the bundle of nanotubes. Necking is observed for a pristine zigzag MWCNT but brittle fracture is reported for armchair and chiral tubes for axially stretched tubes. In most of the cases, the outermost layer of the MWCNTs breaks first and then other shells break gradually. Highest critical buckling strain is observed for a zigzag MWCNT and it is buckled with kink formation. Shell wall buckling is noticed for armchair MWCNT which is depressed at the middle while compressing. Chiral tube is always twisted on buckling and at the same time shell walls are also buckled or depressed at the middle. Due to the presence of van der Waals force, the tubes inside a bundle or inside an MWCNT act as one unit.

KEYWORDS

- **Carbon nanotubes**
- **Chemical bonding**
- **Density of states**
- **Nanoelectromechanical systems**
- **Van Der Waals force**

REFERENCES

1. Alder, B. J. and Wainwright, T. E. Phase Transition for a Hard Sphere System. *Journal of Chemical Physics*, **27**(5)**,** 1208–1209 (1957).

2. Alder, B. J. and Wainwright T. E. Studies in Molecular Dynamics. I. General Method. *Journal of Chemical Physics*, **31**(2), 459–466 (1959).

3. Allen, M. P. and Tildesly, D. J. *Computer Simulation of Liquids*. Oxford University press, Oxford, New York (1989).

4. Andrews, R., Jacques, D., Rao, A. M., Rantell, T., Derbyshire, F., Chen, Y., Chen, J., and Haddon, R. C. Nanotube composite carbon fibers. *Applied Physics Letters*, **75**(9), 1329 (1999).

5. Arroyo, M. and Belytschko, T. Nonlinear Mechanical Response and Rippling of Thick Multiwalled Carbon Nanotubes. *Physical Review Letters*, **91** (21), 215505 (2003).

6. Ashby, M. F. Modelling of materials problem. J*ournal of Computer-Aided Materials Design*, **3**(1–3), 95–99 (1996).

7. Bakshi, S. R., Lahiri, D., and Agarwal, A. Carbon nanotube reinforced metal matrix composites. *International Materials Review*, **55**(1), 41–64 (2010).

8. Baughman, R. H., Cui, C. X., Zakhidov, A. A., Iqbal, Z., Barisci, J. N., Spinks, G. M. et. al., *Carbon Nanotube Actuators. Science*, **284**(5418), 1340–1344 (1999).

9. Berendsen, H. J. C., Postma, J. P. M., van Gunsteren, W. F., DiNola, A., and Haak, J. R. Molecular dynamics with coupling to an external bath. *Journal of Chemical Physics*, **81**(8), 3684–3690 (1984).

10. Bernholc, J., Brabec, C., Buongiorno, M., Maiti, A. and Yakobson. B. I. Theory of growth and mechanical properties of nanotubes. *Applied Physics A*, **67**(1), 39–46 (1998).

11. Bondavalli, P. and Feugnet, G. Gas sensing with carbon nanotubes, *Nanotechnology*, DOI: 10.1117/2.1201207.004311, (2012).

12. Brenner, D. W., Shenderova, O. A., Harrison, J. A., Stuart, S. J., Ni, B., and Sinnot, S. B. A second-generation reactive empirical bond order (REBO) potential energy expression for hydrocarbons. *Journal of Physics: Condensed Matter*, **14**(4), 783–802 (2002).

13. Brenner code, code for MD simulation, URL: http://www.eng.fsu.edu/~dommelen/ research/nano/brenner/ (accessed 10th February, 2013).

14. Buehler, M. J. Atomistic Modeling of Material Failure. S*pringer Science and Business Media*, New York, USA (2008).

15. Chopra, N. G., Benedict, L. X., Crespi, V. H., Cohen, M. L., Louie, S. G., and Zettl, A. Fully collapsed carbon nanotubes. *Nature*, **377**(6545), 135 (1995).

16. Chou, T. W., Gao, L., Thostenson, E. T., Zhang, Z., and Byun, J. H. An Assessment of the Science and Technology of Carbon Nanotube Fibers and Composites. *Composites Science and Technology*, **70**(1), 1–19 (2010).

17. Chung, D. S., Park, S. H., Lee, H. W., Choi, J. H., Cha, S. N., and Kim, J. W. et. al., Carbon nanotube electron emitters with a gated structure using backside exposure processes. *Applied Physics Letter*, **80**(21), 4045 (2002).

18. Cumings, J. and Zettl, A. Localization and nonlinear resistance in telescopically extended nanotubes. *Physical Review Letter*, **93**(8), 086801 (2004).

19. Cumings, J. and Zettl, A. Low-friction nanoscale linear bearing realized from multiwall carbon nanotubes. *Science*, **289**(5479), 602–604 (2000).

20. Demczyk, B. G., Wang, Y. M., Cumings, J., Hetman, M., Han, W., Zettl, A. and Ritchie, R. O. Direct mechanical measurement of the tensile strength and elastic Modulus of multiwalled carbon nanotubes. *Materials Science and Engineering A*, **334**(1–3), 173–178 (2002).

21. Druzhinina, T. S., Hoeppener, S., and Schubert, S. Microwave-Assisted Fabrication of carbon nanotube AFM tips. *Nano Letters*, **10**(10), 4009–4012 (2010).

22. Edwards, B. C. and Westling, E. A. *The Space Elevator: A Revolutionary Earth-to-*

23. Fennimore, A. M., Yuzvinsky, T. D., Han, W. Q., Fuhrer, M. S., Cumings, J., and Zettl ,A. Rotational actuators based on carbon nanotubes. *Nature*, **424**, 408–410(2003).

24. Frankland, S. J. V., Harik, V. M., Odegard, G. M., Brenner, D. W., and Gates, T. S. The stress–strain behavior of polymer–nanotube composites from molecular dynamics simulation. *Composites Science and Technology*, **63**(11), 1655–1661 (2003).

25. Gannon, C. J., Cherukuri, P., Yakobson, B. I., Cognet, L. Kanzius, J. S., Kittrell, C. et. al., Carbon nanotube-enhanced thermal destruction of cancer cells in a noninvasive radiofrequency field. *Cancer*, **110**(12), 2654–2665 (2007).

26. Hafner, J. H., Cheung, C. L., Woolley, A. T., and Lieber, C. M. Structural and functional imaging with carbon nanotube AFM probes. *Progress in Biophysics and Molecular Biology*, **77**(1), 73–110 (2001).

27. Huang, Y. Y. and Terentjev, E. M. Dispersion of Carbon Nanotubes: Mixing, Sonication,

28. Stabilization, and Composite Properties. *Polymers*, **4**(1), 275–295 (2012).

29. Joshi, U. A., Joshi, P., Harsha, S. P., and sharma, S. C. Evaluation of the Mechanical Properties of Carbon Nanotube Based Composites by Finite Element Analysis. International. *Journal of Engineering Science and Technology*, **2**(5), 1098–1107 (2010).

30. Kearns, J. C. and Shambaugh, R. L. Polypropylene fibers reinforced with carbon nanotubes. *J. Appl. Polym. Sci.*, **86**(8), 2079–2084 (2002).

31. Kim, P. and Lieber, C. M. Nanotube Nanotweezers. *Science*, **286**(5447), 2148–2150 (1999).

32. Kis, A., Jensen, K., Aloni, S., Mickelson, W., and Zettl, A. Interlayer forces and ultra low sliding friction in multiwalled carbon nanotubes. *Physical Review Letters*, **97**(2), 025501 (2006).

33. Koziol, V., Vilatela, J., Moisala, A., Motta, M., Cunniff, P., Sennett, M., and Windle, A.. High-performance carbon nanotube fiber. *Science*, **318**(5858), 1892–1895 (2007).

34. Krishnan, A., Dujardin, E., Ebbesen, T. W., Yianilos, P. N., and Treacy, M. M. J. Young's modulus of single-walled nanotubes. *Physical Review B*, **58**(20), 14013–14019 (1998).

35. Kukich, D. Investigation of carbon nanotube composites for structural health monitoring.

36. http://phys.org/news/2012-09-carbon-nanotube-composites-health.html (2012).

37. Kuronuma, Y., Shindo, Y., Takeda, T., and Narita, F. Fracture behavior of cracked carbon nanotube-based polymer composites: Experiments and finite element simulations. *Fatigue and Fracture of Engineering Materials and Structure*, **33**(2), 87–93 (2010).

38. Liew, K. M., He, X. Q., and Wong, C. H. On *the* study of elastic and plastic properties of multi-walled carbon nanotubes under axial tension using molecular dynamics simulation. *Acta Materialia*, **52**(9), 2521–2527 (2004).

39. Liu, Z., Sun, X., Nakayama, N., and Dai, H. Supramolecular Chemistry on Water-Soluble Carbon Nanotubes for Drug Loading and Delivery. *ACS Nano*, 1(1), 50–56 (2007).

40. Lourie, O. and Wagner, H. D. Transmission electron microscopy observations of fracture of single-wall carbon nanotubes under axial tension. *Applied Physics Letters*, **73**(24), 3527 (1998).

41. Lourie, O., Cox, D. M., and Wagner, H. D. Buckling and Collapse of Embedded Carbon Nanotubes. *Physical Review Letters*, **81**(8), 1638–1641 (1998).

42. Mielke, S. L., Troya, D., Zhang, S., Li, J. L., Xiao, S., Car, R., Ruoff, R. S., Schatz, G. C., and Belytschko, T. The role of vacancy defects and holes in the fracture of carbon nanotubes. *Chemical Physics Letters*, **390**(4–6), 413–420 (2004).

43. Mora, R. J., Vilatela, J. J. and Windle, A. Properties of composites of carbon nanotube fibres. *Composites Science and Technology*, **69**(10), 1558 (2009).

44. Nakayama, Y. and Akita, S. Field-emission device with carbon nanotubes for a flat panel display. *Synthetic Metals*, **117** (1–3), 207–210 (2001).

45. Nguyen, C. V., Chao, K. J., Stevens, R. M. D., Delzeit, L., Cassell, A., Han, J. and Meyyappan, M. Carbon nanotube tip probes: stability and lateral resolution in scanning probe microscopy and application to surface science in semiconductors. *Nanotechnology*, **12**(3), 363–367 (2001).

46. Nichols, J. A., Saito, H., Deck, C., and Bandaru, P. A. Artificial introduction of defects into vertically aligned multiwalled carbon nanotube ensembles: Application to electrochemical sensors. *Journal of Applied Physics*, **102**(6), 064306 (2007).

47. O'Connell, M. J., Bachilo, S. M., Huffman, C. B., Moore, V. C., Strano, M. S., Haroz, E. H., Rialon, K. L., Boul, P. J., Noon, W. H., Kittrell, C., Ma, J., Hauge, R. H.,Weisman, R. B., and Smalley, R. E. *Band Gap Fluorescence from Individual Single-Walled Carbon Nanotubes Science*, **297**(5581), 593–596 (2002).

48. Odegard, G. M., Gates, T. S., Wise, K. E., Park, C., and Siochi, E. J. Constitutive modeling of nanotube–reinforced polymer composites. *Composite Science and Technology*, **63**(11), 1671–1687 (2003).

49. Ouyang, M., Huang, J. L., Cheung, C. L., and Lieber, C. M. Energy gaps in "metallic" single-walled carbon nanotubes. *Science*, **292**(5517), 702–705 (2001).

50. Pantano, A., Boyce, M. C., and Parks, D. M. Mechanics of Axial Compression of Single and Multi-Wall Carbon Nanotubes. *Journal of Engineering Material and Technology, Trans. ASME*, **126**(3), 279–284 (2004).

51. Poncharal, P., Wang, Z. L., Ugarte, D., and de Heer, W. A. Electrostatic Defections and Electromechanical Resonances of Carbon Nanotubes. *Science*, **283**(5407), 1513–1516 (1999).

52. Qian, D. and Liu, W. K., Subramoney, S., and Ruoff, R. S. Effect of interlayer potential on mechanical deformation of multiwalled carbon nanotubes. *Journal of Nanoscience and Nanotechnology*, **3**(1–2), 185–191 (2003).

53. Rahman, A. Correlations in the Motion of Atoms in Liquid Argon. *Physical Review A*, **136**(2A), A405–A411 (1964).

54. Reich, S., Thomsen, C., and Ordejon, P. Electronic band structure of isolated and bundled carbon nanotubes. *Physical Review B*, **65**(15), 155411 (2002).

55. Ruan, S., Gao, P. and Yu, T.X. Ultra-strong gel-spun UHMMPE fibers reinforced using multiwalled carbon nanotubes. *Polymer*, **47**(5), 1604–1611 (2006).

56. Rueckes, T., Kim, K., Joselevich, E., Tseng, G. Y., Cheung, C. L., and Lieber, C. M. Carbon Nanotube-Based Nonvolatile Random Access Memory for Molecular Computing. *Science*, **289**(5476), 94–97 (2000).

57. Ruoff, R. S., Calabri, L., Ding, W., and Pugno, N. M. Experimental tests on fracture strength of nanotubes. *Reviews on Advanced Materials Science*, **10**, 110–117 (2005).

58. Salvetat, J. P., Kulik, A. J., Bonard, J. M., Briggs, G. A. D., Stöckli, T., M´et´enier, K., Bonnamy, S., Béguin, F., Burnham, N. A., and Forró L. Elastic modulus of ordered and disordered multiwalled carbon nanotubes. *Advanced Materials*, **11**(2), 161–165 (1999).

59. Salvetat, J. P., Briggs, G. A. D., Bonard, J. M., Bacsa, R. R., Kulik, A. J., and Stöckli, T., Burnham, N. A., and Forró, L. Elastic and shear moduli of single-walled carbon nanotube ropes. *Physical Review Letters*, **82**(5), 944–947 (1999).

60. Sears, A. and Batra, R. C. Buckling of multiwalled carbon nanotubes under axial compression. *Physical Review B*, **73**(8), 085410 (2006).

61. Seo, M. K., Byun, J. H., and Park, S. J. Studies on Morphologies and Mechanical Properties of Multi-walled Carbon Nanotubes/Epoxy Matrix Composites. *Bulletin of the Korean Chemical Society*, **31**(5), 1237–1240 (2010).

62. Shima, H. Buckling of Carbon Nanotubes: A State of the Art Review. *Materials*, **5**(1), 47–84 (2012).

63. Srivastava, D, Menon, M., and Cho, K. Nano-plasticity of Single-Wall Carbon Nanotubes under Uniaxial Compression. *Physical Review Letters*, **83**(15), 2973–2976. (1999).

64. Suehiro, J., Imakiire, H., Hidaka, S., Ding, W., Zhou, G., Imasaka, K., and Hara, M. Schottky-type response of carbon nanotube NO_2 gas sensor fabricated onto aluminum electrodes by dielectrophoresis. *Sensors and Actuators B: Chemical*, **114**(2), 943–949 (2006).

65. Talukdar, K. and Mitra, A. K. The role of potential functions in the mechanical behavior of the single wall carbon nanotubes. *International Journal of Nanoscience*, **11**(3), 1240009-1-5 (2012).

66. Tang, C., Guo, W., and Chen, C. Structural and mechanical properties of partially unzipped carbon nanotubes. *Physicsl Review B*, **83**(7), 075410 (2011).

67. Treacy, M. M. J., Ebbesen, T.W., and Gibson, J. M. Exceptional high Young's modulus. observed for individual carbon nanotubes. *Nature*, **381**(6584), 678–680 (1996).

68. Troya, D., Mielke, S. L., and Schatz, G. C. Carbon nanotube fracture – differences between quantum mechanical mechanisms and those of empirical potentials. *Chemical Physics Letters*, **382**(1–2), 133–141 (2003).

69. Wildoer, J. W. G., Venema, L. C., Rinzler, A. G., Smalley, R. E., and Dekker, C. Electronic structure of atomically resolved carbon nanotubes. *Nature*, 391 (6662), 59–62 (1998).

70. Wong, E. W., Sheehan, P. E., and Lieber, C. M. Nanobeam mechanics: Elasticity, strength and toughness of nanorods and nanotubes. *Science*, **277**(5334), 1971–1975 (1997).

71. Wu, X. and Luo, R. Mechanical properties investigation of carbon/carbon composites fabricated by a fast densification process. *Materials and Design*, **32**(4), 2361–2364 (2011).

72. Xia, Z. H., Guduru, P. R., and Curtin, W. A. Enhancing Mechanical Properties of Multiwall Carbon Nanotubes via sp^3Interwall Bridging. *Physical Review Letters*, **98**(24), 245501 (2007).

73. Xiao, J. R., Lapatnikov, S. L., Gama, B. A., and Gillespie, Jr. J. W. Nanomechanics on the deformation of single- and multi-walled carbon nanotubes under radial pressure.

74. *Material Science and Engineering A*, **416**(1–2), 192–204 (2006).

75. Yakobson, B. I., Brabec, C. J., and Bernholc, J. Nanomechanics of Carbon Tubes: Instabilities beyond Linear Response. *Physical Review Letters*, **76**(14), 2511–2514 (1996).

76. Yao, X., Han, Q., and Xin, H. Bending buckling behaviors of single- and multi-walled carbon nanotubes. *Computational Materials Science*, **43**(4), 579–590 (2008)

77. Yu, M. F., Lourie, O., Dyer, M. J., Molony, K., Kelly, T. F., and Ruoff, R. S. Strength and Breaking Mechanism of Multiwalled Carbon Nanotubes under Tensile Load. *Science*, **287**(5453), 637–640 (2000).

78. Yu, M. F., Files, B. S., Arepally, S., and Ruoff, R. S.. Tensile Loading of Ropes of Single Wall Carbon Nanotubes and their Mechanical Properties. *Physical Review Letters*, **84**(24), 5552–5555 (2000).

79. Yu, M. F., Yakobson, B. I., and Ruoff, R. S. Controlled sliding and pullout of nested shells in individual multiwalled carbon nanotubes. *Journal of Physical Chemistry B*, **104**(37), 8764–8767 (2000).

80. Zhang, Y. Y., Wang, C. M., and Tan, V. B. C. Examining the effects of wall numbers on buckling behavior and mechanical properties of multiwalled carbon nanotubes via molecular dynamics simulations. *Journal of Applied Physics*, **103**(5), 053505 (2008).

81. Zhang, P., Lamert, P. E., and Crespi, V. H. Plastic Deformations of Carbon Nanotubes. *Physical Review Letters*, **81**(24), 5346–5349 (1998).
82. Zou, J., Huang, X., Arroyo, M., and Zhang, S. Effective coarse-grained simulations of super-thick multi-walled carbon nanotubes under torsion. *Journal of Applied Physics*, **105**(3), 033516 (2009).

CHAPTER 11

RECENT ADVANCES IN NANOMEDICINE: APPLICATIONS IN DIAGNOSIS AND THERAPEUTICS

SANDHYA GOPALAKRISHNAN, KANNAN VAIDYANATHAN, and NANDAKUMAR KALARIKKAL

CONTENTS

11.1 INTRODUCTION

Nanoscience is a new branch of technology that studies molecules and chemical structures at the nanoscale. Nanomaterials are not a novel invention, but they are well-known materials, which display properties in the nanorange and not outside it. Measurements performed at that level are also included as nanoscience. Nanomedicine is the application of nanotechnology to medicine. It is a field which has developed in the recent years as a result of developments in a variety of fields like nanoscience, nanotechnology, biology, and medicine. It is the result of a dedicated team of scientists, engineers, and physicians. The ultimate objective of nanoscience is to improve the quality of life and solve problems encountered in routine clinical practice, by developing new drugs, diagnostic agents, and novel drug delivery systems (Balogh, 2009).

There are mainly three powerful technologies which form the basis for nanomedicine:

1. Nanomaterials and devices
2. Advances in molecular medicine via genomics, proteomics, and artificially engineered microbes
3. Molecular machine systems like nanorobots, which can diagnose and at the same time treat diseases.

This chapter covers the latest advances in the field of nanomedicine. We shall discuss the different nanomaterials having medical applications and the criteria of nanoparticles for use as therapeutic agents. We go on to discuss the various tools for detecting nanoparticles and the diagnostic role of nanoparticles, nanogenomics as well as the newer role of nanoparticles as theragnostic agents. We end the chapter with a discussion on regulatory issues surrounding nanomedicine.

The upcoming field of nanomedicine relies on the development of novel nanomaterials. Because of its tremendous potential and ability to reach at sub-cellular level, nanomaterials are being increasingly investigated as therapeutic agents. The fine properties of nanomaterials also make them superior diagnostic agents. The applications of nanomaterials in medicine are an exciting area of research. Toxicity of nanomaterials also needs to be explored before they are employed for human use. In this chapter, we compile the latest advances in the field of nanomedicine such as different nanomedicine tools, nanomaterials, and their potential uses in nanomedicine.

11.2 NANOMATERIALS

11.2.1 PHYSICAL PROPERTIES OF NANOPARTICLES

The most obvious property of nanomaterials is their size. In addition, composition, shape, and architecture of the material also play important roles in determining their properties. The collective property of a nanoparticle is determined by the individual properties of the constituents (that is, the composition of the nanoparticle) and their interaction with each other and the immediate environment (Ventola, 2012a). The effect of size on spherical particles with respect to circulation, extravasation, and distribution in vivo are well known. However, the importance of particle shape has only recently begun to emerge. For example, studies in mouse models have shown that disc-shaped nanoparticles have higher in vivo targeting specificity to endothelial cells than spherical particles of similar size. The effect of shape has potential applications in diagnostic imaging and targeted drug delivery (Tao et. al., 2011).

The molecular and catalytic properties of nanoparticles can be modulated by the combination of nanotechnology and biotechnology tools. This has lead to the development of a large number of sensors that can detect a broad range of analytes like metal ions, other small molecules, nucleic acids, and proteins. Metallic nanoparticles possess distance-dependent optical properties and are useful for designing colorimetric sensors. Another group of materials which are discussed later called semiconductor nanoparticles or quantum dots (QDs) are much superior in their properties to traditional organic fluorophores. Magnetic nanoparticles, also discussed below, are shown to be useful as smart magnetic resonance imaging (MRI) contrast agents. Carbon nanotubes show near-IR emission properties, and thus, are useful for in vivo sensing (Lu and Liu, 2009).

11.2.2   ADVANTAGES OF NANOMATERIALS

Decreasing particle size results in increased surface interactions, which leads to increased solubility. When nanoparticles are used as drugs, this helps their absorption, irrespective of whether they are lipophobic or lipophilic. Liposomes, polymers, surface compatibilized nanoparticles, and dendrimers are examples of such agents. Further, chemical modifications of their surfaces increase receptor-specific targeting (Fahmy 2005). Nanoscale devices have different biodistribution, pharmacodynamic, and pharmacokinetic behavior compared to traditional drugs (Li, 2010). They have unusual pharmacodynamics and pharmacokinetics which traditional drugs do not possess. They include stability in body fluids, lack of toxicity, appropriate biodistribution, predictable, and ideal pharmacokinetic behavior without significant immunogenicity. The agents are capable of functional interactions like binding to receptors, blocking, or internalization, initiating signaling events and so on, which are inherent in their design. The effect of body's immune system is also different. They enter tissues and cells, interact with proteins and DNA at a molecular level and directly and indirectly modulate the immune system by novel mechanisms.

11.2.3   PROBLEMS WITH NANOMATERIALS

On the flip side, they are capable of activating proinflammatory cytokines, chemokines, and, adhesion molecules. This leads to recruitment of inflammatory cells like basophils, macrophages, dendritic cells, and T cells. Overall influence on body's immune defense, Th1/Th2 balance, and other non-immunologic functions occur. As a result, increases in incidence of autoimmune, allergic and even neoplastic diseases occurs (Chang, 2010). Toxicity profiling is therefore an important aspect of nanodrug research.

11.2.4 SYNTHESIS OF NANOPARTICLES

Engineered scaffolds supplements or replaces injured, missing, or defective tissue or organs. Scaffolds that mimic the structure and function of the native extracellular matrix (ECM) are a popular area of research. There are three major processing techniques; these are self-assembly, phase separation, and the most popular technique, electrospinning. All these techniques can produce fibers that rival the size of those found in the native ECM. Electrospinning is the most popular methods since it is simple, versatile, and scalable. It can reproduce many of the complex aspects that characterize the native ECM. Many novel electrospinning strategies are available and further continue to be evolved. These include alterations of scaffold composition and architecture, addition and encapsulation of cells, drugs, and growth factors within the scaffold (Ayres et. al., 2010). Natural ECM contains fibers in both micro and nano-scales and provides a structural scaffold which helps localization, migration, proliferation, and differentiation. The technique helps further functionalizing polymeric nanofibres by using them for drug delivery devices and improving their design to improve control of delivery (Ashammakhi et. al., 2009).

11.3 APPLICATIONS OF NANOMATERIALS

Nanomaterials can be used as contrast agents in MRI, optical imaging, and photoacoustic imaging (Ryvolova et. al., 2012, Tong et. al., 2012). Nanoformulations can be used as drug carriers, as for example in cancer therapy. Multiple therapeutic agents like chemotherapy, antiangiogenic, and gene therapy agents can be simultaneously delivered by nanocarriers to tumor sites to enhance the effectiveness of therapy. Different therapeutic modalities such as chemotherapy and hyperthermia can be co-administered to take advantage of synergistic effects. Examples of nanoformulations are liposomes, metallic and polymeric nanoparticles, dendrimers, carbon nanotubes, and quantum dots; these are used as multifunctional platforms for cancer theranostics (Fernandez-Fernandez et. al., 2011, Kumar and Khan, 2010).

We shall now see the common agents used as nano-drugs and some of their important properties.

11.3.1 VIRAL NANOPARTICLES

Viral nanoparticles (VNPs) are naturally occurring nanomaterials. They are biocompatible and biodegradable. Genetic and chemical protocols are used for their synthesis. Since their introduction 20 years back, they have evolved rapidly. They are strategically modified for their functionalization with imaging reagents, targeting ligands, and a variety of drugs (Steinmetz, 2010).

11.3.2 INORGANIC NANOPARTICLES

Many inorganic nanomaterials (INMs) and nanoparticles (NPs) are used as chemotherapeutic agents, imaging molecules or antiseptics. Examples include metals, silica derivatives, dendrimers, and organic-inorganic, and bioinorganic hybrids. Gold NPs are important in imaging, as carrier molecules for various drugs, and for thermotherapy of biological targets. Many gold NPs are used in the form of nanoshells, nanorods, and so on as anti-cancer agents. Metal NP contrast agents are used in MRI and other imaging techniques like ultrasound imaging. Hollow and porous INMs are used as therapeutic agents and in gene delivery. They also have potential uses as diagnostic imaging tools and in photothermal therapy. Silver NPs have antimicrobial activity. Inorganic nanohybrids are used in targeted imaging and therapy, as therapeutic agents and in gene delivery, and even in regenerative medicine. Dendrimers are used as therapeutic agents, gene carriers, diagnostic contrast agents, and as sensors for metal ions (Sekhon and Kamboj, 2010a).

Fluorescent QDs have a variety of applications like as cell labeling tools, biosensors, and other in vivo imaging and other diagnostic uses. Biocompatible QD conjugates are used in treatment of cancer. Magnetic nanowires are used as biosensors and as nucleic acid sensors. Magnetic nanoparticles (MNPs) are important for MRI, as therapeutic agents, and in cell labeling. Iron oxide nanoparticles are used for hyperthermic treatment of cancer tissues. The MNPs play a crucial role in developing techniques to diagnose, monitor, and treat a wide range of common diseases (Sekhon and Kamboj, 2010b).

11.3.3   MESOPOROUS SILICA NANOMATERIALS (MSN)

The MSN is a type of silica nanoparticle with wide applications in imaging, as biosensors, and in drug delivery (Mai and Meng, 2013). They can be synthesized by simple sol-gel or spray drying methods. Ordered MSN boosts in vivo and in vitro dissolution of hydrophobic drugs. These bioceramics exhibit two important features, they can regenerate osseous tissues and they are able to act as controlled release systems. Nanodrugs loaded onto MSNs can be locally released in a controlled fashion. The MSNs can be chemically modified so that adsorption of biomolecules like peptides, proteins, or growth factors is possible. Smart biomaterials (drug is released under an external stimulus) might even be possible with MSNs (Vallet-Regi, 2010; Rosenholm et. al., 2010; Rother et. al., 2011). Functionalized nanoparticles can be made to attach to specific ligands, known as smart nanoparticles (Friedman et. al., 2013; Lehner et. al., 2013).

11.3.4   QUANTUM DOTS

The QDs (semiconducting nanoparticles) have a variety of promising applications. They are used as optical and electro-optical devices, and also have computing applications. The QDs along with polymers form nanocomposites, with a wide variety of nanomedical applications (Mansur, 2010). New generations of QD probes are also promising for imaging cellular processes at the single-molecule level and in the field of personalized medicine (Tomczak et. al., 2012, Zrazhevskiy et. al., 2010).

11.3.5   UPCONVERSION NANOPARTICLES

Upconversion nanoparticles (UCNs) are an emerging class of luminescent nanoparticles with high signal to noise ratio and superior photo-stability. Potential applications are in the field of imaging, detection and therapeutic tools. The major features are deep tissue penetration and low photodamage to cells (Ang et. al., 2011, Wang et. al., 2011).

11.3.6   CARBON NANOTUBES (CNT)

The CNT are extremely stiff and strong and they have unique properties, which make them ideal agents in nanomedicine (Wujcik and Monty, 2013). Functionalized CNTs act as nanocarriers for therapeutic agents, peptides, and nucleic acids. Carbohydrate functionalized CNTs are used to detect microbes, to bind specific lectins, to deliver glycomimetic 1chemotherapeutic agents into cells and as biosensors (Gorityala et. al., 2010). Polyethylene glycol (PEG) modified CNTs are used in preclinical models in the treatment of cancer, neurological disorders, vaccination, and imaging (Bottini et. al., 2011). They are used as novel nanocarriers in drug delivery systems and biomedical applications. Literature reveals that CNTs are versatile carriers for controlled and targeted drug delivery, especially for cancer cells (Beg et. al., 2011). However, some studies show that they may induce oxidative stress, inflammation, apoptosis, and cytotoxic effects on lungs. (Kayat et. al., 2011).

Single-walled carbon nanotubes (SWNTs) are used for noninvasive and high sensitivity detection and as imaging agents *in vitro* and *in vivo*. Functionalized SWNTs are used as carriers to deliver various anticancer drugs, proteins, and nucleic acids. Their advantages include that the drug can be delivered to specific sites thus maximizing their bioavailability (Liang and Chen, 2010). Carbon nanodiamonds are among the most biocompatible materials even among other carbon nanomaterials. Additional features of carbon nanodiamonds like lack of cytotoxicity, strong and stable photoluminescence, very small size, large surface area and capability to be easily functionalized with biomolecules make them suitable agents (Xing and Dai, 2009).

11.3.7   GOLD AND SILVER NANOPARTICLES

Gold nanoparticles (GNPs) are now being incorporated into polymer and lipid based NPs to build multifunctional devices. Gold nanostructures have been used to build multifunctional platforms with imaging, targeting, and therapeutics with great benefits (Chitrani, 2010). The GNP fluorescence-based activatable probes are designed to increase fluorescence intensity in fluorescence-based assays and detection techniques (Swierczewska et. al., 2011). There is a trend toward viral-based hybrid systems to furnish viral nanoparticles with advanced features, for function beyond a delivery vehicle. (Portney et. al., 2009). At nanoscale, silver exhibits remarkable physical, chemical, and biological properties including strong antibacterial activity. Nanosilver is used for treatment of wounds and burns. Use of nanosilver is becoming more and more widespread in medicine (Chen and Schluesener, 2007). Two of the best examples of commercial nanopharmaceuticals as topical products are ActicoatTM and Silveron wound dressings. These dressings use nanocrystalline silver for topical application to dermal wounds (Atiyeh et. al., 2005).

11.3.8   CHITOSAN

Chitosan is a polysaccharide that forms nanoparticles through ionic gelation. Chitosan nanoparticles have received much attention since early 2000 as potential nanocarriers (Gonçalves et. al., 2012). Chitosan is commonly found in the hard skeleton of shellfish. It has been incorporated into a number of nanoparticle formulations and has played a critical role in improving delivery of many drugs since there is better bioavailability, drug stability, ability to delivery at the exact site (targeted delivery), and increased cytostatic activity over a long duration. Chitosan-modified graphene is used for DNA mutation analysis (Alwarappan et. al., 2012).

11.3.9   MULTIMODULAR NANOASSEMBLY

Majority of nanotherapeutics/nanodiagnostics being developed accommodate single- or multiple- functionalities on the same entity. There are many heterogeneous biological barriers that can prevent therapeutic and imaging agents from reaching their intended targets in sufficient concentrations. Hence there is an emerging requirement to develop a multimodular nanoassembly, where there is synergism of individual components. The multistage nanovectors (MSVs) were introduced in 2008 as the first system of this type comprising several nanocomponents or "stages". Each stage designed to negotiate one or more biological barriers. For example, Stage 1 mesoporous silicon particles (S1MPs) were rationally designed and fabricated in a nonspherical geometry to enable superior blood margination and to increase cell surface adhesion. S1MPs can efficiently transport nanoparticles that are loaded into their porous structure and to protect them during transport from the administration site to the disease lesion (Godin et. al., 2011).

11.3.10   POLYHEDRAL OLIGOMERIC SILSESQUIOXANE (POSS)

The POSS have a distinctive nanocage structure consisting of an inner inorganic framework and an outer shell of organic functional groups. Biomedical applications of POSS include the development of biomedical devices, tissue engineering, drug delivery, dentistry, and biosensors. The POSS nanocomposites in combination with other nanostructures including silver nanoparticles and QD nanocrystals are available. Functionalization confers antimicrobial efficacy to POSS. Polymer nanocomposites provides a biocompatible surface coating for QD nanocrystals, which increase the efficacy of the materials for different biomedical and biotechnological applications (Ghanbari et. al., 2011).

11.3.11   GLYCONANOPARTICLE

The term glyconanoparticle was coined in 2001 to denote nanoparticles constructed with gold. These were used as tools in carbohydrate-based interaction studies and to interfere in biological process where carbohydrates are involved. Gold inorganic core can be replaced by a wide variety of materials, thus a range of glyconanoparticles of various optical, electronic, mechanical, and magnetic properties can be formed. Their sizes can be modulated and the surface can be engineered to modify multivalence and insert multifunctionality (Marradi et. al., 2010, Garcia et. al., 2010).

11.3.12   CYCLODEXTRINS

Cyclodextrins are cyclic oligosaccharides. They are usually synthesized from starch by enzymatic action and are used as nebulizers. Cyclodextrins are used in conjunction with nanoparticles for oral delivery of proteins (Kanwar et. al., 2011). Cyclodextrin-based nanosponges are used as a drug delivery system, since they are safe and bio-degradable. They improve the water-soluble of otherwise water-insoluble drugs (Trotta et. al., 2012).

11.3.13   FERRITIN

Ferritin is one of the storage forms of iron. Ferritin level indicates the iron content in the body. It is mainly seen as a globular intracellular protein and is ubiquitous in distribution. In material science, ferritin is used as a precursor for iron nanoparticles in carbon nanotube synthesis. Native and synthetic ferritins are used as bionanoparticles (Domanguez et. al., 2010).

11.3.14   FIRST-GENERATION NANOPARTICLES

First-generation nanoparticles are pharmaceutical drug delivery carriers with added drug tolerability, circulation half-life, and efficacy. Different organs, tissues and sub-cellular compartments, and some disease states are characterized by their pH levels. When exposed to changes in pH, pH-responsive NPs respond with physical as well as chemical changes to their material structure and surface characteristics favoring drug release (Gao et. al., 2010).

11.3.15   NANORODS

Noble metal nanorods are a useful tool for tracking binding events in different applications, for example assembly, biosensing, in vivo targeting and imaging, and single-molecule detection by surface-enhanced Raman spectroscopy (Mannelli et. al., 2010).

11.3.16   NANOWIRES

Potential applications of nanowires include as nanosensors and nanocarriers. They can be readily functionalized with various biochemical through different linkage chemistries. This integration of nanowires and biomolecules leads to novel hybrid systems with catalytic properties of biomaterials and other characteristics of nanowires. Applications include direct real-time label-free electrical detection of biomolecular interactions. Nanowires can perform several tasks simultaneously. Multisegment nanowires designed for nanomedicine applications can couple the selective targeting, therapy, and imaging functions (Wang, 2009).

11.3.17 DENDRITIC POLYMERS

Dendritic polyglycerols (PGs) exhibit good chemical stability and inertness under biological conditions and are highly biocompatible. Oligoglycerols are used as cosmetics, food industries, pharmaceuticals, polymers, and polymer additives (Calderon et. al., 2010). Dendrimers are used as nanocontainers to conjugate and encapsulate drugs or imaging moieties (Tian et. al., 2013). Dendrimers are used as high-contrast MRI imaging agents (Menjoge et. al., 2010).

11.3.18 DENDRITIC CELLS

Dendritic cells (DCs) are potent antigen-presenting cells capable of initiating a primary immune response. The DCs have an important role in various diseases and conditions involving the immune system, particularly in cancer and some autoimmune diseases. Targeting NPs to DCs provides a promising strategy for developing an efficient balanced and protective immune response, since they can modulate the immune response and are useful as effective vaccine adjuvants for infectious disease and cancer therapy (Klippstein and Pozo, 2010).

11.3.19 MAGNETIC NANOPARTICLES (MNPS)

The MNPs for cancer therapy and diagnosis have been developed on the basis of their unique physico-chemical properties not present in other materials. Their versatility finds many applications like cell and macromolecule separation and/or purification, diagnostic immunoassays, therapeutic targeted drug delivery, controlled release, gene therapy, and MRI (Babincova et. al., 2009). The use of MNPs for biological and biomedical applications is one of the most attractive fields of nanotechnology today because of their unique magnetic properties and the potential to function at cellular and molecular level of biological interactions (Chanana et. al., 2009, Frimpong et. al., 2010, Sandhu et. al., 2010, Banerjee et. al., 2010).

11.3.20 PACA BIODEGRADABLE POLYMERS

The NPs developed from poly (alkyl cyanoacrylate) (PACA) biodegradable polymers have opened new and exciting perspectives in the field of drug delivery. They have ideal characteristics as drug carriers in biomedical applications. They are available as nanospheres and nanocapsules (either oil- or water-containing) as well as long-circulating and ligand-decorated NPs (Nicolas et. al., 2009).

11.3.21 IRON BASED NANOPARTICLES

Superparamagnetic iron oxide nanoparticles (SPIONs) have attracted a great deal of interest in biomedical research and clinical applications over the past decades. The SPIONs have in vivo applications such as hyperthermia (HT), magnetic drug targeting (MDT), MRI, and gene delivery (GD). The SPIONs have precise control of the particle's shape, size, and size distribution (Lin et. al., 2008, Figuerola et. al., 2010).

The SPIONs act synergistically with agents which amplify reactive oxygen species stress and this is turn improves efficacy of medications like anti-cancer drugs (Huang et. al., 2013).

11.3.22   *LAYERED DOUBLE HYDROXIDES (LDHs)*

The LDHs are used in delivery of therapeutic molecules like peptides, proteins, and nucleic acids to mammalian cells their properties are ideal for gene and drug delivery (Ladewig et. al., 2009).

11.4   NANOMEDICINE TOOLS

Progress in nanomedicine is possible only with the help of advanced tools for imaging and manipulating biological systems at the nanoscale. The development of the science has coincided with the development of a variety of microscopic techniques. The imaging techniques can be categorized based on the size scale of detection. Atomic force microscopy (AFM), transmission electron microscopy (TEM), and scanning electron microscopy (SEM) are useful for measuring at nanoscale. Confocal laser scanning microscopy (CLSM), MRI, and superconducting quantum interference devices are useful for measurement at micro to macro-scale. Combined techniques like AFM-CLSM, correlative light, and electron microscopy (CLEM) and SEM spectroscopy are used with additional advantages (Jin et. al., 2010). Using the technique of localization microscopy, individual molecules can be analyzed (Owen et. al., 2012).

11.4.1   *ATOMIC FORCE MICROSCOPY (AFM)*

The AFM is to analyze the structure, properties, and functions of microbes at high resolution. Additionally, they can be used to monitor their remodeling upon interaction with drugs. Single-molecule force spectroscopy helps to localize cell wall constituents and to study their mechanics and interactions. They also help to understand microbe-drug and microbe-host interactions, which are necessary requisites while developing new antimicrobial agents (Alsteens et. al., 2011, El-Kirat-Chatel et. al., 2012). Further, their application can be extended to the study of drug resistance, blood component analysis, studying the toxicity of nano-drugs (nanotoxicology), and biocompatibility of nanomaterials. (Iaminskii et. al., 2011). The AFM can study the structural and physical properties of live biological systems at an atomic scale (Shi et. al., 2012). It has an open architectural framework that allows it to be integrated with other techniques, tools, and operating environments. It can be used to create and characterize nanocarriers and implantable vehicles for controlled delivery. It has helped to understand the pathogenesis of various diseases, including protein misfolding diseases (e.g., Alzheimer' s disease, cancer, diabetes) and lifestyle and environmental diseases (Lal and Arnsdorf, 2010).

11.4.2   *NUCLEAR MAGNETIC RESONANCE (NMR) SPECTROSCOPY*

The NMR spectroscopy is considered as an advanced profiling technique which can assess even minor fluctuations in the level of various endogenous metabolites

(Kurhanewicz et. al., 2008, Golman et. al., 2003). The NMR can be combined with multivariate statistics, the so-called metabonomics approach. This approach is used for diagnosis of human diseases, human nutrition, pharmaceutical research, and environmental toxicology. Metabonomics is a new tool in nanomedicine since it gives a deeper understanding of the mechanisms by which nanomaterials modify biological responses (Duarte, 2011).

11.4.3   QUARTZ CRYSTAL MICROBALANCE (QCM)

The QCM is a highly sensitive instrument that determines the nature of binding interactions in real time within a label free environment. Recent advances in this technology have made possible increased stability, sensitivity, and high throughput capacity, which make it an ideal candidate to the study of nanomedicine. For example, QCM is used to study the activation of the complement system and induction of apoptosis (Hunter, 2009).

11.4.4   CRYO-ELECTRON TOMOGRAPHY

Many NPs cannot be studied by conventional techniques like x-ray crystallography or nuclear magnetic resonance spectroscopy, because of their large size or heterogeneity of the sample specimen. Cryo-electron tomography is an emerging technique which provides high-resolution structural information of these NPs. The technique enables the visualization and quantification of heterogeneous nanomaterials (Lengvel et. al., 2008).

11.4.5   SURFACE PLASMONS

Free electrons in a noble metal nanoparticle can be resonantly excited. This result in their collective oscillation, the phenomenon referred to as surface plasmon. Eventuaaly this leads to absorption of light, generation of heat, transfer energy, and re-radiation of incident photons. Plasmon resonant NPs are used in quantitative biology and nanomedicine. Examples are (1) Dual-functional nanoplasmonic optical antennae for label-free biosensors and nanoplasmonic gene switches, and (2) Integrated photo-acoustic-photothermal contrast agents (Lee and Lee, 2010).

11.4.6   SINGLE PHOTON EMISSION COMPUTED TOMOGRAPHY (SPECT)/POSITRON EMISSION TOMOGRAPHY (PET)

Many methods of labeling liposomes with both diagnostic and therapeutic radionuclides are available. Some of these methods can be used to track nanometer-sized liposomes in the body in a quantitative fashion (Phillips et. al., 2009). Examples are SPECT and PET imaging. Liposomes can be labeled with single photon or dual photon positron emission radionuclides and the uptake can be tracked providing an excellent tool for developing liposome-based drug delivery agents. They are also useful in cancer therapy. Not only liposomes, they can also be useful for the imaging and tracking of other NPs (Phillips et. al., 2009).

11.4.7 OTHER TOOLS

Other important tools include Infra-Red (IR) Spectroscopy (Ghosh et. al., 2012), Scanning Probe Microscopy (SPM) (Brodusch et. al., 2012), Near Field Scanning Optical Microscopy (NSOM) (Rasmussen and Deckert, 2005), Single Molecule Fluorescence Microscopy (Juffmann et. al., 2012), Photo-Activated Localizing Microscopy (PALM) (Quirin et. al., 2012), inductively coupled plasma-mass spectrometry (ICP-MS) (Yang et. al., 2012), graphite furnace atomic absorption spectrometry (GF-AAS) (Cunha et. al., 2012), isotherm N_2 absorption/desorption measurements (BET) (Sheykhan et. al., 2011), field flow fractionation (FFF) (Roda et. al., 2009), energy dispersive X-ray spectroscopy (EDX) (Iannuccelli et. al., 2013), and so on.

11.5 DIAGNOSTICS

Medical devices using nanomaterials and/or nanostructures could be micro- or macro-sized devices, or truly nanosized active systems. Examples of micro- and/or macro-sized devices are diagnostic arrays that use nonbleaching QDs as fluorescence markers and microfluidic devices that contain nanoscopic structures in their inner surface to improve detection or catalytic processes. Nanoscience permits the integration of biological and physical systems at the nanoscale allowing the fabrication of real nanoscale devices or in situ fabrication of nanostructures (Ventola, 2012b). Some outstanding achievements of nanomedicine are analytical applications with greatly improved diagnostic accuracy and convenience (Greehalgh and Turos 2009, Wong et al. 2006), tissue engineering using autologous donor cells (Guo 2009), cancer cell–specific markers that allow surgeons to visually identify the tissue that needs to be removed during the procedure (Jiang et. al., 2004, Nguyen et. al., 2010), replacement of surgical brachytherapy procedures by a single injection (Khan, 2010), and development of personalized and disease-specific intelligent nanodrugs (Kawasaki and Player, 2005; Dutta, 2009).

The detection of mismatched base pairs in DNA plays a crucial role in the diagnosis of genetic-related diseases and conditions, especially for early stage treatment. Advancements in micro- and nanotechnological techniques, like fabrication techniques, and new nanomaterials, have lead to the development of highly sensitive and specific sensors making them attractive for the detection of small sequence variations. In addition, the integration of sensors with sample preparation and fluidic processes lead to development of rapid, multiplexed DNA detection essential for Point of care clinical diagnostics (Wei et. al., 2010). Nonmetallic polysaccharide based NPs are used in drug delivery and molecular imaging. Glyconanoparticles (GNPs) have diagnostic applications (Marradi et. al., 2011). Glycosylated NPs (that is GNPs having sugar residues on the surface) like glycosylated gold NPs, glycosylated QDs, and GNPs self-assembled from amphiphilic glycopolymers are used for biomedical applications such as bioassays and targeted drug delivery (Dong, 2011). Alterations of blood rheology (hemorheology) are important for the early diagnosis, prognosis and prevention of diseases like myocardial infarction, stroke, inflammation, thromboembolism, trauma, malignancies, and sickle cell anemia. Real-time in vivo assessment of multiple hemorheological parameters over longer periods of time is possible by lable-free photoacoustic (PA) and photothermal (PT) flow cytometry (Galanzha et. al., 2011).

11.5.1   CELL LABELING TECHNIQUES

Cell based therapeutics are an important new technique. This had led to the better understanding of the migration and proliferation mechanisms of implanted cells, since it provides a means to track molecules. They are easily detected by appropriate imaging modalities in living subjects at a high spatial and temporal resolution. These features precise differentiation from host cells. Advantage over traditional histological methods include that they permit non-invasive, real-time tracking *in vivo* (Bhirde et. al., 2011).

11.5.2   NANOINFORMATICS

Nanoinformatics would accelerate the introduction of nano-related research and applications into clinical medicine, leading to an era of "translational nanoinformatics" (Maojo et. al., 2010).

11.5.3   NANOSENSORS

High sensitivity nanosensors pushes detection limits of biomarkers to a very high level. The unique properties of nanomaterials are exploited to design biomarker diagnostics. Recognition with very high-sensitivity is achieved by signal and target amplification. The applications are in early detection of disease and improved patient care (Swierczewska et. al., 2012). There are many novel nanosensors with the potential to improve our understanding of cellular processes on the molecular level. For example, the hybrid sensor consisting of gold or silver NPs with an attached reporter species (Kneipp et. al., 2009).

11.5.4   THERANOSTICS

Theranostics is a new field of nanomedicine which combines diagnostic and therapeutic modalities (Luk et. al., 2012, Prabhu and Patravale, 2012).

An ideal theranostic agent should have following properties
(1) Selectively and rapidly accumulating in the tissue of need
(2) Reporting biochemical and morphological characteristics of the area
(3) Delivering an effective therapeutic agent
(4) Safety and degradation into nontoxic products (Jokerst and Gambhir, 2011, Nystrom and Wooley, 2011, Lammers et. al., 2010).

Multiple therapeutic agents such as chemotherapy, antiangiogenic, and gene therapy agents can be simultaneously delivered by nanocarriers to tumor sites to enhance the effectiveness of therapy and to take advantage of synergistic effects. Examples of nanoformulations that can be used as multifunctional platforms for cancer theranostics are liposomes, metallic and polymeric NPs, carbon nanotubes, dendrimers, and QDs (Fernandez-Fernandez et. al., 2011).

Other agents are polymerizable lipid molecules (like phospholipids) (Puri and Blumenthal, 2011), multifunctional nanocomposite NPs based on mesoporous silica nanostructures (MSNs) (Lee et. al., 2011), MNPs (Yoo et. al., 2011), and so on. Inorganic NPs< 20 nm (that is, sizes close to those of biomolecules) have unique physical and chemical properties which are not observed in bulk forms. (Murray et. al., 2000). Magnetic NPs have properties like greater size control, surface functionalization and specific binding capacities. In addition, the magnetic fields have deep tissue penetration and the ability to enhance MRI sensitivity and magnetic heating efficiency. These properties make magnetic NPs promising candidates for successful future theranostics. Common agents used are iron oxide NPs, gold-iron oxide NPs, metallic iron NPs, and Fe-based alloy NPs, such as iron-cobalt (FeCo), and iron-platinum NPs (Ho et. al., 2011).

11.6   NANOPHARMACEUTICALS

The greatest impact of nanotechnology in medicine has been the development of novel therapeutic agents. Nanopharmaceuticals can mean either drug produced using nanotechnology or nanocarriers. A variety of nanotechniques can be used to create drugs with better properties like increased bio-availability and stability. They may also be administered through easier routes (nasal, ophthalmic, and so on) and with better results. Nano-carriers enable precise targeting of drugs to specific sites like cancer tissues. Wang et al suggest that nanodrugs could be used in future to target cancer stem cells as well as in eliminating drug resistance (Wang et. al., 2013). Poly (lactic-co-glycolic acid) NPs (PLGA NP) are a new class of NPs with applications in the treatment of drug resistant breast cancer (Bharali et. al., 2013) as well as treat diseases of the central nervous system, since they can cross blood-brain barrier (Tosi et. al., 2013).

11.6.1   NANOCARRIERS

Examples of nanocarriers are inorganic NPs, lipid aggregates, liposomes, and synthetic polymeric systems, like micelles, or nanotubes, vesicles, and polymeric vesicles (Tanner et. al., 2011). The targeted delivery of therapeutic peptide by nanocarriers systems requires the knowledge of interactions of nanomaterials with the internal body environment, peptide release, and stability of peptides used as drugs. Therapeutic application of nanoencapsulated peptides are increasing exponentially and >1000 peptides in nanoencapsulated form are in different clinical/trial phase (Yadav et. al., 2011).

Drug delivery to the brain has been conventionally a difficult issue. The reason behind the failure of conventional delivery systems in reaching the brain is the blood brain barrier (BBB). However using nanocarrierrs, drugs can cross the BBB. These have a special role in treatment of autoimmune diseases and cancer (Jatariu et. al., 2009).

11.6.2   TRANSDERMAL DRUG DELIVERY

Drug delivery into the skin has inherent difficulties since as a natural barrier it has a very low permeation rate. Solid lipid nanoparticles (SLNs) have been used for drug delivery into the skin (Kristl et. al., 2010). Nanocarriers (microemulsions, liposomes and micro, and NPs) are promising dermal and transdermal drug delivery systems

(Schroeter et. al., 2010). The most important example is transdermal estrogen therapy (Valenzuela and Simon, 2012).

11.6.3   LIPID DRUG DELIVERY SYSTEMS

The SLNs, nanostructured lipid carriers (NLCs), and lipid drug conjugate (LDCs) NPs are novel lipid drug delivery systems (Attama, 2011). Liposomes and other vesicular systems have wide applications in drug delivery and the discovery of new nanomedicines (Elizondo et. al., 2011, Allen and Cullis, 2012, Peterson et. al., 2012). Therapeutic liposomes and other lipid excipient based therapeutic products can activate complements and contribute to a hypersensitivity syndrome known as C activation-related pseudoallergy (CARPA). Mostly mild and transient, in an occasional patient it can be severe or even lethal. The CARPA may be a safety issue in cardiac patients (Szebeni et. al., 2011). Liposomes and other vesicles as mucosal and transcutaneous adjuvants are attractive alternatives to parenteral vaccination (Romero and Morilla, 2011). Ethosomes are specially tailored vesicular carriers able to efficiently deliver various molecules with different physicochemical properties into deep skin layers and across the skin (Ainbinder et. al., 2010). Inorganic NPs, mainly zinc oxide (ZnO) and titanium dioxide (TiO_2) can be used for sunscreen applications. These particles have the properties of attenuation of the ultraviolet light (by absorption and scattering) and formation of free radicals (that is, photo toxicity), which are necessary for sunscreens design. However, particle penetration into skin can lead to possible adverse effects associated with interaction between NPs and skin living cells (Popov et. al., 2010).

11.6.4   CAVEOLAE

Caveolae are special types of lipid rafts, they are small invaginations of plasma membranes rich in proteins and lipids like cholesterol and sphingolipids. They function in signal transduction, endocytosis, oncogenesis and uptake of bacteria and viruses (Anderson, 1998; Frank and Lisanti, 2004; Li et. al., 2005; Pelkmans, 2005). Limited penetration across the vascular endothelium and uptake by the reticuloendothelial system (RES) substantially impede effectiveness of nanoparticle delivery into tissues. Use of active transendothelial transport pathways, like caveolae, is an effective solution to both target and deliver NPs (Chrastina et. al., 2011).

11.6.5   PROTEIN BASED DRUG DELIVERY SYSTEMS

Protein-based nanomedicine platforms (self-assembled protein subunits of the same protein or a combination of proteins) are ideal for drug-delivery platforms due to their biocompatibility, biodegradability, and low toxicity. Many proteins have been used and characterized for therapeutic-delivery systems, including the ferritin/apoferritin protein cage, plant-derived viral capsids, albumin, soy and whey protein, collagen, and gelatin (Maham et. al., 2009). Cyclodextrins are used in conjunction with NPs for oral delivery of proteins. The use of absorption enhancers like cyclodextrins, bile salts, and other surfactants facilitates bio-availability into the system (Kanwar et. al., 2011).

11.6.6   NUCLEIC ACID LIGANDS (APTAMERS)

Nucleic acid ligands (aptamers), are a class of macromolecules with many novel nanobiomedical applications. They have many special properties like high affinity and specificity for their target, a versatile selection process, ability to be easily synthesized chemically and small physical size. These properties make them attractive molecules for targeting diseases or as drugs (Levy-Nissenbaum et. al., 2008).

11.6.7   PULMONARY DELIVERY

The lung is an attractive target for drug delivery. The most important reason is that it is a noninvasive procedure (inhalation aerosols). Metabolically also there are advantages, there is no first-pass metabolism, direct delivery to the site of action, and large surface area for absorption. Colloidal carriers (that is, nanocarrier systems) in pulmonary drug delivery offer many advantages including uniform distribution of drug dose among the alveoli, achievement of improved solubility, a sustained drug release and reduced dosage and increased patient compliance, decreased toxicity, and the potential of drug internalization by cells (Mansour et. al., 2009, Dombu and Betbeder, 2012).

11.6.8   CELL MEDIATED DRUG DELIVERY

Immune cells can be exploited as Trojan horses for drug delivery. Immunocytes laden with drugs can cross the blood-brain or blood-tumor barriers to facilitate treatments for infectious diseases, cancer and other inflammatory diseases. Advantages include targeted drug transport, prolonged circulation times, and reduction in cell and tissue toxicities (Batrakova et. al., 2011).

11.6.9   MICROBIAL ORIGIN FOR DRUG DELIVERY

Many bacterial species contain intracellular nano- and micro-compartments consisting of self-assembling proteins that form protein-only shells. The self assembling properties of these particles can be used for directing shell assembling and enzyme packaging. Eventually they act as a platform for drug delivery (Corchero and Cedano, 2011). Virus-based nanotechnology takes advantage of the natural circulatory and targeting properties of viruses, in order to design drugs and vaccines that specifically target tissues of interest *in vivo*. Cowpea mosaic virus (CPMV) and flock house virus (FHV) nanoparticle-based strategies hold great promise for the design of targeted drugs and structure-based vaccine approaches (Destito et. al., 2009).

11.6.10   MULTIFUNCTIONAL NANODRUGS

Polyethylene glycol (PEG) modified carbon nanotubes (CNTs) are used in preclinical models in the treatment of cancer, neurological disorders, vaccination, and imaging and so are classified as multifunctional nanodrugs (Bottini et. al., 2011). Mononuclear phagocyte system is a major constraint to nano-carrier based drug delivery system. The PEG overcomes this to a great extent. Covalent addition of PEG, PEG to a protein produces PEGylated protein. PEGylation also correlates with reduced hemolysis, thrombogenicity, complement activation and protein adsorption, attributed to its uncharged and hydrophilic nature. Hydrophilicity and cell-repelling delivery systems are not always beneficial and hence different alternatives are needed (Cavadas et. al., 2011). Studies by Kim et al show that DNA NPs may be given PEG coating which enable their transfer across mucosal barriers and therefore better gene transfer (Kim et. al., 2013). Hybrid NPs consisting of PEG shell, an inorganic core and a lipid monolayer or bi-layer between the two, have the properties of NPs and liposomes and have a wide variety of properties like longer circulation time, higher drug loading efficiency, increased stability, bio-compatibility, sustained release properties, and so on (Tan et. al., 2013).

Polymer-protein conjugates like PEGylated enzymes and cytokines, polymeric drugs, and sequestrants are the first generation "nanomedicines". Later on, a wider range of fatal debilitating diseases (For example, viral infections, arthritis, multiple sclerosis and hormone disorders) have been targeted via intravenous (I.V.), subcutaneous (S.C.), or oral routes of administration. Many polymeric materials formed the second-generation polymer therapeutics (Gaspar and Duncan, 2009).

11.6.11   NANOCERIA

Cerium oxide NPs (nanoceria) have significant pharmacological properties including antioxidant properties. They react with superoxide and hydrogen peroxide. They might also fight chronic inflammation and oxidative stress induced diseases including cancer and neurodegeneration (Celardo et. al., 2011). For example, cerium oxide NPs prevent metastasis of melanoma cells *in vivo* (Alili et. al., 2013).

11.6.12   NANOSCALE METAL ORGANIC FRAMEWORKS (NMOF)

The NMOFs possess several potential advantages over conventional nanomedicines like structural and chemical diversity, high loading capacity, and intrinsic biodegradability. Under mild conditions NMOFs are crystalline or amorphous and particle size, composition and morphology can be easily tuned to optimize final properties (Della Rocca et. al., 2011).

11.6.13   ANTI-CANCER DRUGS

Gemcitabine is a nucleoside analog anti-cancer drug effective against a variety of solid tumors, however its use is limited by the fact that it is destroyed by deoxycytidine deaminase following in vivo administration. Polymeric NPs are used for gemcitabine drug delivery (Celia et. al., 2011). Polymeric micelles have been under extensive investigation during the past years as drug delivery systems, particularly for anticancer drugs (Talelli et. al., 2010). Most nanomedicines achieve selective tumor accumulation *via* the enhanced permeability and retention (EPR) effect or a combination of the EPR effect and active targeting to cellular receptors. The fundamental physicochemical properties of a nanomedicine (its size, charge, hydrophobicity, and so on) can significantly affect its distribution to cancer tissue, transport across blood vessel walls, and retention in tumors. Nanoparticle characteristics such as stability in the blood and cancerous tissue, ability to dissociate into covalently bound components, tumor cell uptake, and cytotoxicity contribute to efficacy once the nanoparticle has reached the tumor's interstitial space (Adiseshaiah et. al., 2010).

Anticancer drug resistance almost invariably emerges and poses major obstacles towards curative therapy of various human malignancies. In recent years, multiple NP-based therapeutic systems have been developed that were rationally designed to overcome drug resistance (Shapira et. al., 2011).

11.6.14   INFECTIOUS DISEASES

Nanomaterials provide added benefits due to their small size, allowing for an increased ability to surpass most physiologic barriers and reach their targets. The high surface area-to-volume ratio gives increased potential to interact with pathogen membranes and cell walls (Blecher et. al., 2011). The main drawbacks for conventional antimicrobial agents are the development of multiple drug resistance and adverse side effects. Drug resistance is another major problem. Several classes of antimicrobial NPs and nanosized carriers for antibiotics delivery have proven their effectiveness for treating infectious diseases and antibiotic resistant conditions, in vitro as well as in animal models (Huh and Kwon, 2011).

11.6.15   PHOTODYNAMIC THERAPY

Photodynamic therapy (PDT) is a technique, which has developed over the last century. It is used for the treatment of various diseases such as cancer and macular degeneration of the eye. In this technique, initially there is delivery of a photosensitizing drug, which is followed by light irradiation. Activated photosensitizers are formed, which transfer their energy to molecular oxygen. This results in the generation of reactive oxygen species which can cause apoptosis or necrosis of diseased cells. Nanosized carriers may also be developed for photosensitizers, which can improve the efficiency of photodynamic activity. Thus many side effects associated with classic photodynamic therapy can be overcome (Paszko et. al., 2011).

11.6.16   ANTIVIRAL DRUGS

Nanoparticulate-based systems might change the release kinetics of antiviral drugs, increase their bioavailability, improve efficacy, reduce drug side effects, and reduce treatment costs. In addition, they help the delivery of antiviral drugs to specific target sites and viral reservoirs in the body. In situations where high drug doses are needed, where drugs are expensive and situations, where patient compliance are important, these features become very significant (Lembo and Cavalli, 2010). Nanocarriers like polymeric, inorganic, and SLNPs, liposomes, dendrimers, cyclodextrins, and cell-based nanoformulations have been studied for delivery of drugs intended for HIV prevention or therapy. For anti-HIV drugs to be effective, proper biodistribution must be achieved, and adequate drug concentrations must be maintained at those sites for longer durations. Nanocarriers also provide a means to overcome cellular and anatomical barriers to drug delivery (Mallipeddi and Rohan, 2010).

11.6.17   GENE DELIVERY

Site specific vascular GD for therapeutic is theoretically possible for genetic disorders, however there is low transfection rates due to suboptimal delivery. A combination of ultrasonic microbubble technology with magnetic nanoparticle enhanced gene transfer could make it successful (Mannell et. al., 2012). High-intensity nanosecond pulses, known as nanosecond electroporation (nsEP), are used in electroporation GD systems (Sundararajan, 2009). Amphiphilic block copolymers of polyethylene oxide and polypropylene oxide are commercially available pharmaceutical excipients (known as Pluronic or poloxamer), especially in GD applications with adenovirus and lentivirus (Kabanov et. al., 2005).

11.6.18   NANOGELS

Nanogels are swollen nanosized networks composed of hydrophilic or amphiphilic polymer chains. They are used as carriers for chemotherapeutic agents. They can be designed to spontaneously form biologically active molecules. Polyelectrolyte nanogels can also be made, which can readily incorporate oppositely charged low-molecular-mass drugs and biomacromolecules such as oligo- and polynucleotides (siRNA, DNA) and proteins. Multiple chemical functionalities can be employed in the nanogels to introduce imaging labels and to allow targeted drug delivery. Recent studies suggest that nanogels have a very promising future in biomedical applications (Kabanov and Vinogradov, 2009).

11.6.19   MUCINS

Mucins help as lubricating agents and protects tissues from pathogens and enzymatic attack. MUC1, MUC4, and MUC16 (also known as cancer antigen 125) are the notable mucins. Currently, several approaches are being examined that target mucins for immunization or nanomedicine using mucin-specific antibodies (Constantinou et. al., 2011).

11.6.20   CELL ENCAPSULATION THERAPY

Cell encapsulation therapy (CET) provides an attractive means to transplant cells without the need for immunosuppression. The cells are immunoisolated, which allows selective permeation of nutrients and therapeutics while isolating the cells from hostile immune components (Randall et. al., 2011).

11.7   NEWER DIRECTIONS

Nanomedicine offers promise for the treatment and prevention of various diseases, including infectious diseases. They act as potent free radical scavengers and antioxidants and as anti-inflammatory agents (Elswaifi et. al., 2009). They may therefore be used in a variety of diseases, which are thought to be mediated by oxidative stress.

11.7.1 NANOGENOMICS

The emergence of "-omic" technologies like genomics, transcriptomics, proteomics, metabolomics as well as advances in systems biology are leading towards a personalized approach to medicine. This has been closely linked with advances in nanomedicine. However, this transition will depend heavily upon the evolutionary development of a systems biology approach to clinical medicine based upon "-omic" technology analysis and integration (Sakamoto et. al., 2010). The DNA and RNA can be used to construct artificial nanodevices with strong potential for future biomedical applications. The DNA nanodevices are used as biosensors, to detect and report various natural proteins mRNA or microRNAs. Applications of DNA nanodevices also extend to controlled release and drug delivery, nanocontainers for drugs, switchable hydrogels ,and so on (Simmel, 2007). The RNA can be designed to produce a variety of different nanostructures. The RNA has a flexible structure and possesses catalytic functions that are similar to proteins. Diversity with RNA technology provides a platform for identifying viable building blocks for various applications. Since they are heat-stable, RNA technology is also used for the production of multivalent nanostructures with defined stoichiometry (Guo, 2010, Davis et. al., 2009). The NPs are used as vectors for gene therapy. Liu and Zhang have described several effective strategies to increase the transfection efficiency of NPs (Liu and Zhang, 2011). The successful delivery of nucleic acids to particular target sites is the challenge that is being addressed using a variety of viral and nonviral delivery systems, each with distinct advantages and disadvantages. Nonviral vectors have the advantage of safety and flexibility over viral vectors, but they lack in efficiency. Dendrimers have the ability to interact with various forms of nucleic acids such as plasmid DNA, antisense oligos, and RNA to form complexes that protect the nucleic acid from degradation (Dutta et. al., 2010).

11.7.2   RNA INTERFERENCE AND siRNA

The RNA interference (RNAi) has been named as "breakthrough technology" in 2002 by Science magazine (Couzin, 2002). In a short time span this technology has conquered life sciences and numerous therapeutic approaches and is now well underway to become an important novel class of RNA based therapeutics (Keller, 2009). Harnessing RNA interference using small RNA-based drugs has great potential to develop drugs designed to knock down expression of any pathological gene, thereby greatly expanding the range of possible drug targets (Peer, 2011). The NPs made from synthetic polymers have been developed to deliver small interfering RNA (siRNA). For efficient siRNA delivery, these NPs need to encapsulate siRNA, actively target specific sites and release siRNA intracellularly (Gao et. al., 2010). The ability to specifically silence genes using RNA interference (RNAi) has wide therapeutic applications for the treatment of disease or the augmentation of tissue formation (Tokatlian et. al., 2010). The size of the carrier is important as carriers <100 nm in diameter have been reported to have higher accumulation levels in cancer tissues, liver cells and inflamed tissue, larger NPs are taken up by cell components of the reticulo-endothelial system (RES). Carriers can be surface modified with hydrophilic materials, such as PEG, which reduces this uptake and increases their circulation time (Schroeder et. al., 2010).

Small interfering RNA (siRNA) technology holds great promise as a therapeutic intervention for targeted gene silencing in cancer and other diseases. The major limitations against the use of siRNA as a therapeutic tool are its degradation by serum RNases and DNases, relatively low cellular uptake and rapid renal clearance following systemic administration. Nuclease-resistant chemically modified siRNAs and variety of synthetic and natural biodegradable lipids and polymers have been developed to systemically deliver siRNA with different efficacy and safety profiles. Cationic liposomes are very effective carriers for these siRNA (Ozpolat et. al., 2010).

11.7.3 BIOLOGICAL PORES AND PORE FORMING PEPTIDES

Biological protein pores and pore-forming peptides have diverse and essential functions that range from maintaining cell homeostasis and participating in cell signaling to activating or killing cells. Applications are single-molecule sensing to drug delivery and targeted killing of malignant cells, single strand DNA sequencing and biocompatible visual detection systems (Majd et. al., 2010).

11.7.4 PROTEIN ENGINEERING

Protein engineering is the technique of synthesizing novel proteins, by either rational design or directed evolution. In rational design, knowledge of protein structure and function are used to make necessary changes. In directed evolution, proteins undergo random mutagenesis and better variants are selected. More detailed knowledge of protein structure and function, and development of high-throughput techniques have led to the synthesis of many new proteins. The relation between protein engineering and nanotechnology is a synergistic one. Nanotechnology provides optimal vectors for various biomolecules, while some proteins can be used as active ligands to help NPs loaded with chemotherapeutic or other drugs to reach particular sites in the body. This combination of proteins and NPs is mainly achieved by absorption, bioconjugation, and encapsulation. A clear understanding of nanoparticle-protein interactions could make possible the design of precise and versatile hybrid nanosystems. This allows control of pharmacokinetics, activity, and their safety (Di Marco et. al., 2010). Cell-targeting peptides (CTPs) and cell-penetrating peptides (CPPs) assist in the delivery of NPs, liposomes, and other nanocarriers (Juliano et. al., 2009).

11.7.5 GENE THERAPY AND NANOTECHNOLOGY

Nanomedicine based on the use of adenovirus vectors for therapeutic GD shows broad potential, as in metastatic cancers or cardiovascular diseases. Adenoviral vector targeting upon contact with the bloodstream interacts with a variety of host proteins which minimizes the efficacy of the virus and induces toxicity (Duffy et. al., 2012). We are now getting more understanding about the interaction of adenovirus with the body immune system; this knowledge would be helpful in developing better gene therapy tools in future.

11.7.6   APTAMER BIOLOGY

Aptamers are agents that bind proteins with affinities and specificities that have therapeutic utility. Aptamers are susceptible to nuclease-mediated degradation and cannot easily cross biological barriers, but amenable to modifications, for example, they could be modified with virus-like particles (Wu et. al., 2011). Virus-like particles (VLP) resemble viruses, however, they are non-infectious and do not contain any viral genetic material (Santi et. al., 2006). They are used for development of vaccines (Akahata et. al., 2010) and in gene therapy (Petry et. al., 2003).

11.8   TOXICITY OF NANOPARTICLES

Nanomaterials are very useful agents for the treatment of a variety of diseases, and in diagnostic applications. However, it is important that their toxicity profile be monitored carefully before their introduction as clinical drugs. Numerous in vitro and in vivo studies have shown that carbon nanotubes and/or associated contaminants may induce oxidative stress and prominent pulmonary inflammation. Since carbon nanotubes can be readily functionalized they have a place in the treatment and monitoring of cancer, infectious diseases, and other diseases (Shvedova et. al., 2009, Bussy et. al., 2012). Nanomaterials can enter the body through the lungs or other organs via oral route and as medicines and affect different organs and tissues causing cytotoxic effects (Singh and Nalwa, 2007). Intravenous and subcutaneous injections of nanoparticulate carriers deliver exogenous NPs directly into the human body without passing through the normal absorption process. The carriers are also responsible for cytotoxicity and interaction with biological macromolecules. Insoluble nanoparticulate carriers may accumulate in human tissues or organs. Considering all these aspects, toxicological studies for biosafety evaluation of these nanomaterials are extremely important (Zhao and Castranova, 2011).

Nano-sized materials and nano-scaled processes have wide industrial applications and are increasingly found in numerous consumer products. The small size of NPs may affect normal physiological functions of cells and cause cytotoxicity. There may also be unknown risks to human health and the environment (Shvedova and Kagan, 2010). The NPs interact with proteins, a variety of membranes, cellular organelles, DNA, and so on, and establish a series of nanoparticle/biological interfaces (Nel et. al., 2009). The effects of these interactions need to be studied carefully.

11.9   REGULATION OF NANOMEDICINE

As the science and technology of nanomedicine are progressing, many ethical and legal issues are emerging. We need to proactively address these issues in order to minimize its adverse impacts on the environment and public health and also to avoid a public backlash. A variety of different regulatory bodies are present in US and other Western countries. The general opinion among scientists and lawmakers is that none of the issues surrounding nanomedicine and nanomaterials is unique and general regulations, which are applicable to any therapeutic agent would apply to nanomedicines as well.

More *in vivo* animal experiments and ex vivo laboratory analyses are needed before introducing the products into clinical trials. Questions of social justice, public access to healthcare and the use of nanotechnology for physical enhancement are also becoming increasingly important (Resnik and Tinkle, 2007, Lanone and Boczkowski, 2006, Chan, 2006, Bawa, 2011, Kuiken, 2010).

11.10   INDIAN SCENARIO

In India, Nanoscience has rapidly progressed in the last 5 years or so. The Department of Science and Technology (DST) launched "Nano Mission" in 2007 to "foster, promote and develop all aspects of nanoscience and nanotechnology, which have the potential to benefit the country". The objectives of Nano Mission include "promoting basic research, infrastructure development, public private partnership and nano applications and technology development centers, human resource development and international collaborations". Further details can be found in the website, http,// www. dst.gov.in/about_us/ar07-08/nano-mission.htm. Department of Biotechnology (DBT), Indian Council of Medical Research (ICMR), and Department of Scientific and Industrial Research (DSIR) have also supported research on nanotechnology in India in a big way.

A large number of universities, companies, and researchers are also working on various areas of nanomedicine (nanodiagnosis, nano-cardiology, nano-neurology, nano-oncology, nano-orthopedics, nano-ophthalmology, nano-surgery, and infectious diseases), development of novel nanodrugs, nanocarriers, tissue engineering, and so on. The country has witnessed tremendous amount of excitement in the field, and the developments as well as liberal government funding have prompted many scientists, doctors, and other professionals to switch their career to nanotechnology and nanomedicine. Much has been achieved, but much more remains to be done.

11.11   CONCLUSION

Undoubtedly nanomedicine is the future of diagnostics and therapeutics. There are a large number of nanomaterials, which have been developed over the years; many of them have special properties which can be exploited as diagnostic or therapeutic agents. The increasing use of these compounds raises the doubt of toxicity. We need to evaluate nanomaterials for their toxicity before introducing them into human use.

KEYWORDS

- **Inorganic nanomaterials**
- **Mesoporous silica**
- **Nanomaterials**
- **Nanoscience**
- **Viral nanoparticles**

REFERENCES

1. Adiseshaiah, P. P., Hall, J. B., and McNeil, S. E. Nanomaterial standards for efficacy and toxicity assessment. *Wiley Interdiscip Rev Nanomed Nanobiotechnol.*, **2**(1), 99–112 (Jan-Feb, 2010).
2. Ainbinder, D., Paolino, D., Fresta, M., and Touitou, E. Drug delivery applications with ethosomes. *J Biomed Nanotechnol.*, **6**(5), 558–568 (Oct, 2010).
3. Akahata, W., Yang, Z. Y., Andersen, H., Sun, S., Holdaway, H. A., Kong, W. P., Lewis, M. G., Higgs, S., Rossmann, M. G., Rao, S., and Nabel, G. J. A virus-like particle vaccine for epidemic Chikungunya virus protects nonhuman primates against infection. *Nat Med.*, **16**(3), 334–338 (Mar, 2010) doi, 10.1038/nm.2105. Epub 2010 Jan 28.
4. Alili, L., Sack, M., von, Montfort, C., Giri, S., Das, S., Carroll, K. S., Zanger, K., Seal, S., and Brenneisen, P. Downregulation of Tumor Growth and Invasion by Redox-Active Nanoparticles. *Antioxid Redox Signal*, **24** (Jan, 2013) [Epub ahead of print].
5. Alsteens, D., Dupres, V., Andre, G., and Dufrene, Y. F. Frontiers in microbial nanoscopy. *Nanomedicine (Lond).*, **6**(2), 395–403 (Feb, 2011).
6. Alwarappan, S., Cissell, K., Dixit, S., Mohapatra, S., and Li, C. Z. Chitosan-Modified Graphene Electrodes for DNA Mutation Analysis. *J Electroanal Chem (Lausanne Switz).*, **686**, 69–72 (Oct 15, 2012).
7. Anderson, R. G. The caveolae membrane system. *Annu. Rev. Biochem.*, **67**, 199–225 (1998) doi 10.1146/annurev.biochem.67.1.199.
8. Ang, L. Y., Lim, M. E., Ong, L. C., and Zhang, Y. Applications of upconversion nanoparticles in imaging, detection and therapy. *Nanomedicine (Lond).*, **6**(7), 1273–1288 (Sep, 2011).
9. Ashammakhi, N., Wimpenny, I., Nikkola, L., and Yang, Y. Electrospinning, methods and development of biodegradable nanofibres for drug release. *J Biomed Nanotechnol.*, **5**(1), 1–19 (Feb, 2009).
10. Atiyeh, B. S., S. N. Hayek, and Gunn, S. W. New technologies for burn wound closure and healing—Review of the literature. *Burns,* **31**, 944–956 (2005).
11. Attama, A. A. SLN, NLC, LDC, state of the art in drug and active delivery. *Recent Pat Drug Deliv Formul.*, **5**(3), 178–187 (Sep, 2011).
12. Ayres, C. E., Jha, B. S., Sell, S. A., Bowlin, G. L., and Simpson, D. G. Nanotechnology in the design of soft tissue scaffolds, innovations in structure and function. *Wiley Interdiscip Rev Nanomed Nanobiotechnol.*, **2**(1), 20–34 (Jan–Feb, 2010).
13. Babincova, M. and Babinec, P. Magnetic drug delivery and targeting, principles and applications. *Biomed Pap Med Fac Univ Palacky Olomouc Czech Repub.*, **153**(4), 243–250 (Dec, 2009).
14. Balogh, L. P. The future of nanomedicine and the future of Nanomedicine, NBM (Editorial). Nanomedicine, Nanotechnol. *Biol. Med.*, **5**, 1 (2009).
15. Banerjee, R., Katsenovich, Y., Lagos, L., McIntosh, M., Zhang, X., and Li, C. Z. Nanomedicine, MNPs and their biomedical applications. *Curr Med Chem.*, **17**(27), 3120–3141 (2010).
16. Batrakova, E. V., Gendelman, H. E., and Kabanov, A. V. Cell-mediated drug delivery. *Expert Opin Drug Deliv.*, **8**(4), 415–433(Apr, 2011) Epub 2011 Feb 24.
17. Bawa, R. Regulating nanomedicine - can the FDA handle it? *Curr Drug Deliv.*, **8**(3), 227–234 (May, 2011).
18. Beg, S., Rizwan, M., Sheikh, A. M., Hasnain, M. S., Anwer, K., and Kohli, K. Advancement in carbon nanotubes, basics, biomedical applications and toxicity. *J Pharm*

Pharmacol., **63**(2), 141–163 (Feb, 2011). doi, 10.1111/j.2042–7158.2010.01167.x. Epub 2010 Nov 16.

19. Bharali, D. J., Yalcin, M., Davis, P. J., and Mousa, S. A. Tetraiodothyroacetic acid-conjugated PLGA nanoparticles, ananomedicine approach to treat drug-resistant breast cancer. *Nanomedicine (Lond).*, **28** (Feb, 2013) [Epub ahead of print].

20. Bhirde, A., Xie, J., Swierczewska, M., and Chen, X. Nanoparticles for cell labeling. *Nanoscale*, **3**(1), 142–153 (Jan, 2011) Epub 2010 Oct 11.

21. Blecher, K., Nasir, A., and Friedman, A. The growing role of nanotechnology in combating infectious disease. *Virulence*, **2**(5), 395–401 (Sep–Oct, 2011) Epub 2011 Sep 1.

22. Bottini, M., Rosato, N., and Bottini, N. PEG-modified carbon nanotubes in biomedicine, current status and challenges ahead. *Biomacromolecules*, **12**(10), 3381–3393 (Oct 10, 2011) Epub 2011 Sep 21.

23. Bottini, M., Rosato, N., and Bottini, N. PEG-modified carbon nanotubes in biomedicine, current status and challenges ahead. *Biomacromolecules*, **12**(10), 3381–3393 (Oct 10, 2011) Epub 2011 Sep 21.

24. Brodusch, N., Trudeau, M., Michaud, P., Rodrigue, L., Boselli, J., and Gauvin, R. Contribution of a new generation field-emission scanning electron microscope in the understanding of a 2099 Al-Li alloy. *Microsc Microanal.*, **18**(6), 1393–1409 (Dec, 2012) doi, 10.1017/S143192761200150X. Epub 2012 Oct 29.

25. Bussy, C., Ali-Boucetta, H., and Kostarelos, K. Safety Considerations for Graphene, Lessons Learnt from Carbon Nanotubes. *Acc Chem Res.*, (Nov 20, 2012) [Epub ahead of print].

26. Calderon, M., Quadir, M. A., Sharma, S. K., and Haag, R. Dendritic polyglycerols for biomedical applications. *Adv Mater.*, **22**(2), 190–218 (Jan 12, 2010).

27. Cavadas, M. Gonzalez-Fernandez, A, and Franco, R. Pathogen-mimetic stealth nanocarriers for drug delivery, a future possibility. *Nanomedicine*, **7**(6), 730–743 (Dec, 2011) Epub 2011 May 19.

28. Celardo, I., Pedersen, J. Z., Traversa, E., and Ghibelli, L. Pharmacological potential of cerium oxide nanoparticles. *Nanoscale*, **3**(4), 1411–1420 (Apr, 2011) Epub 2011 Mar 2.

29. Celia, C., Cosco, D., Paolino, D., and Fresta, M. Gemcitabine-loaded innovative nanocarriers vs GEMZAR, biodistribution, pharmacokinetic features and in vivo antitumor activity. *Expert Opin Drug Deliv.*, **8**(12), 1609–1629 (Dec, 2011) Epub 2011 Nov 5.

30. Chanana, M., Mao, Z., and Wang, D. Using polymers to make up magnetic nanoparticles for biomedicine. *J Biomed Nanotechnol.*, **5**(6), 652–668 (Dec, 2009).

31. Chang, C. The immune effects of naturally occurring and synthetic nanoparticles. *J Autoimmun*, **34**(3), J234–246 (May, 2010) Epub 2009 Dec 7.

32. Chan, V. S. Nanomedicine, An unresolved regulatory issue. *Regul Toxicol Pharmacol.*, **46**(3), 218–224 (Dec, 2006) Epub 2006 Nov 1.

33. Chen, X. and Schluesener, H. J. Nanosilver, a nanoproduct in medical application. *Toxicol Lett.*, **176**(1), 1–12 (Jan 4, 2008) Epub 2007 Oct 16.

34. Chithrani, D. B. Intracellular uptake, transport, and processing of gold nanostructures. *Mol Membr Biol.*, **27**(7), 299–311(Oct, 2010) Epub 2010 Oct 7.

35. Chrastina, A., Massey, K. A., and Schnitzer, J. E. Overcoming in vivo barriers to targeted nanodelivery. *Wiley Interdiscip Rev Nanomed Nanobiotechnol.*, **3**(4), 421–437 (Jul–Aug, 2011) doi, 10.1002/wnan.143. Epub 2011 Apr 27.

36. Constantinou, P. E., Danysh, B. P., Dharmaraj, N., and Carson, D. D. Transmembrane mucins as novel therapeutic targets. *Expert Rev Endocrinol Metab.*, **6**(6), 835–848 (Nov, 2011).

37. Corchero, J. L. and Cedano, J. Self-assembling, protein-based intracellular bacterial organelles, emerging vehicles for encapsulating, targeting and delivering therapeutical cargoes. *Microb Cell Fact.*, **10**, 92 (Nov 3, 2011).

38. Couzin, J. Breakthrough of the year. Small RNAs make big splash. *Science*, **298**(5602), 2296–2297 (Dec 20, 2002).

39. Cunha, F. A., Sousa, R. A., Harding, D. P., Cadore, S., Almeida, L. F., and Araújo, M. C. Automatic microemulsion preparation for metals determination in fuel samples using a flow-batch analyzer and graphite furnace atomic absorption spectrometry. *Anal Chim Acta.*, **727**, 34–40 (May 21, 2012) doi, 10.1016/j.aca.2012.03.014. Epub 2012 Mar 29.

40. Davis, D., Akhtar, U., Keaster, B., Grozinger, K., Washington, L., Kelsey, S., Sarkar, A., and DeLong, R. K. Challenges and potential for RNA nanoparticles (RNPs). *J Biomed Nanotechnol.*, **5**(1), 36–44 (Feb, 2009).

41. Della, Rocca J., Liu, D., and Lin, W. Nanoscale metal-organic frameworks for biomedical imaging and drug delivery. *Acc Chem Res.*, **44**(10), 957–968 (Oct 18, 2011) Epub 2011 Jun 7.

42. Destito, G., Schneemann, A., and Manchester, M. Biomedical nanotechnology using virus-based nanoparticles. *Curr Top Microbiol Immunol.*, **327**, 95–122 (2009).

43. Di, Marco M., Shamsuddin, S., Razak, K. A., Aziz, A. A., Devaux, C., Borghi, E., Levy, L., and Sadun, C. Overview of the main methods used to combine proteins with nanosystems, absorption, bioconjugation, and encapsulation. *Int J Nanomedicine.*, **5**, 37–49 (Feb 2, 2010).

44. Dombu, C. Y. and Betbeder, D. Airway delivery of peptides and proteins using nanoparticles. *Biomaterials*, **34**(2), 516–525 (Jan, 2013) doi, 10.1016/j.biomaterials.2012.08.070. Epub 2012 Oct 6.

45. Dominiquez-Vera, J. M., Fernandez, B., and Galvez, N., Native and synthetic ferritins for nanobiomedical applications, recent advances and new perspectives. *Future Med Chem.*, **2**(4), 609–618 (Apr 2010).

46. Dong, C. M. Glyconanoparticles for biomedical applications. *Comb Chem High Throughput Screen.*, **14**(3), 173–181 (Mar 1, 2011).

47. Duarte, I. F. Following dynamic biological processes through NMR-based metabonomics: a new tool in nanomedicine. *J Control Release*, **153**(1), 34–39 (Jul 15, 2011) Epub 2011 Mar 22.

48. Duffy, M. R., Parker, A. L., Bradshaw, A. C., and Baker, A. H. Manipulation of adenovirus interactions with host factors for gene therapy applications. *Nanomedicine (Lond).*, **7**(2), 271–288 (Feb, 2012).

49. Dutta, T., Jain, N. K., McMillan, N. A., and Parekh, H. S. Dendrimer nanocarriers as versatile vectors in gene delivery. *Nanomedicine*, **6**(1), 25–34 (Feb, 2010) Epub 2009 May 18.

50. Dutta, T., Burgess, M., McMillan, N. A. J., and Parekh, H. S. Dendrosome-based delivery of siRNA against E6 and E7 oncogenes in cervical cancer, Nanomedicine, Nanotechnol. *Biol. Med.*, DOI, 10.1016/j.nano.2009.12.001. 2009.

51. Elizondo, E., Moreno, E., Cabrera, I., CÃ³rdoba, A., Sala, S., Veciana, J., and Ventosa, N. Liposomes and other vesicular systems, structural characteristics, methods of preparation, and use in nanomedicine. *Prog Mol Biol Transl Sci.*, **104**, 1–52 (2011).

52. El-Kirat-Chatel, S. and Dufrêne, Y. F. Nanoscale imaging of the Candida-macrophage interaction using correlated fluorescence-atomic force microscopy. *ACS Nano*, **6**(12), 10792–10799 (Dec 21, 2012) doi, 10.1021/nn304116f. Epub 2012 Nov 12.

53. Elswaifi, S. F., Palmieri, J. R., Hockey, K. S., and Rzigalinski, B. A. Antioxidant nanoparticles for control of infectious disease. *Infect Disord Drug Targets*, **9**(4), 445–452 (Aug, 2009).

54. Fahmy, T. M., Fong, P. M., Goyal, A., and Saltzman, W. M. Targeted for drug delivery. *Nano Today*, p 18–26 (August 18, 2005).

55. Fernandez-Fernandez, A., Manchanda, R., McGoron, A. J. Theranostic applications of nanomaterials in cancer, drug delivery, image-guided therapy, and multifunctional platforms. *Appl Biochem Biotechnol.*, **165**(7–8), 1628–1651 (Dec, 2011) Epub 2011 Sep 27.

56. Figuerola, A., Di, Corato, R., Manna, L., and Pellegrino, T. From iron oxide nanoparticles towards advanced iron-based inorganic materials designed for biomedical applications. *Pharmacol Res.*, **62**(2), 126–143 (Aug, 2010) Epub 2010 Jan 4.

57. Frank, P. and Lisanti, M. Caveolin-1 and caveolae in atherosclerosis, differential roles in fatty streak formation and neointimal hyperplasia. *Curr Opin Lipidol.*, **15**(5), 523–529 (2004) doi,10.1097/00041433–200410000–00005.

58. Friedman, A. D., Claypool, S. E., and Liu, R. The Smart Targeting of Nanoparticles. *Curr Pharm Des.*, (Mar 4, 2013) [Epub ahead of print].

59. Frimpong, R. A. and Hilt, J. Z. Magnetic nanoparticles in biomedicine, synthesis, functionalization and applications. *Nanomedicine (Lond).*, **5**(9), 1401–1414 (Nov, 2010).

60. Galanzha, E. I., and Zharov, V. P. *In vivo* photoacoustic and photothermal cytometry for monitoring multiple blood rheology parameters. *Cytometry A*, **79**(10), 746–757 (Oct, 2011) doi, 10.1002/cyto.a.21133. Epub 2011 Aug 30.

61. Gao, W., Chan, J. M., and Farokhzad, O. C. pH-Responsive nanoparticles for drug delivery. *Mol Pharm.*, **7**(6), 1913–1920 (Dec 6, 2010) Epub 2010 Oct 27.

62. Gao, W., Xiao, Z., Radovic-Moreno, A., Shi, J., Langer, R., and Farokhzad, O. C. Progress in siRNA delivery using multifunctional nanoparticles. *Methods Mol Biol.*, **629**, 53–67 (2010).

63. Garcia, I., Marradi, M., and Penades, S. Glyconanoparticles, multifunctional nanomaterials for biomedical applications. *Nanomedicine (Lond).*, **5**(5), 777–792 (Jul, 2010).

64. Gaspar, R. and Duncan, R. Polymeric carriers, preclinical safety and the regulatory implications for design and development of polymer therapeutics. *Adv Drug Deliv Rev.*, **61**(13), 1220–1231 (Nov 12, 2009) Epub 2009 Aug 12.

65. Ghanbari, H., Cousins, B. G., and Seifalian, A. M. A nanocage for nanomedicine, polyhedral oligomeric silsesquioxane (POSS). *Macromol Rapid Commun.*, **32**(14), 1032–1046 (Jul 15, 2011) doi, 10.1002/marc.201100126. Epub 2011 May 19.

66. Ghosh, D., Choudhury, S. T., Ghosh, S., Mandal, A. K., Sarkar, S, Ghosh, A., Saha, K. D., and Das, N. Nanocapsulated curcumin, oral chemopreventive formulation against diethylnitrosamine induced hepatocellular carcinoma in rat. *Chem Biol Interact.*, **195**(3), 206–214 (Feb 5, 2012) doi, 10.1016/j.cbi.2011.12.004. Epub 2011 Dec 16.

67. Godin, B., Tasciotti, E., Liu, X., Serda, R. E., and Ferrari, M. Multistage nanovectors, from concept to novel imaging contrast agents and therapeutics. *Acc Chem Res.*, **44**(10), 979–989 (Oct 18, 2011) Epub 2011 Sep 8.

68. Golman, K., Ardenkjaer-Larsen, J. H., Petersson, J. S., Mansson, S., Leunbach, and I. *Molecular imaging with endogenous substances. Proc Natl Acad Sci.*, U S A. **100**(18), 10435–10439 (2003).

69. Gonçalves, M. C., Mertins, O., Pohlmann, A. R., Silveira, N. P., and Guterres, S. S. Chitosan coated liposomes as an innovative nanocarrier for drugs. *J Biomed Nanotechnol.*, **8**(2), 240–250 (Apr, 2012).

70. Gorityala, B. K., Ma, J., Wang, X., Chen, P., and Liu, X. W. Carbohydrate functionalized carbon nanotubes and their applications. *Chem Soc Rev.*, **39**(8), 2925–2934 (Aug, 2010) Epub 2010 Jun 28.

71. Greenhalgh, K. and Turos, E. *In vivo* studies of polyacrylate nanoparticle emulsions for topical and systemic applications. *Nanomed. Nanotech. Bio. Med.*, **5**, 46–54 (2006).

72. Guo, P. The emerging field of RNA nanotechnology. *Nat Nanotechnol*, **5**(12), 833–842 (Dec, 2010) Epub 2010 Nov 21.

73. Guo, J. S., Leung, K. K. G., Su, H. X., Yuan, Q., Wang, L., Chu, T. H., Zhang, W. M., Pu, J. K. S., Ng, G. K. P., Wong, W. M., Dai, X., and Wu, W. L. Self-assembling peptide nanofi ber scaffold promotes the reconstruction of acutely injured brain. Nanomedicine, Nanotechnol. *Biol. Med.*, **5**, 345–351 (2009).

74. Ho, D., Sun, X., and Sun, S. Monodisperse magnetic nanoparticles for theranostic applications. *Acc Chem Res.*, **44**(10), 875–882 (Oct 18, 2011) Epub 2011 Jun 10.

75. http,// www.dst.gov.in/about_us/ar07–08/nano-mission.htm.

76. Huang, G., Chen, H., Dong, Y., Luo, X., Yu, H., Moore, Z., Bey, E. A., Boothman, D. A., and Gao, J. *Superparamagnetic Iron Oxide Nanoparticles, Amplifying ROS Stress to Improve Anticancer Drug Efficacy*, **3**(2), 116–126 (2013) doi, 10.7150/thno.5411. Epub 2013 Feb 1.

77. Huh, A. J. and Kwon, Y. J. Nanoantibiotics, a new paradigm for treating infectious diseases using nanomaterials in the antibiotics resistant era. *J Control Release*, **156**(2), 128–145 (Dec 10, 2011) Epub 2011 Jul 6.

78. Hunter, A. C. Application of the quartz crystal microbalance to nanomedicine. *J Biomed Nanotechnol.*, **5**(6), 669–675 (Dec, 2009).

79. Iaminskii, I. V., Gorelkin, P. V., and Dubrovin, E. V. Nanoanalytics for medicine. *Biofizika*, **56**(5), 955–960 (Sep-Oct, 2011).

80. Iannuccelli, V., Coppi, G., Romagnoli, M., Sergi, S., and Leo, E. *In vivo* detection of lipid-based nano and microparticles in the outermost human stratum corneum by EDX analysis. *Int J Pharm.*, (Mar 11, 2013) pii, S0378–5173(13)00211–1. doi, 10.1016/j.ijpharm.2013.03.002. [Epub ahead of print].

81. Jatariu, A., Peptu, C., Popa, M., and Indrei, A. Micro and nanoparticles medical applications. *Rev Med Chir Soc Med Nat Iasi.*, **113**(4), 1160–1169 (Oct–Dec, 2009).

82. Jiang, T., Olson, E. S., Nguyen, Q. T., Roy, M., Jennings, P. A., and Tsien, R. Y. Tumor imaging by means of proteolytic activation of cell-penetrating peptides. *Proc. Natl. Acad. Sci.*, USA. **101**(51), 17867–17872 (2004).

83. Jin, S. E., Bae, J. W., and Hong, S., Multiscale observation of biological interactions of nanocarriers, from nano to macro. *Microsc Res Tech.*, **73**(9), 813–823 (Sep, 2010).

84. Jokerst, J. V. and Gambhir, S. S. Molecular imaging with theranostic nanoparticles. *Acc Chem Res.*, **44**(10), 1050–1060 (Oct 18, 2011) Epub 2011 Sep 15.

85. Juffmann, T., Milic, A., Müllneritsch, M., Asenbaum, P., Tsukernik, A., Tüxen, J., Mayor, M., Cheshnovsky, O., and Arndt, M. Real-time single-molecule imaging of quantum interference. *Nat Nanotechnol.*, **7**(5), 297–300 (Mar 25, 2012) doi, 10.1038/nnano.2012.34.

86. Juliano, R. L., Alam, R., Dixit, V., and Kang, H. M. Cell-targeting and cell-penetrating peptides for delivery of therapeutic and imaging agents. *Wiley Interdiscip Rev Nanomed Nanobiotechnol.*, **1**(3), 324–335 (May–Jun, 2009).

87. Kabanov, A., Zhu, J., and Alakhov, V. Pluronic block copolymers for gene delivery. *Adv Genet.*, **53**, 231–261 (2005).

88. Kabanov, A. V., and Vinogradov, S. V. Nanogels as pharmaceutical carriers, finite networks of infinite capabilities. *Angew Chem Int Ed Engl.*, **48**(30), 5418–5429 (2009).

89. Kang, H. G., Tokumasu, F., Clarke, M., Zhou, Z., Tang, J., Nguyen, T., and Hwang, J. Probing dynamic fluorescence properties of single and clustered quantum dots toward quantitative biomedical imaging of cells. *Wiley Interdiscip Rev Nanomed Nanobiotechnol.*, **2**(1), 48–58 (Jan-Feb, 2010).

90. Kanwar, J. R., Long, B. M., and Kanwar, R. K. The use of cyclodextrins nanoparticles for oral delivery. *Curr Med Chem.*, **18**(14), 2079–2085 (2011).

91. Kawasaki, E. S. and Player, A. Nanotechnology, nanomedicine, and the development of new, effective therapies for cancer, Nanomedicine, Nanotechnol. *Biol. Med.*, **1**, 101–109 (2005).

92. Kayat, J., Gajbhiye, V., Tekade, R. K., and Jain, N. K. Pulmonary toxicity of carbon nanotubes, a systematic report. *Nanomedicine*, **7**(1), 40–49 (Feb, 2011) Epub 2010 Jul 8.

93. Keller, M. Nanomedicinal delivery approaches for therapeutic siRNA. *Int J Pharm.*, **379**(2), 210–211 (Sep 11, 2009) Epub 2009 Apr 5.

94. Khan, M. K., Minc, L. D., Nigavekar, S. S., Kariapper, M. S. T., Nair, B. M., Schipper, M., and Cook, A. C.

95. Klippstein, R. and Pozo, D. Nanotechnology-based manipulation of dendritic cells for enhanced immunotherapy strategies. *Nanomedicine*, **6**(4), 523–529 (Aug, 2010) Epub 2010 Jan 18.

96. Kim, A. J., Boylan, N. J., Suk, J. S., Hwangbo, M., Yu, T., Schuster, B. S., Cebotaru, L., Lesniak, W. G., Oh, J. S., Adstamongkonkul, P., Choi, A. Y., Kannan, R. M., and Hanes, J. Use of Single-Site-Functionalized PEG Dendrons to Prepare Gene Vectors that Penetrate Human Mucus Barriers. *Angew Chem Int Ed Engl.*, (Mar 4, 2013) doi, 10.1002/anie.201208556. [Epub ahead of print].

97. Kneipp, J., Kneipp, H., Wittig, B., and Kneipp, K. Novel optical nanosensors for probing and imaging live cells. *Nanomedicine*, **6**(2), 214–226 (Apr, 2010) Epub 2009 Aug 20.

98. Kristl, J., Teskac, K., and Grabnar, P. A. Current view on nanosized solid lipid carriers for drug delivery to the skin. *J Biomed Nanotechnol.*, **6**(5), 529–542 (Oct, 2010).

99. Kuiken, T. Nanomedicine and ethics is there anything new or unique, *Wiley Interdiscip Rev Nanomed Nanobiotechnol.*, (Apr 20, 2010) [Epub ahead of print].

100. Kumart, S. A. and Khan, M. I. Heterofunctional nanomaterials, fabrication, properties and applications in nanobiotechnology. *J Nanosci Nanotechnol.*, **10**(7), 4124–4134 (Jul, 2010).

101. Kurhanewicz, J., Bok, R., Nelson, S. J., and Vigneron, D. B. *Current and potential applications of clinical 13C MR spectroscopy. J Nucl Med.*, **49**(3), 341–344 (2008).

102. Ladewig, K., Xu, Z. P., and Lu, G. Q. Layered double hydroxide nanoparticles in gene and drug delivery. *Expert Opin Drug Deliv.*, **6**(9), 907–922 (Sep, 2009).

103. Lal, R. and Arnsdorf, M. F. Multidimensional atomic force microscopy for drug discovery, a versatile tool for defining targets, designing therapeutics and monitoring their efficacy. *Life Sci.*, **86**(15–16), 545–562 (Apr 10, 2010) Epub 2009 Apr 26.)

104. Lammers, T., Kiessling, F., Hennink, W. E., and Storm, G. Nanotheranostics and image guided drug delivery, current concepts and future directions. *Mol Pharm.*, **7**(6), 1899–1912 (Dec 6, 2010) Epub 2010 Oct 6.

105. Lanone, S. and Boczkowski, J. Biomedical applications and potential health risks of nanomaterials, molecular mechanisms. *Curr Mol Med.*, **6**(6), 651–663 (Sep, 2006).

106. Lee, J. E., Lee, N., Kim, T., Kim, J., and Hyeon, T. Multifunctional mesoporous silica nanocomposite nanoparticles for theranostic applications. *Acc Chem Res.*, **44**(10), 893–902 (Oct 18, 2011) Epub 2011 Aug 17.

107. Lee, S. E. and Lee, L. P. Biomolecular plasmonics for quantitative biology and nanomedicine. *Curr Opin Biotechnol.*, **21**(4), 489–497 (Aug, 2010).

108. Lehner, R., Wang, X., Marsch, S., and Hunziker, P. Intelligent Nanomaterials for Medicine, Carrier platforms and targeting strategies in the context of clinical application. *Nanomedicine*, (Feb 19, 2013) pii, S1549–9634(13)00033-6. doi, 10.1016/j.nano.2013.01.012. [Epub ahead of print].

109. Lembo, D. and Cavalli, R. Nanoparticulate delivery systems for antiviral drugs. *Antivir Chem Chemother*, **21**(2), 53–70 (2010).

110. Lengyel, J. S., Milne, J. L., and Subramaniam, S. Electron tomography in nanoparticle imaging and analysis. *Nanomedicine (Lond).*, **3**(1), 125–131 (Feb, 2008).
111. Levy-Nissenbaum, E., Radovic-Moreno, A. F., Wang, A. Z., Langer, R., and Farokhzad, O. C. Nanotechnology and aptamers, applications in drug delivery. *Trends Biotechnol.*, **26**(8), 442–449 (Aug, 2008) Epub 2008 Jun 19.
112. Li, X., Everson, W., and Smart, E. Caveolae, lipid rafts, and vascular disease. *Trends Cardiovasc Med.*, **15**(3), 92–96 (2005) doi,10.1016/j.tcm.2005.04.001.
113. Liang, F. and Chen, B. A review on biomedical applications of single walled carbon nanotubes. *Curr Med Chem.*, **17**(1), 10–24 (2010).
114. Lin, M. M., Kim, do K., El, Haj A. J., and Dobson, J. Development of superparamagnetic iron oxide nanoparticles (SPIONS) for translation to clinical applications. *IEEE Trans Nanobioscience.*, **7**(4), 298–305 (Dec, 2008).
115. Liu, C. and Zhang, N. Nanoparticles in gene therapy principles, prospects, and challenges. *Prog Mol Biol Transl Sci.*, **104**, 509–562 (2011).
116. Lu, Y. and Liu, J. Catalyst-functionalized nanomaterials. *Wiley Interdiscip Rev Nanomed Nanobiotechnol.*, **1**(1), 35–46 (Jan–Feb, 2009).
117. Luk, B. T., Fang, R. H., and Zhang, L. Lipid and polymer based nanostructures for cancer theranostics. *Theranostics*, **2**(12), 1117–1126 (2012) doi, 10.7150/thno.4381. Epub 2012 Dec 10.
118. Maham, A., Tang, Z., Wu, H., Wang, J., and Lin, Y. Protein-based nanomedicine platforms for drug delivery. *Small*, **5**(15), 1706–1721 (Aug 3, 2009).
119. Majd, S., Yusko, E. C., Billeh, Y. N., Macrae, M. X., Yang, J., and Mayer, M. Applications of biological pores in nanomedicine, sensing, and nanoelectronics. *Curr Opin Biotechnol.*, **21**(4), 439–476 (Aug, 2010) Epub 2010 Jun 18.
120. Mai, W. X. and Meng, H. Mesoporous silica nanoparticles, A multifunctional nano therapeutic system. *Integr Biol (Camb).*, **5**(1), 19–28 (Jan, 2013) doi, 10.1039/c2ib20137b.
121. Mallipeddi, R. and Rohan, L. C. Progress in antiretroviral drug delivery using nanotechnology. *Int J Nanomedicine.*, **5**, 533–547 (Aug 9, 2010).
122. Mannell, H., Pircher, J., Fochler, F., Stampnik, Y., Räthel, T., Gleich, B., Plank, C., Mykhaylyk, O., Dahmani, C., Wörnle, M., Ribeiro, A., Pohl, U., and Krötz, F. Site directed vascular gene delivery *in vivo* by ultrasonic destruction of magnetic nanoparticle coated microbubbles. *Nanomedicine*, (Apr 2, 2012) [Epub ahead of print].
123. Mannelli, I. and Marco, M. P. Recent advances in analytical and bioanalysis applications of noble metal nanorods. *Anal Bioanal Chem.*, **398**(6), 2451–2469 (Nov, 2010) Epub 2010 Jul 20.
124. Mansour, H. M., Rhee, Y. S., and Wu, X. Nanomedicine in pulmonary delivery. *Int J Nanomedicine.*, **4**, 299–319 (2009) Epub 2009 Dec 29.
125. Mansur, H. S. Quantum dots and nanocomposites. *Wiley Interdiscip Rev Nanomed Nanobiotechnol.*, **2**(2), 113–129(Mar–Apr, 2010).
126. Maojo, V., Martin-Sanchez, F., Kulikowski, C., Rodriguez-Paton, A., and Fritts, M. Nanoinformatics and DNA-based computing, catalyzing nanomedicine. *Pediatr Res.*, **67**(5), 481–489 (May, 2010).
127. Marradi, M., Garcia, I., and Penades, S. Carbohydrate.based nanoparticles for potential applications in medicine. *Prog Mol Biol Transl Sci.*, **104**, 141–173 (2011).
128. Marradi, M., Martin-Lomas, M., and Penades, S. Glyconanoparticles polyvalent tools to study carbohydrate based interactions. *Adv Carbohydr Chem Biochem.*, **64**, 211–290 (2010).

129. Menjoge, A. R., Kannan, R. M., and Tomalia, D. A. Dendrimer-based drug and imaging conjugates, design considerations for nanomedical applications. *Drug Discov Today.*, **15**(5–6), 171–185 (Mar, 2010) Epub 2010 Jan 29.

130. Murray, C. B., Kagan, C. R., and Bawendi, M. G. Synthesis and characterization of monodisperse nanocrystals and close packed nanocrystal assemblies. *Annu Rev Mater Sci.*, **30**, 545–610 (2000).

131. Musacchio, T. and Torchilin, V. P. Recent developments in lipid based pharmaceutical nanocarriers. *Front Biosci.*, **16**, 1388–1412 (Jan 1, 2011).

132. Nel, A. E., Madler, L., Velegol, D., Xia, T., Hoek, E. M., Somasundaran, P., Klaessig, F., Castranova, V., and Thompson, M. Understanding biophysicochemical interactions at the nano-bio interface. *Nat Mater.*, **8**(7), 543–557 (Jul, 2009) Epub 2009 Jun 14.

133. Nguyen, Q. T., Olson, E. S., Aguilera, T. A., Jiang, T., Scadeng, M., Ellies, L. G., and Tsien, R. Y. Surgery with molecular fluorescence imaging using activatable cell penetrating peptides decreases residual cancer and improves survival. *Proc. Natl. Acad. Sci.*, USA. **107**(9), 4317–4322 (Mar 2, 2010) Epub 2010 Feb 16.

134. Nicolas, J. and Couvreur, P. Synthesis of poly (alkyl cyanoacrylate)-based colloidal nanomedicines. *Wiley Interdiscip Rev Nanomed Nanobiotechnol.*, **1**(1), 111–127 (Jan-Feb, 2009).

135. Nystrom, A. M. and Wooley, K. L. The importance of chemistry in creating well-defined nanoscopic embedded therapeutics, devices capable of the dual functions of imaging and therapy. *Acc Chem Res.*, **44**(10), 969–978 (Oct 18, 2011) Epub 2011 Jun 15.

136. Owen, D. M., Sauer, M., and Gaus, K. Fluorescence localization microscopy, The transition from concept to biological research tool. *Commun Integr Biol.*, **5**(4), 345–349 (Jul 1, 2012) doi, 10.4161/cib.20348.

137. Ozpolat, B., Sood, A. K., and Lopez-Berestein, G. Nanomedicine based approaches for the delivery of siRNA in cancer. *J Intern Med.*, **267**(1), 44–53 (Jan, 2010).

138. Paszko, E., Ehrhardt, C., Senge, M. O., Kelleher, D. P., and Reynolds, J. V. Nanodrug applications in photodynamic therapy. *Photodiagnosis Photodyn Ther.*, **8**(1), 14–29 (Mar, 2011) Epub 2010 Dec 28.

139. Peer, D. and Lieberman, J. Special delivery, targeted therapy with small RNAs. *Gene Ther.*, **18**(12), 1127–1133 (Dec, 2011) doi, 10.1038/gt.2011.56. Epub 2011 Apr 14.

140. Pelkmans, L. Secrets of caveolae and lipid raft-mediated endocytosis revealed by mammalian viruses. *Biochim Biophys Acta.*, **1746**(3), 295–304 (2005).doi,10.1016/j.bbamcr.2005.06. 009.

141. Petry, H., Goldmann, C., Ast, O., and Lüke, W. The use of virus-like particles for gene transfer. *Current Opinion in Molecular Therapeutics*, **5**(5), 524–528 (October, 2003).

142. Phillips, W. T., Goins, B. A., and Bao, A. Radioactive liposomes. *Wiley Interdiscip Rev Nanomed Nanobiotechnol.*, **1**(1), 69–83 (Jan-Feb, 2009).

143. Popov, A. P., Zvyagin, A. V., Lademann, J., Roberts, M. S., Sanchez, W., Priezzhev, A. V., and MyllylÃ, R. Designing inorganic light-protective skin nanotechnology products. *J Biomed Nanotechnol.*, **6**(5), 432–451 (Oct, 2010).

144. Portney, N. G., Destito, G., Manchester, M., and Ozkan, M. Hybrid assembly of CPMV viruses and surface characteristics of different mutants. *Curr Top Microbiol Immunol.*, **327**, 59–69 (2009).

145. Prabhu, P. and Patravale, V. The upcoming field of theranostic nanomedicine, an overview. *J Biomed Nanotechnol.*, **8**(6), 859–882 (Dec, 2012).

146. Puri, A. and Blumenthal, R. Polymeric lipid assemblies as novel theranostic tools. *Acc Chem Res.*, **44**(10), 1071–1079 (Oct 18, 2011) Epub 2011 Sep 15.

147. Quirin, S., Pavani, S. R., and Piestun, R. Optimal 3D single-molecule localization for super resolution microscopy with aberrations and engineered point spread functions. *Proc Natl Acad Sci.*, U S A. **109**(3), 675–679 (Jan 17, 2012) doi, 10.1073/pnas.1109011108. Epub 2011 Dec 30.

148. Ramachandran, S. and Lal, R. Scope of atomic force microscopy in the advancement of nanomedicine. *Indian J Exp Biol.*, **48**(10), 1020–1036 (Oct, 2010).

149. Randall, C. L., Kalinin, Y. V., Jamal, M., Shah, A., and Gracias, D. H. Self-folding immunoprotective cell encapsulation devices. *Nanomedicine.*, **7**(6), 686–689 (Dec, 2011) Epub 2011 Sep 21.

150. Rasmussen, A. and Deckert, V. New dimension in nano-imaging, breaking through the diffraction limit with scanning near-fieldoptical microscopy. *Anal Bioanal Chem.*, **381**(1), 165–172 (Jan, 2005) Epub 2004 Nov 13.

151. Resnik, D. B. and Tinkle, S. S. Ethics in nanomedicine. *Nanomedicine (Lond).*, **2**(3), 345–350 (Jun, 2007).

152. Roda, B., Zattoni, A., Reschiglian, P., Moon, M. H., Mirasoli, M., Michelini, E., and Roda, A. Field-flow fractionation in bioanalysis, A review of recent trends. *Anal Chim Acta.*, **635**(2), 132–143 (Mar 9, 2009) doi, 10.1016/j.aca.2009.01.015. Epub 2009 Jan 18. Review.

153. Romero, E. L. and Morilla, M. J. Topical and mucosal liposomes for vaccine delivery. *Wiley Interdiscip Rev Nanomed Nanobiotechnol.*, **3**(4), 356–375 (Jul–Aug, 2011) doi, 10.1002/wnan.131. Epub 2011 Feb 25.

154. Rosenholm, J. M., Sahlgren, C., and Linden, M. Towards multifunctional, targeted drug delivery systems using mesoporous silica nanoparticles-opportunities & challenges. *Nanoscale.*, **2**(10), 1870–1883 (Oct, 2010) Epub 2010 Aug 23.

155. Ross, J., Hunter, Victor, R., and Preedy. *Nanomedicine in health and disease.*, Science Publishers, CRC Press (Taylor and Francis Group). Enfield, New Hampshire. USA. (2011).

156. Rother, D., Sen, T., East, D., and Bruce, I. J. Silicon, silica and its surface patterning activation with alkoxy and amino-silanes for nanomedical applications. *Nanomedicine (Lond).*, **6**(2), 281–300 (Feb, 2011).

157. Ryvolova, M., Chomoucka, J., Drbohlavova, J., Kopel, P., Babula, P., Hynek, D., Adam, V., Eckschlager, T., Hubalek, J., Stiborova, M., Kaiser, J., and Kizek, R. Modern micro and nanoparticle-based imaging techniques. *Sensors (Basel).*, **12**(11), 14792–14820 (Nov 2, 2012) doi, 10.3390/s121114792.

158. Sakamoto, J. H., van de. Ven, A. L., Godin, B., Blanco, E., Serda, R. E., Grattoni, A., Ziemys, A., Bouamrani, A., Hu, T., Ranganathan, S. I., De, Rosa. E., Martinez, J. O., Smid, C. A., Buchanan, R. M., and Lee, S. Y.,

159. Srinivasan, S., Landry, M., Meyn, A., Tasciotti, E., Liu, X., Decuzzi, P., and Ferrari, M. Enabling individualized therapy through nanotechnology. *Pharmacol Res.*, **62**(2), 57–89 (Aug, 2010) Epub 2010 Jan 5.

160. Sandhu, A., Handa, H., and Abe, M. Synthesis and applications of magnetic nanoparticles for biorecognition and point of care medical diagnostics. *Nanotechnology*, **21**(44), 442001 (Nov 5, 2010) Epub 2010 Oct 8.

161. Santi, L., Huang, Z., Mason, H. Virus like particles production in green plants. *Methods*, **40**(1), 66–76 (September, 2006) doi,10.1016/j.ymeth.2006.05.020.

162. Schroeder, A., Levins, C. G., Cortez, C., Langer, R., and Anderson, D. G. Lipid-based nanotherapeutics for siRNA delivery. *J Intern Med.*, **267**(1), 9–21 (Jan, 2010).

163. Schroeter, A., Engelbrecht, T., Neubert, R. H., and Goebel, A. S. New nanosized technologies for dermal and transdermal drug delivery. A review. *J Biomed Nanotechnol.*, **6**(5), 511–528 (Oct, 2010).

164. Sekhon, B. S. and Kamboj, S. R. Inorganic nanomedicine-part 1. *Nanomedicine*, **6**(4), 516–522 (Aug, 2010) Epub 2010 Apr 22.
165. Sekhon, B. S. and Kamboj, S. R. Inorganic nanomedicine-part 2. *Nanomedicine*, **6**(5), 612–618 (Oct, 2010) Epub 2010 Apr 22.
166. Shapira, A., Livney, Y. D., Broxterman, H. J., and Assaraf, Y. G. Nanomedicine for targeted cancer therapy, towards the overcoming of drug resistance. *Drug Resist Updat.*, **14**(3), 150–163 (Jun, 2011) Epub 2011 Feb 16.
167. Sheykhan, M., Heydari, A., Mamani, L., and Badiei, A. The synthesis and spectroscopic characterization of nano calcium fluorapatite using tetra-butylammonium fluoride. *Spectrochim Acta A Mol Biomol Spectrosc.*, **83**(1), 379–383 (Dec, 2011) doi, 10.1016/j.saa.2011.08.049. Epub 2011 Aug 31.
168. Shi, X., Zhang, X., Xia, T., and Fang, X. Living cell study at the single-molecule and single-cell levels by atomic force microscopy. *Nanomedicine (Lond).*, **7**(10), 1625–1637 (Oct, 2012) doi, 10.2217/nnm.12.130.
169. Shvedova, A. A. and Kagan, V. E. The role of nanotoxicology in realizing the helping without harm paradigm of nanomedicine, lessons from studies of pulmonary effects of single-walled carbon nanotubes. *J Intern Med.*, **267**(1), 106–118 (Jan, 2010).
170. Shvedova, A. A., Kisin, E. R., Porter, D., Schulte, P., Kagan, V. E., Fadeel, B., and Castranova, V. Mechanisms of pulmonary toxicity and medical applications of carbon nanotubes, Two faces of Janus. *Pharmacol Ther.*, **121**(2), 192–204 (Feb, 2009) Epub 2008 Dec 6.
171. Simmel, F. C. Towards biomedical applications for nucleic acid nanodevices. *Nanomedicine (Lond).*, **2**(6), 817–830 (Dec, 2007).
172. Singh, S. and Nalwa, H. S. Nanotechnology and health safety-toxicity and risk assessments of nanostructured materials on human health. *J Nanosci Nanotechnol.*, **7**(9), 3048–3070 (Sep, 2007).
173. Steinmetz, N. F. Viral nanoparticles as platforms for next-generation therapeutics and imaging devices. *Nanomedicine*, **6**(5), 634–641 (Oct, 2010) Epub 2010 Apr 28.
174. Sundararajan, R. Nanosecond electroporation, another look. *Mol Biotechnol.*, **41**(1), 69–82 (Jan, 2009) Epub 2008 Sep 26.
175. Swierczewska, M., Lee, S., and Chen, X. The design and application of fluorophore-gold nanoparticle activatable probes. *Phys Chem Chem Phys.*, **13**(21), 9929–9941 (Jun 7, 2011) Epub 2011 Mar 7.
176. Swierczewska, M., Liu, G., Lee, S., and Chen, X. High-sensitivity nanosensors for biomarker detection. *Chem Soc Rev.*, **41**(7), 2641–2655 (Mar 12, 2012) Epub 2011 Dec 20.
177. Szebeni, J., Muggia, F., Gabizon, A., and Barenholz, Y. Activation of complement by therapeutic liposomes and other lipid excipient-based therapeutic products, prediction and prevention. *Adv Drug Deliv Rev.*, **63**(12), 1020–1030 (Sep 16, 2011) Epub 2011 Jul 14.
178. Talelli, M., Rijcken, C. J., van Nostrum, C. F., Storm, G., and Hennink, W. E. Micelles based on HPMA copolymers. *Adv Drug Deliv Rev.*, **62**(2), 231–239 (Feb 17, 2010) Epub 2009 Dec 22.
179. Tan, S., Li, X., Guo, Y., and Zhang, Z. Lipid-enveloped hybrid nanoparticles for drug delivery. *Nanoscale.*, **5**(3), 860–872 (Feb 7, 2013) doi, 10.1039/c2nr32880a. Epub 2013 Jan 4.
180. Tanner, P., Baumann, P., Enea, R., Onaca, O., Palivan, C., and Meier, W. Polymeric vesicles, from drug carriers to nanoreactors and artificial organelles. *Acc Chem Res.*, **44**(10), 1039–1049 (Oct 18, 2011) Epub 2011 May 24.
181. Tao, L., Hu, W., Liu, Y., Huang, G., Sumer, B. D., and Gao, J. Shape-specific polymeric nanomedicine, emerging opportunities and challenges. *Exp Biol Med (Maywood).*, **236**(1), 20–29 (Jan, 2011).

182. Tian, W. D. and Ma, Y. Q. Theoretical and computational studies of dendrimers as delivery vectors. *Chem Soc Rev.*, **42**(2), 705–727 (Jan 21, 2013) doi, 10.1039/c2cs35306g.

183. Tokatlian, T. and Segura, T. siRNA applications in nanomedicine. *Wiley Interdiscip Rev Nanomed Nanobiotechnol.*, **2**(3), 305–315 (May–Jun, 2010).

184. Tomczak, N., Janczewski, D., Dorokhin, D., Han, M. Y., and Vancso, G. J. Enabling biomedical research with designer quantum dots. *Methods Mol Biol.*, **811**, 245–265 (2012).

185. Tong, S., Cradick, T. J., Ma, Y., Dai, Z., and Bao, G. Engineering imaging probes and molecular machines for nanomedicine. *Sci China Life Sci.*, **55**(10), 843–861 (Oct, 2012) doi, 10.1007/s11427–012–4380–1. Epub 2012 Oct 31.

186. Tosi, G., Bortot, B., Ruozi, B., Dolcetta, D., Vandelli, M. A., Forni, F., and Severini, G. M. Potential use of polymeric nanoparticles for drug delivery across the blood-brain barrier. *Curr Med Chem.*, (Feb 21, 2013) [Epub ahead of print].

187. Trotta, F., Zanetti, M., and Cavalli, R. Cyclodextrin-based nanosponges as drug carriers. *Beilstein J Org Chem.*, **8**, 2091–2099 (2012) doi, 10.3762/bjoc.8.235. Epub 2012 Nov 29.

188. Valenzuela, P. and Simon, J. A. Nanoparticle delivery for transdermal HRT. *Maturitas*, (Feb 8, 2012) [Epub ahead of print].

189. Vallet-Regi, M. Nanostructured mesoporous silica matrices in nanomedicine. *J Intern Med.*, **267**(1), 22–43 (Jan, 2010).

190. Ventola, C. L. The nanomedicine revolution, part 1, emerging concepts. *P T.*, **37**(9), 512–525 (Sep, 2012).

191. Ventola, C. L. The nanomedicine revolution, part 2, current and future clinical applications. *P T.*, **37**(10), 582–591 (Oct, 2012).

192. Lesniak, W. G. and Balogh, L. P. Fabrication of {198Au0} radioactive composite nanodevices by radiation polymerization and their use for nanobrachytherapy. Nanomedicine, Nanotechnol. *Biol. Med.*, **4**, 57–69 (2008).

193. Wang, J. Biomolecule-functionalized nanowires, from nanosensors to nanocarriers. *Chemphyschem*, **10**(11), 1748–1755 (Aug 3, 2009).

194. Wang, K., Wu, X., Wang, J., and Huang, J. Cancer stem cell theory, therapeutic implications for nanomedicine. *Int J Nanomedicine*, **8**, 899–908 (2013) doi, 10.2147/IJN.S38641. Epub 2013 Feb 28.

195. Wang, M., Abbineni, G., Clevenger, A., Mao, C., and Xu, S. Upconversion nanoparticles, synthesis, surface modification and biological applications. *Nanomedicine*, **7**(6), 710–729 (Dec, 2011) Epub 2011 Mar 17.

196. Wei, F., Lillehoj, P. B., and Ho, C. M. DNA diagnostics, nanotechnology-enhanced electrochemical detection of nucleic acids. *Pediatr Res.*, **67**(5), 458–468 (May, 2010).

197. Wong, K. K. Y., Tian, J., Ho, C. M., Lok, C. N., Che, C. M., Chiu, J. F., and Tam, P. K. H. Topical delivery of silver nanoparticles reduces systemic inflammation of burn and promotes wound healing. *Nanomedicine Nanotechnol. Biol. Med.*, **2**, 306 (2006).

198. Wujcik, E. K. and Monty, C. N. Nanotechnology for implantable sensors, carbon nanotubes and graphene in medicine. *Wiley Interdiscip Rev Nanomed Nanobiotechnol.*, (Feb 28, 2013) doi, 10.1002/wnan.1213 [Epub ahead of print].

199. Wu, Z., Tang, L. J., Zhang, X. B., Jiang, J. H., and Tan, W. Aptamer-modified nanodrug delivery systems. *ACS Nano*, **5**(10), 7696–7699 (Oct 25, 2011).

200. Xing, Y. and Dai, L. Nanodiamonds for nanomedicine. *Nanomedicine (Lond).*, **4**(2), 207–218 (Feb, 2009).

201. Yadav, S. C., Kumari, A., and Yadav, R. Development of peptide and protein nanotherapeutics by nanoencapsulation and nanobioconjugation. *Peptides*, **32**(1), 173–187 (Jan, 2011) Epub 2010 Oct 8.

202. Yang, M. H., Lin, C. H., Chang, L. W., and Lin, P. Application of ICP-MS for the study of disposition and toxicity of metal-based nanomaterials. *Methods Mol Biol.*, **926**, 345–359 (2012) doi, 10.1007/978–1–62703–002–1_22.

203. Yoo, D., Lee, J. H., Shin, T. H., and Cheon, J. Theranostic magnetic nanoparticles. *Acc Chem Res.*, **44**(10), 863–874 (Oct 18, 2011) Epub 2011 Aug 8.

204. Zhao, J. and Castranova, V. Toxicology of nanomaterials used in nanomedicine. *J Toxicol Environ Health B Crit Rev.*, **14**(8), 593–632 (Nov, 2011).

205. Zrazhevskiy, P., Sena, M., and Gao, X. Designing multifunctional quantum dots for bio-imaging, detection, and drug delivery. *Chem Soc Rev.*, **39**(11), 4326–4354 (Nov, 2010) Epub 2010 Aug 9.

CHAPTER 12

ROLE OF NANOGROOVES ON THE PERFORMANCE OF ULTRA-FINE GRAINED TITANIUM AS A BIO-IMPLANT

P. PRIYANKA, K. SUJITH, A. M. ASHA, MANITHA NAIR, S. V. NAIR, R. BISWAS, T. N. KIM, and A. BALAKRISHNAN

CONTENTS

12.1   INTRODUCTION

The rejection of the implants by the body and the need for replacement after a few years has exaggerated not only the cost, but also requires complicated surgical procedures. This has demanded the use of novel technologies that incorporate nanostructured metal implants with enhanced healing time and biocompatibility. The development of fine grained nanomaterials has gained increased interest due to the possibility of obtaining improved mechanical properties compared to conventional coarse-grained materials. Among the metals and alloys known, stainless steel, Co–Cr alloys, titanium, and its alloys are the most widely used for the making of bio-implants (Mudali, 2003, Zheng, 2011). Of these, titanium and its alloys have shown to have high strength-to-weight ratio, good corrosion resistance, excellent physio-mechanical, and biocompatibility properties hence favoring its use in orthopedics (Callister, 1994, Oliveira, 1998, Aragon, 1972, Bagambisa and Joos, 1990).

Chemical stability, mechanical properties, and biocompatibility of the metallic implants are the basic requirements for its successful application in orthopedic replacements. There are many factors like corrosion and strength, which affects the durability of orthopedic devices that are made of metals and alloys. Various alloys like stainless steels (SS), Co–Cr alloys and titanium alloys are widely used for generating durable bio-implants. Frequent incidences of failures of stainless steel implants in the body raise the need for a next generation implant material. Titanium and its alloy have been widely researched because of its excellent corrosion resistance and mechanical properties. However, its susceptibility to body fluids under load bearing conditions mobilizes toxic alloying ions from the surface into the body. Further processing of bio-medical grade Ti alloy can be expensive if these issues are not addressed. One of the solutions to this problem is by strengthening pure Ti by grain refinement, which can lead to a significant improvement in mechanical, electrochemical, and biocompatible properties. This overview discusses the performance of ultra fine grained titanium as a bio-implant by comparing it to its coarser grained and alloy counterparts. The *in vivo* and *in vitro* responses have been discussed in detail and analyzed to understand its underlying mechanism.

12.2   TITANIUM AS AN IMPLANT

Titanium forms a stable film of titanium oxide on its surface. These materials are accepted by the body environment because of this passive bio-inert oxide layer. However, in order to improve its physio-chemical properties and enhance its performance as an implant, bulk alloying, ion implantation and surface modification through bio ceramic coatings has been widely researched and applied (Webster, 1999, Chu, 2006, Cui, 2008, Jayaram, 2004, Xavier, 2003, Zhou, 2005). Of the titanium alloys that provide sufficient strength and corrosion resistance Ti-6Al-4V, are widely used in generating bio implants (Callister, 1994, Oliveira, 1998, Aragon, 1972, Bagambisa and Joos, 1990, Dobbs, 1982, Kim, 2008). However, when these metallic alloys react with the body plasma it becomes susceptible to corrosion. Various factors like pH, applied load, wear, and interaction with biological fluids result in roughening of surface, weakening of restoration, and liberation of elements from these metallic alloys

(Adya, 2005, Smetana, 1987). The released metal ions are ingested by macrophages that mediate the release of cytokines causing an inflammatory response leading to implant fracture and failure (Balakrishnan, 2008, Eastwood, 1998). One of the approaches in addressing this problem is by nanostructuring the implants, for example, nanostructured plate and conic screw implants for osteosynthesis and spine fixation respectively (Valiev, 2004, Prajapati, 2000). These kinds of approaches generates lighter and stronger implants and can exhibit improved biocompatibility by further coating it with nanomaterials or by incorporating nanomaterials as secondary phases. This also increases implants' potency against bacterial and fungal infections (Lekkas, 2008 Curtis, 1994). Nanomaterials are capable of accelerating cell growth, adhesion, spread, and migration after implantation. According to Linder et. al., titanium implants could achieve osseointegration without fibrous tissue growth at bone-implant interface. It was shown that the level of integration and the efficiency of healing were dependent on the bone-to-implant contact area. The higher contact area resulted in enhanced implant integration into the host bone. Incorporating growth factors and biomolecules can also be exploited onto implants to enhance growth and integration (Carbone, 2006, Goldspink, 1992). Nanomaterials can extend the durability and provide good physio-mechanical integrity although the cost (Bindu, 2009) and large scale production limits its commercialization (Linder, 1983).

Commercially produced titanium (CP-Ti) shows lower mechanical strength (Kim, 2008), but good biocompatibility, and thereby modification in this metal by alloying and surface treatment is of interest for surgical implants. If the strength of pure titanium could be increased by reducing the grain size significantly to nano-regime, it could be used as a novel substitute for alloy implants. Materials with the finer grain size were reported to show improved biocompatibility over normal materials. Materials for bio-implant can be classified based on the grain sizes that is, Ultra-Fine Grained (UFG) materials with grain sizes smaller than 1 μm and coarse grain materials greater than 1 μm (Sanosh, 2010). Titanium parts with the ultra-fine grain structure feature can be produced either through powder metallurgy (PM) route (Panigrahi, 2005, Panigrahi, 2006, Min, 2010) or by Severe Plastic Deformation (SPD) process (Shin, 2003, Zhernakov, 2001, Valiev, 2004, Chon, 2004, Curtis, 1978). The finest microstructures for load bearing applications in titanium can be obtained by introducing a very large deformation in the sample through a complex strain path using SPD techniques such as Equal Channel Angular Pressing (ECAP) or severe rolling. The present overview focuses on using ECAP for developing ultra fine grained titanium and evaluate its performance as a substitute bio-implant for Ti alloys.

12.3 PROCESSING OF ULTRA FINE GRAINED TITANIUM

The ECAP is a promising technique for obtaining bulk UFG materials (Valiev, 1997, Alexandrov, 1998, Stolyarov, 1999). Though, several parameters are considered for grain refining and enhancing the physiomechanical properties of a metal during ECAP, one of the important parameters is the stability of metals to withstand larger strains imposed during the deformation process, which is carried out at temperature higher than 0.2 T_m (T_m = Melting temperature) (Burow, 2008, Valiev, 2006, Balyanov,

2004). In this process, a metal billet is pressed through a die having two channels that are equal in cross-section and intersecting at an angle Φ. The deformation pressings are repeated for definite number of passes, where each pass refines the grain size. The grain size refining during the pressing imposes a large amount of plastic strain on the sample, since the shearing does not change the dimensions of the sample. Different ECAP routes are developed by rotating the metal billet around its longitudinal axis, between two consequent passes. Four rotation routes are used to impose ECAP (Rack, 2006, Ferrasse, 1997, Prangnell, 2000, Iwahashi, 1997, 1998 a, b, c, and d, Ishi, 1998, Langdon, 1999, Segal, 1995, Furukawa, 1998) on the metal.

These four ECAP routes are

i)   Route A, where the billet is not rotated
ii) Route B_c, where the billet is rotated 90° clockwise
iii) Route B_A, where the billet is rotated 90° clockwise and counter clockwise alternatively and
iv) Route C, where the billet is rotated 180°. The channel-intersection angle, Φ, plays an important role in determining the shear strain during each pass of the billet through the die (Nemoto, 1999).

$$\gamma = 2\cot\left(\frac{\phi}{2}\right) \tag{1}$$

It has been observed that lower values of Φ, exhibit higher shear strain from each pass and therefore is more effective in refining grains (Ishi, 1998). The values reported for Φ generally ranges from 90° to 157.5° (Ferrasse, 1997; Ishi, 1998).

To apply ECAP to hexagonal close packing (HCP) metals, such as titanium and its alloys, is challenging because the malleability of these materials is normally inferior to that of cubic phased metals. This structural behavior is mainly attributed to the fact that slip occurs primarily only on basal or prism planes along the close packed (a) direction. For instance, it has been shown (Iwahashi, 1996), that a slip in titanium tends to occur along the <11–20> direction primarily on {10–10} planes and less frequently on the {0001} plane. As this slip does not induce a plastic strain along the c axis of the crystal, deformation twinning or c + a slip on pyramidal planes tends to accommodate the plastic strain imposed by conventional deformation processes (Yoo, 1981, Paton, 1970, Akhtar, 1975). The twinning planes in Ti metal are {10–12}, {11–21}, and {11–22} at ambient temperatures and {10–11} at temperatures above 400°C (Minonishi, 1982). Furthermore, twins resulting from these deformations increase the slips along the <11–23> direction (that is, c + a slip) on {10–11} or {11–22} planes (Paton, 1970, Akhtar, 1975). Several studies have been reported on a nanostructural development and mechanical properties of Ti by grain refinement through different ECAP routes (Koiwa, 2006, Stolyarov, 2001, Delo, 1999). It has been shown that the deformation of titanium during ECAP involves not only the slip of dislocations but also deformation twinning (Shin, 2003). Among the different ECAP routes, route B_c, where the sample is rotated 90° clockwise between each pass, is considered to be the most effective in grain size refinement to nanoscale (Koiwa, 2006). The deformation temperature was found to affect grain refinement significantly as well (Stolyarov, 2001, Delo, 1999). In the present study, commercially available ASTM Grade-2 pure titanium (CP-Ti) rod was normalized in an electric furnace at 705°C for 30 min and cooled in air. Subsequently, this CP-Ti was severely deformed via route B$_c$ (as mentioned earlier) by annealing at 600°C for 10 min.

12.4   PHYSICAL AND MECHANICAL PROPERTIES OF ULTRA FINE GRAINED TITANIUM

It was seen that by imposing such a high deformation onto CP-Ti, the average grain size was found to be reduced from 15μm to 238 nm (hereby this will be referred as ultra fine grained titanium (UFG-Ti)) (see Figure 1). It is also interesting to note that after ECAP the grains showed high sheared patterns. The UFG-Ti sample also showed significant improvement in strength (684 MPa, ~ 60% higher) than the strength of the CP-Ti (418 MPa) sample and comparable to that of Ti–6Al–4V alloy (730 MPa) (Kim, 2008, Bindu, 2009). The hardness of UFG-Ti (~3.3 GPa) was found to be 3 times higher than CP-Ti (~1.0 GPa) (Balakrishnan, 2008). The hardening and strengthening effect was mainly resulting from the reduced grain size and micro-lattice strains. X-ray Diffractometry (XRD) showed peak broadening in UFG-Ti indicating reduced grain size and micro-lattice strains during ECAP (Figure 2).

The UFG-Ti showed peak shifts towards lower diffraction angles compared to CP-Ti indicating residual compression on UFG-Ti surface. Peak broadening was also seen in UFG-Ti indicating reduction in the crystallite size to the nanolevel and micro-lattice strains during ECAP. The presence of high residual compression was confirmed by XRD peak shifts and calculated as ~ 200 MPa using low incident beam angle diffraction (LIBAD) method (Kim, 2008). The reduction in UFG-Ti grain size could be attributed to large shear deformation, which occurs inside the material (Sanosh, 2010). The technique employed here (Route B$_c$) uses four passes, where each pass of CP-Ti through the die introduces a very intense plastic strain into the material. During the first pass, the dislocations are introduced to form cells and sub-grain bands and this deformation occurs at a very fast rate. On further passes through the die, the orientations across these boundaries increase as more dislocations move from the sub-grain interiors and become absorbed in the sub-grain walls, thereby refining the nanostructure (Carbone, 2006). Such a display of strength and hardness by UFG-Ti is encouraging considering the fact that, it could ideally replace Ti–6Al–4V alloy for load bearing applications. However, the interest was to evaluate the biocompatibility of such a metallic system for implant purposes.

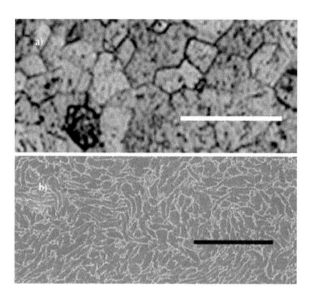

FIGURE 1   Microstructure of a) CP-Ti and b) UFG-Ti. Bar represents 50 μm (Sanosh, 2010).

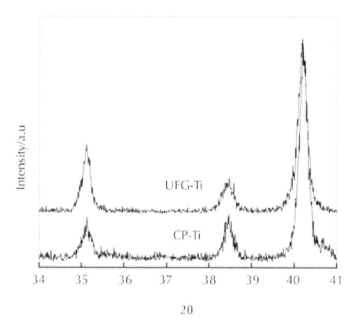

FIGURE 2   The XRD patterns of UFG-Ti and CP- Ti.

Contact angle measurements were performed using distilled water to understand the wetting properties of the UFG-Ti surfaces .The intention of performing contact angle tests was to get an insight into the wetting behavior of the surfaces, which would influence the cell adhesion and proliferation. It was seen that the contact angles for UFG-Ti exhibited lower values (~69°, good wettability) compared to their coarser grained counterparts (CP-Ti~ 75°) (Figure 3). The increased wettability was attributed to the high surface energy generated onto CP-Ti during the ECAP. It is known from the expression $\Delta E = \gamma \Delta A$ (ΔE is the free energy, γ is the surface tension, and ΔA is the increase in the surface area) that the increase in free energy, increases the surface area of interaction. The atomic densities at the grain boundary regions would be relatively lower and a large number of atoms would be energetically unstable, compared to the inside of the grain. When the deformed material is recrystallized, some nanovoids or nanogrooves remain at the triple grain junctions, which would not have been healed completely by the solid-state mass flow. This leads to a drastic increase in the grain boundary energy and tends to release the energy in every possible way. This accounts for the increase in wettability or the hydrophilicity of the material as seen in UFG-Ti (Lenel, 1979, German, 1996, Kilpadi, 1994, Mosmann, 1983). This nature of the increased wettability seen in UFG- Ti inspired to extrapolate its performance in *in vitro* and *in vivo* environment.

a)   b)

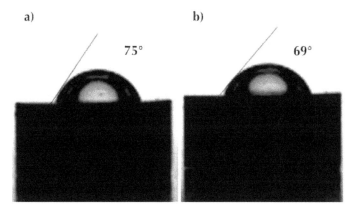

75°   69°

FIGURE 3   Wettability tests: Contact angle measurements a) CP-Ti and b) UFG-Ti (Bindu, 2009).

12.5   *IN VITRO* RESPONSE OF UFG-TI

Controlled experiments of cell adhesion, proliferation, and cytotoxicity, conducted on a 96 well plate for each set of samples (n = 3) using mouse fibroblast cell line 3T3, showed significant increase in the absorbance value for UFG- Ti in MTT assay Ferrari, 1990, Chon, 2004) (Figure 4). A significant increase was also observed in the absorbance values of UFG- Ti samples post 5 days of cell seeding, however no statistical difference was observed in cytotoxicities (Kim, 2008) (Figure.4). It was observed that under similar surface roughness conditions (~150 nm), UFG-Ti exhibited improved cyto-compatibility (cell adhesion and proliferation, *in vitro*) compared to CP-Ti and also Ti-6Al-4V alloy (Figure 5). It is hypothesized that when these implants come in contact with body fluids and cells, the resultant solid-liquid contact tends to lower the energy, thus lowering the contact angle of the liquid. This affects the adsorption rate of the extracellular proteins like fibronectin on implants followed by significant increase in cell adhesion through cell membrane integrin proteins. This would also initiate intracellular signaling affecting cytoskeleton changes and actin microfilament assembly, the focal adhesion plates, thus influencing cell spreading (Kim, 2008, Bindu, 2009). In order to prove that, the immunolabelling of fibronectin was done, which showed 50% increase in the intensity of signal with UFG-Ti over CP-Ti (Figure 6).

As proposed by Walboomers et al., (Walboomers, 1998), surface discontinuities of about 1 μm or less stimulates mechano-receptive responses in cells, implying dynamics of cytoskeleton. A relative increase in the adherence and the proliferation in the UFG-Ti could be attributed due to (a) The increased surface free energy and (b) the presence of a number of nano size grooves (Bindu, 2009). A schematic presentation of possible nanogrooves formation has been shown in Figure 7. The high energy grain boundaries and nanogrooves provide active sites for the osteogenic process. It was found that the cell growth rate was better when the dimensions of the grooves were similar to that of cell dimensions. Jansen and co-workers (Walboomers, 1998) observed that the cell proliferation in smaller size grooves (~1 μm) was significantly higher compared to the larger size grooves (~5 and ~10 μm). The front edges of the cells (the lamellipodia) contain actin micro spikes, which could be influenced by the surface discontinuities. Several recent *in vitro* studies in different metal systems have shown similar effect of grain size. For instance, UFG-Zr showed improved post induction mineralization when supplemented with osteogenic induction medium compared to their coarser grained counterparts (Saldana, 2010). The influence of grain boundaries on enhanced osteoblast cell like attachments in ultrafine-grained NiTi alloy was also reported recently (Iwahashi, 1998). These studies showed that indeed nanomaterials exhibit not only to improved mechanical properties, but also enhanced biocompatibility.

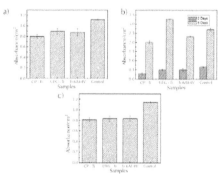

FIGURE 4   Cellular response (mouse fibroblast cell line 3T3) on various materials–MTT assay.

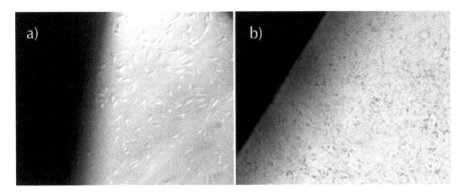

FIGURE 5   Phase contrast micrographs (10 x) showing non-cytotoxicity of 3T3 mouse fibroblast cells in contact with UFG-Ti after (a) 2 days (b) 5 days (Bindu 2009).

12.6   *IN VIVO* RESPONSE OF UFG-TI

While considering the inflammatory response elicited by foreign materials, the response is very much dependent on the surface properties that stimulate the inflammatory cells, like macrophages. Therefore, *in vivo* bio compatibility of UFG-Ti was done (See Table 1) by quantitatively assessing inflammatory cells, especially macrophages surrounding the implant site, using image analysis and immunohistochemistry. While considering the inflammatory response elicited by foreign materials, the response is very much dependent on the surface properties that stimulate the inflammatory cells, like macrophages. For the *in vivo* study, metal implants were implanted subcutaneously in Wistar rats. Tissues surrounding the implants were retrieved at 30 days post implantation and immunohistochemical assay was carried out to detect the number of macrophages in the peri-implant tissue. To quantify the number of macrophages, images were taken from five different fields around the implant site labeled with FITC (Fluorescence Isothio-cyanate) and analyzed using image analysis software (Bindu, 2009, Balakrishnan, 2008). It was observed that the number of macrophages surrounding the implant was similar compared to CP-Ti. Interactions at the material-macrophage interface triggers macrophage activation leading to the production of pro-inflammatory signaling molecules responsible for the extended immune response (Bindu, 2009). The number of inflammatory cells, especially macrophages in the peri-implant tissue was less in UFG-Ti compared to CP-Ti, indicating that UFG-Ti is less immunogenic. (See Table 1).

TABLE 1   Quantification and comparison of average number of macrophages in UFG-Ti, CP-Ti, and control in tested rats

Quantification of Macrophages								
Samples/Trials	1	2	3	4	5	6	7	8
Control	6.4	6.0	8.0	7.4	9.0	8.6	8.2	5.6
CP-Ti	11.6	13.0	11.6	9.8	10.63	13	10.2	10.8
UFG-Ti	9.4	8.2	9.8	10.0	11.4	8.8	8.4	8.6

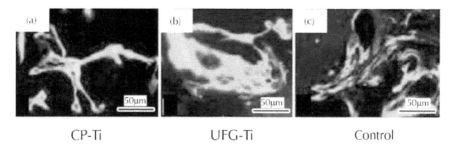

CP-Ti   UFG-Ti   Control

FIGURE 6   Detection of fibronectin (green) produced on the surfaces of (a) CP-Ti (b) UFG-Ti (c) Control (Bindu, 2009).

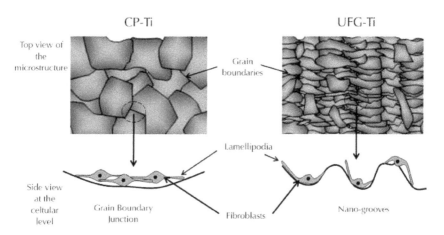

Figure 7   Schematic representation of cellular response in presence of grooves on the polished surfaces of (a) CP-Ti (b) UFG-Ti.

12.7   ELECTROCHEMICAL BEHAVIOR OF UFG –TI

The biocompatibility and corrosion resistance of the titanium metal is the result of passive TiO_2 thin films of thickness 2–6 nm formed on the titanium surface (Stothard, 1996, Esposito, 1999, Lopez, 2003, Strietzel, 1998, Dearnley, 2003). Many studies have been reported (Velten, 2002, Mabilleau, 2006, Clark, 1982, Contu, 2002, Ibris, 2002) to understand the corrosion mechanism of pure titanium in different biological media. Various surface modifications like anodic oxidation treatments (Choubey, 2005), electrochemical treatments (Kim, 2007), sandblasting (Masmoudi, 2006), carbide coatings (Jiang, 2006, Martinez, 2002), laser nitriding (He, 2006), electrolytic polishing (Carpene, 2005, Trehan, 1972), and so on, on CP-Ti implants have been carried out to improve their corrosion resistance. The above techniques have shown to improve the corrosion resistance of CP-Ti, but as mentioned and shown earlier (See 12.2), the strength of these implants still remains an issue of concern. It introduces potentiodynamic polarization studies that show UFG-Ti exhibiting higher corrosion resistance compared to coarse grained CP-Ti (Balakrishnan, 2008, Stolyarov, 2004) in simulated body fluid (SBF). The results (see Table 2) showed that UFG–Ti exhibited a quite increase in corrosion resistance from its coarse grained counterparts. It is a quite known fact that surface roughness is of considerable importance in creating pits, which facilitate pitting corrosion. To minimize this effect, the surface roughness was kept constant to base the further evaluation on grain size. It was observed that the corrosion potential value (E_{corr} ~ -340 mV) for the UFG-Ti was more negative than that of CP-Ti (E_{corr} ~ -125 mV). This lower E_{corr} of UFG-Ti was attributed to the inhibition of the cathodic reaction that takes place due to quicker passivation, due to which a displacement of this E_{corr} more negative value occurs.

It is hypothesized that by reducing the grain size in case of UFG-Ti an increased electron activity (Guilherme, 2005, Li, 2006, Fiebig, 2011) results, which in turn results in a decrease of the electron work function. As a result, the surface becomes more electrochemically reactive giving rise to an increase in passivation ability resulting in rapid formation of a mechanically strong and stable passive film. Electrochemical Impedance Spectroscopy (EIS) studies by Shukla et.al, (Shukla, 2005) showed that corrosion of CP-Ti in simulated body fluid was mainly due to dissolution of weak passive oxide layer by Cl^- ions resulting in localized corrosion. The local break down of these passive layer zones can lead to increase in pitting corrosion (Prokofiev, 2010, Valor, 2007, Gonzalez, 2011). However, for UFG-Ti, ECAP results in high-density grain boundaries, which may help to increase the interfacial adherence of the passive film. This is possible by the formation of oxide protrusions penetrating into grain boundaries (Figure 8), similar to oxide pegs of thick oxide scales on some ferrous alloys developed at grain boundaries, which increases corrosion resistance significantly. Interestingly, energy dispersive x-ray (EDX) analysis on UFG-Ti surface showed that this oxide layer had a chemical affinity towards Ca^{2+} and Na^+ ions in the SBF (Kim, 2005, Liu, 2004) and combine selectively to form amorphous Na-Ca-Ti-O compounds.

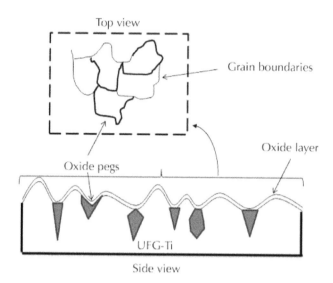

FIGURE 8   Schematic representation of (side and top view) oxide peg protrusions penetrating into grain boundaries.

TABLE 2   Electrochemical responses of CP-Ti and UFG-Ti

Sample	E_{corr}	i_{corr}	*Corrosion
	(mV)	(μA/cm^2)	Rate, mils per year (mpy)
CP-Ti	-125 ± 8	0.75 ± .06	1. 1 ± .08
UFG-Ti	-340 ± 7	0.06 ± .04	0.12 ± .03

*The corrosion values represented here have been calculated using Tafel extrapolation plots.

12.8   CONCLUSION

The effect of micro/nano-structure on the strength of the implant and the cell-substrate interaction is reported. The UFG-Ti prepared by ECAP route showed improved strength, better biocompatibility in terms of wettability, cell adhesion and proliferation than conventional CP-Ti. Higher cell proliferation was observed on UFG-Ti, which was attributed to high surface energy and the presence of nano sized grooves. These two factors were primarily responsible for mimicking the extracellular structures that aid significantly in protein adsorption. Although, the fibronectin adsorption has been phenotypically characterized by the cell adhesion and migration, the molecular mechanism between the titanium surface and biological system are yet to be fully understood. Potentiodynamic polarization studies shows UFG-Ti exhibits higher corrosion resistance compared to coarse grained CP-Ti. However, the scope of this study can be further improved by comparative studies of UFG-Ti with other coated/surface treated substrates considering factors like cost, mechanical and corrosion properties and not the least, the biological response of the substrate. In addition to the above, stress shielding studies needs to be conducted. This can address the issues related to long-term complications from stress shielding and various other factors resulting in implant life expectancy of 10 to 15 years. Although from the *in vivo* point of view the present overview limits itself to the subcutaneous tissue, where no load bearing is experienced. The long term data will be very crucial to validate the use of UFG-Ti for implant application. However, with the present data UFG seems to be a promising candidate for bioimplants.

KEYWORDS

- **Grain size**
- **Nanomorphology**
- **Ultra fine grained titanium**
- **Titanium**

ACKNOWLEDGEMENT

Department of Biotechnology (DBT), India is gratefully acknowledged for supporting Dr R. Biswas under Ramalingaswami Fellowship scheme.

REFERENCES

1. Adya, et a). Corrosion in titanium dental implant,: literature review. *Journal of Indian Prosthodontics Society*, **5**, 12--131 (2005).
2. Akht5). Basal slip and twinning in alpha-titanium single crystals. *Metallurgical Transactions A*, **6**, 115--1113 (1975.
3. Alexandrov, et a). Microstructures and properties of nanocomposites obtained through SPTS consolidation of powders. *Metallurgical and Materials Transactions A*, **29**, 223--2260 (1998).
4. Aragon, et a). Corrosion of Ti-6Al-4V in simulated body fluids and bovine plasma. *Journal of Biomedical Materials Research*, **6**, 15--164 (1972.
5. Bagambisa, Joo). Preliminary studies on the phenomenologicaurbehavior of osteoblasts cultured on hydroxyapatite ceramics. *Biomaterials*, **11**, 0--56 (1990.
6. Balakrishnan, et a). Corrosiourbehavior of ultra fine grained titanium in simulated body fluid for implant application. *Trends Biomaterials and Artificial Organs*, **22**, 8--64 (2008.
7. Balyanov, et a). Corrosion resistance of ultra fine-grained Ti. *Scripta Materialia*, **51**, 25--229 (2004.

8. Bindu, et a). An in vitro evaluation of ultra-fine grained titanium implants. *Journal of Materials Science and Technology*, 25, 56--560 (2009.

9. Burow, et a). Martensitic transformations and functional stability in ultra-fine grained NiTi shape memory alloys. *Materials Science Forum*, **852**, 54--586 (2008).

10. Calliste). *Materials Science & Engineerin,: An Introduction*, 6th edn, John Wiley & Sons, Inc., New York, pp. 10--132 (1994).

11. Carbon). Biocompatibility of cluster-assembled nanostructured TiO_2 with primary and cancer cells. *Biomaterials*, **27**, 321--3229 (2006.

12. Carpene, et a). Free-electron laser surface processing of titanium in nitrogen atmosphere. *Applied Surface Science*, **247**, 37--314 (2005.

13. Chon, et a). Effect of oxygen and iron impurities on the hot deformation behavior of CP-Ti at intermediate temperatures. *Metals and Mateials International*, **10**, 57--573 (2004).

14. Choubey, et a). Electrochemical behavior of Ti-based alloys in simulated human body fluid environment. *Trends in Biomaterials and Artificial Organs*, **18**, 4--72 (2005.

15. Chu, et a). *In vivo* study on biocompatibility and bonding strength of Ti/Ti-20 vol.% HA/Ti-40 vol.% HA functionally graded biomaterial with bine tissues in the rabbit. *Materials Science and Engineerig, : A*, **429**, 8--24 (2006.

16. Clark, et a). The effects of proteins on metallic corrosion. *Journal of Biomedical Materials Research*, **16**, 15--134 (1982.

17. Conu , et a). Characterization of implant materials in fetal bovine serum and sodium sulfate by electrochemical impedance spectroscoy. IlyMechanically polished samples. *Journal of Biomedical Materials Research*, **62**, 42--421 (2002.

18. Cui, et a). Improving the biocompatibility of NiTi alloy by chemical treatmens:Anan in vitro evaluation in 3T3 human fibroblast cell. *Materials Science and Engineering C*, **28**, 117--1122 (2008.

19. Curti4). *Mechanical tensing of cells and chromosome arrangemen*e F. Lyall. and A. J. El Ha. (Eds.), Cambridge University Press, Cambride., (1994).

20. Curtis et, a). The control of cell division by tension or diffusion. *Nature*, **274**, 2--53 (1978.

21. Dearnley, et a). The corrosion-weaurbehavior of thermally oxidized CP-Ti and Ti-6Al-4V. *Wear*, **256**, 49--479 (2003.

22. Delo, Semiati). Hot working of Ti-6Al-4V via equal channel angular extrusion. *Metallurgical and Materials Transaction A*, **30**, 243--2481 (1999.

23. Dobbs. Fracture of titanium orthopedic implants. *Journal of Materials Science*, **1**, 238--2404 (1982.

24. Eastwood, et a). Fibroblast responses to mechanical forces. *Proceedings of the Institution of Mechanical Engineers*, **212**, 5--92 (1998).

25. Esposito, et a). Surface analysis of failed oral titanium implants. *Journal of Biomedical Materials Research*, **48**, 59--568 (1999.

26. Ferrari, et a). MTT colorimetric assay for testing macrophage cytotoxic activity in vitro. *Journal of Immunological Methods*, **131**, 15--172 (1990.

27. Ferrasse, et a). Development of a submicrometer-grained microstructure in aluminum 6061 using equal channel angular extrusion. *Journal of Materialh ,Research*, **12**, 123--1261. Microstructure and properties of copper and aluminum alloy 3003 heavily worked by equal channel angular extrusion. *Metallurgical and Materials Transactions A*, **28**, 107--1060 (1997 (a) and (b).

28. Fiebig, et a). Diffusion of Ag and Co in ultrafine-grained α-Ti deformed by equal channel angular pressing. *Journal of Applied Physics*, **110**, 8354– 83522 (2011).

29. Fronkova, et a). Simultaneous detection of endogenous lectins and their binding capacity at the single-cell level—a technical note. *Folia Biologica*, **45**, 17--162 (1999).

30. Furukawa, et a). The shearing characteristics associated with equal-channel angular pressing. *Materials Science and Engineerin,: A*, **257**, 38--332 (1998.

31. Germ6). Sintering Theory and Practice, *Wiley-Interscience*, New York (1996.

32. Goldspink, et a). Gene expression in skeletal muscle in response to stretch and force generation. *The American Journal of Physiology*, **262** R36--R363 (1992).

33. Gonzalez, et a). Corrosion behavior of Ni-Cr based coatings in simulated human body fluid environment. *International Journal of Electrochemical Science*, **6**, 364--3655 (2011.

34. Guilherme, et a). Surface roughness and fatigue performance of commercially pure titanium and Ti-6Al-4V alloy after different polishing protocols. *Journal of Prosthetic Dentistry*, **93**, 38--385 (2005.

35. He, et a). Surface modification of pure titanium treated with B4C at high temperature. *Surface and Coatings Technology*, **200**, 306--3020 (2006.

36. Ibris, Ro2). EIS study of Ti and its alloys in biological media. *Journal of Electroanalytical Chemistry*, **526**, 3--62 (2002).

37. Ishi, et a). Rotation conditions for grain refinement in equal-channel angular pressing. *Metallurgical and Materials Transactions A*, **29**, 201--2013 (1998.

38. Iwahashi, et a). Principle of equal-channel angular pressing for the processing of ultra-fine grained materials. *Scripta Materialia*, **35**, 13--146. An investigation of microstructural evolution during equal-channel angular pressing. *Acta Materialia*, **45**, 473--4741. Microstructural characteristics of ultrafine-grained aluminum produced using equal-channel angular pressing. *Metallurgical and Materials Transaction A* **29**, 225--2252. The process of grain refinement in equal-channel angular pressing. *Acta Materialia*, **46**, 337--3331. Factors influencing the equilibrium grain size in equal-channel angular pressing. *Metallurgical and Materials Transactions A*, **29**, 253--2510 (1996, 1997, 1998(a), (b), and (c)).

39. Jayaraman, et a). Influence of titanium surfaces on attachment of osteoblast-like cells *in vitro*. *Biomaterials*, **25**, 65--631 (2004.

40. Jiang, et a). Enhancement of fatigue and corrosion properties of pure Ti by sandblasting. *Materials Science and Engineerin,: A*, **429**, 0--35 (2006.

41. Kilpadi, Lemo4). Surface energy characterization of unalloyed titanium implant. *Journal of Biomedical Materials Research*, **28**, 149--1425 (199. .

42. Kim, et a). Process and kinetics of bonelike apatite formation on sintered hydroxyapatite in a simulated body fluid. *Biomaterials*, **26**, 436--4373. The effect of spark anodizing treatment of pure titanium metals and titanium alloys on corrosion characterization. *Surface and Coatings Technology*, **201**, 878--8745. In vitro fibroblast response to ultra fine grained titanium produced by severe plastic deformation process. *Journal of Materials Science*, **19**, 53--557 (2005, 2007, and 2008).

43. Koiwa, Numaku6). The snoek effect in ternary BCC alloys. A review. *Solid State Phenomena*, **115**, 7--40 (2006.

44. Langd9). Recent developments in high strain rate superplasticity. *Materials Transactions JIM.*, **40**, 76--722 (1999).

45. Lekkas, et a). Ultrafine particles (UFP) and health effects. Dangerous. Like no other P.? Review and analysis. *Global NEST Journal*, **10**, 49--452 (2008.

46. Len9). *Reviews on powder metallurgy and physical ceramics*, Freund Publishing House, Israel. (1979).

47. 6). Electron work function at grain boundary and the corrosiourbehavior of nanocrystalline metallic materials. *Materials Research Society Symposium Proceedings*, **887**, 27--235 (2006).

48. Linder, et a). Electron microscopic analysis of the bone-titanium interface. *Acta Orthopaedica Scandinavica*, **54**, 5--52 (1983.

49. Liu, et, a). Surface modification of titanium, titanium alloys, and related materials for bio-medical applications. *Materials Science and EngineeringR:. Reports*, **47**, 9--121 (2004.

50. Lopez, et a). Corrosion study of oxidized titanium alloys. *Electrochimica Acta*, **48**, 135--1401 (2003.

51. Mabilleau, et a). Influence of fluoride, hydrogen peroxide and lactic acid on the corrosion resistance of commercially pure titanium. *Acta Biomaterialia*, **2**, 11--129 (2006.

52. Martine, et a). Tribological performance of TiN supported molybdenum and tantalum carbide coatings in abrasion and sliding contact. *Wear*, **9237**,1--6 (2002.

53. Masmoudi, et a). Friction and weaurbehavior of cp Ti and Ti6Al4V following nitric acid passivation. *Applied Surface Science*, **253**, 227--2243 (2006.

54. Min, et a). MAO-Prepared hydroxyapatite coating on ultrafine-grained titanium. *Materials Science Forum,* **64--656**, 218--2171 (2010.

55. Minonishi, et a). {1122} <1123> slip in titanium. *Scripta Metallurgica*, **16**, 47--430 (1982.

56. Mosma3). Rapid colorimetric assay for cellular growth and survival: application to proliferation and cytotoxicity assays. *Journal of Immunological Methods*, **65**, 5--63 (1983.

57. Mudali, et a). Corrosion of bioimplants. *Sadhana*, **28**, 61--637 (2003.

58. Nemoto, et a). Microstructural evolution for superplasticity using equal-channel angular pressing. *Materials Science Forum*, **306**, 9--66 (1999.

59. Oliveira, et a). Preparation and characterization of Ti-Al-Nb alloys for orthopedic implants. *Brazilian Journal of Chemical Engineering*, **15**, 36--333 (1998.

60. Panigrahi, Godkhin6). Sintering of titanium: Effect of particle size. *International Journal of Powder Metallurgy*, **42**, 5--42 (2006).

61. Panigrahi, et a). *Journal of Materials Research*, **20**, 87--836 (2005.

62. Paton, Backof0). Plastic deformation of titanium at elevated temperatures. *Metallurgical Transactions*, **1**, 289--2847 (1970).

63. Prajapati, et a). Mechanical loading regulates protease production by fibroblasts in three-dimensional collagen substrates. *Wound Repair and Regeneration*, **8**, 26--237 (2000.

64. Prangnell, et a). *Investigations and Applications of Severe Plastic Deformati06,* Kluwer Academic Publishers, Dordrecht, **106**, (2000).

65. Prokofiev, et a). Suppression of Ni_4Ti_3 precipitation by grain size refinement in Ni-Rich NiTi shape memory alloys. *Advanced Engineering Materials*, **12**, 77--753 (2010.

66. Rack, Qa6). Titanium alloys for biomedical applications. *Materials Science and Engineering: C.*, **26**, 129--1277 (2006.

67. Saldana, Vilab0). Effects of micrometric titanium particles on osteoblast attachment and cytoskeleton architecture. *Acta Biomaterialia*, **6**, 169--1660 (2010.

68. Sanosh, et a). Vickers and knoop micro-hardnesurbehavior of coarse and ultrafine-grained titanium. *Journal of Material Science and Technology*, **26**, 94--907 (2010.

69. Seg5). Materials processing by simple shear *Materials Science and Engineering:: A*, **197**, 17--164 (1995.

70. Shin, et a). Microstructure development during equal-channel angular pressing of titanium. *Acta Materialia*, **51**, 93--996 (2003.

71. Shukla, et a). Properties of passive film formed on CP titanium, Ti-6Al-4V and Ti-13.4Al-29Nb alloys in simulated human body conditions. *Intermetallics*, **13**, 61--637 (2005.

72. Smetana. Multinucleate foreign-body giant cell formation. *Experimental and Molecular Pathology*, **46**, 28--265 (1987.

73. Stolyarov, et a). A two step SPD processing of ultrafine-grained titanium. *Nanostructured Materials*, **11**, 97--954. Microstructure and properties of pure Ti processed by ECAP and cold extrusion. *Materials Science and Engineerin.: A*, **303**, 2--89. Reduction of friction

coefficient in ultrafine-grained CP Ti. *Materials Science and Engineerin.: A.* **371**, 33--317 (1999, 2001, and 2004.

74. Stothard, et al. Guidance and activation of murine macrophages by nanometric scale topography. *Experimental Cell Research*, **223**, 46--435 (1996.

75. Strietzel, et a). In vitro corrosion of titanium. *Biomaterials*, **19**, 145--1499 (1998.

76. Trehan, Shar2). The electrolytic polishing of aluminium and aluminium alloys in a universal perchloric acid-bath. *NML Technical.,Journal*, **14**, 19--126 (1972.

77. Valib). Structure and mechanical properties of ultrafine grained metals. *Materials Science and Engineering A.*, **24--236**, 9--66. Nanostructuring of metals by severe plastic deformation for advanced properties. *Nature Materials*, **3**, 51--516. Strength and ductility of nanostructured SPD metals. *Metallic materials with high structural efficiency*, **146**, 9--90 (1997, 2004 (a) and (b)).

78. Valiev, Langd6). Principles of equal-channel angular pressing as a processing tool for grain refinement. *Progress in Materials Science*, **51**, 81--981 (2006.

79. Valor, et a). Stochastic modeling of pitting corrosion: A new model for initiation and growth of multiple corrosion pits. *Corrosion Science*, **49**, 59--579 (2007.

80. Velten, et a). Preparation of TiO_2 layers on cp-Ti and Ti6Al4V by thermal and anodic oxidation and by sol-gel coating techniques and their characterization. *Journal of Biomedical Materials Research*, **59**, 8--8. (2002).

81. Walboomers, et al. Growth behavior of fibroblasts on microgrooved polystyrene. *Biomaterials*, **19**, 1861–1868 (1998).

82. Webster, et al. Design and evaluation of nanophase alumina for orthopedic/dental applications. *Nanostructed materials*, **12**, 983–986 (1999).

83. Xavier, et al. Response of rat bone marrow cells to commercially pure titanium submitted to different surface treatments. *Journal of Dentistry*, **31**, 173–180 (2003).

84. Yoo. Slip, twinning, and fracture in hexagonal close-packed metals. *Metallurgical Transactions A*, **12**, 409–418 (1981).

85. Zheng, et al. Enhanced biocompatibility of ultrafine-grained biomedical NiTi alloy with microporous surface. *Applied Surface Science*, **257**, 9086–9093 (2011).

86. Zhernakov, et al. The developing of nanostructured SPD Ti for structural use. *Scripta Materialia*, **44**, 1771–1774 (2001).

87. Zhou, et al. Corrosion resistance and biocompatibility of Ti-Ta alloys for biomedical applications. *Materials Science and Engineering A*, **398**, 28–36 (2005).

88. Zhu, Lowe. Observations and issues on mechanisms of grain refinement during ECAP process. *Materials Science and Engineering A*, **291**, 46–53 (2000).

CHAPTER 13

SUPER-PARAMAGNETIC IRON OXIDE NANOPARTICLES (SPIONS) AS NANO-FLOTILLAS FOR HYPERTHERMIA: A PARADIGM FOR CANCER THERANOSTICS

MADHURI SHARON, GOLDIE OZA, ARVIND GUPTA, and SUNIL PANDEY

CONTENTS

13.1 INTRODUCTION

Nature is the source of inspiration to design nanomagnetic particles, which magnetized the attention of scores of material scientists. The magnetic gizmo, which can act like an artificial robot plays an imperative role in synaphic targeting of the miscreant cell with the aid of antibody. When a magnetic field is applied, all such targeted cells get fried up due to hyperthermal property. The history of magnetism in medicine is very ancient and is credited with amazing anecdotes about the pioneering work of physicians and physicists. The first medical use of magnetite powder was reported in the form of *Lauha Bhasma* or Iron powder in *Ayurveda* by great seer of all ages, *Acharya Charak*, writer of the compendium *Charak Samhita*. One of the most important concerns of *Ayurveda* is Noble Metal *Bhasma* as well as decoctions and extractions of various medicinal herbs. It is a matter of adoration that the ancient teachers were not oblivious of Nanoparticles in the form of *Bhasma*. This ancient Indian wisdom was then disseminated to the whole world, giving rise to different branches of medicine like Allopathy, Homeopathy, Unani, and Siddha. European physician and philosopher, Avicenna in the 10th century A. D. recommended the antidote for accidental ingestion of rust in the form of one magnetite particle taken along with milk. This magnetite was exploited for the attraction of poisonous inert iron and enhancing its excretion through the intestine. Similar *modus operandi* was used for the removal of iron oxide embedded in the eye by the force of magnetization.

A journey from *Ayurveda* to allopathy is self-explicable if we talk about the medicinal characteristics of Iron. But recently the upsurge of nanotechnology has expanded the horizons of its medical value since it can act as an android, which possess plethora of applications such as synaphic targeting, multimodal imaging, acting as a drug-delivery piggyback with real-time monitoring, thus can be termed as "Trojan horse". This has led to its applications in diagnosis and treatment of many catastrophic diseases such as cancer, cardiovascular diseases, Alzheimers' disease, and an unending list continues.

Super Paramagnetic Iron Oxide Nanoparticles (SPIONs) are considered to create a niche in the field of therapeutics and diagnostics with their unique and distinct properties. Due to the reduction of spatial extension of magnetic materials, ferromagnetic particles exhibit novel characteristics since there is a transition from multi-magnetic domain to single magnetic domain below critical size as well as below Curie temperature. Such mono-domain magnetic nanoparticles are considered to be super paramagnetic in nature. In such a small nanoparticle, magnetization can randomly flip direction under the regulation of temperature and the time lag between two flips is known as the Neel Relaxation time, which is named after the Magnetic giant, *Louis Neel*, Nobel prize winner on magnetism, who worked on the modification of Brownian relaxation time (Neel, 1949). When there is no magnetic field, the time used to measure the magnetization of the nanoparticles is much larger than the Neel relaxation time, thus the magnetization appears to be in average zero a state termed to be known as superparamagnetic in nature. The magnitude of magnetic moment is proportional to its volume. These mono-domain superparamagnetic materials are analogous to large magnetic units, which possess magnetic moment of thousands of Bohr magnetons.

The shape of the nanoparticles must be such that the magnetic moments tend to align the longest axis, thus defining the largest shape anisotropic energy.

The SPIONs also possess enhanced magnetic moments as well as proton relaxivities thus is quintessentially exploited as a competent MRI contrast agent. Earlier, Gadolinium and Iron oxide were used as MRI contrast agents, but due to their worse performance in proton relaxation their precision was under constant scrutiny. The SPIONs have come to the rescue of medical scientists, who deciphered that such nanoparticles possess a combination of T1 and T2 relaxivity, long vascular half-life and inhibit any interstitial leakage. This has led to the development of a proficient system for multimodal imaging of tissues, an impossible endeavor coming true.

All the above mesmerizing properties of SPIONs makes it an efficient candidate for Theranaustics (Therapeutics and Diagnostics) acting like an android having capability of molecular actions when interacting with a cell by means of antibody, proved to be an efficient MRI contrast agent causing Hyperthermia under magnetic field leading to actively targeting erred cells via receptor-ligand interaction. SPIONS, nano-scale "magic bullets", are cataclysmic towards diseased cells since they possess enhanced magnetic moments and are superparamagnetic in nature leading to augmentation of Neel relaxation causing Hyperthermia.

The four characteristics proven to be effective as drug-delivery systems *in vivo* are composition, morphology, size, and surface chemistry. An inorganic magnetic core and a biocompatible shell made up of polymers can act as a contrivance encapsulating drugs, which can be made synaphic using antibodies, thus targeting at the diseased site. This zombie thus specifically move towards the pathological site, thus decreasing the dosage and inimical side-effects associated with non-specific uptake of drugs by normal cells. The biocompatible polymers increase the stealthiness of SPIONs thus escaping the reticuloendothelial clearance from the circulation.

The SPIONs are next generation multi-functional nanoparticles involved in probing, monitoring, and treating tissue-devastating disease such as cancer, which is orchestrated by abnormal molecular assemblies and signaling molecules. When the immune surveillance mechanism of the body errs, then the hostility of the unfavorable signals and inefficient tumor suppressing agents leads to the transformation of normal cells. The hallmark of cancer is characterized by evasion of apoptosis, loss of anchorage dependency, over-expression of anti-apoptotic factors, mutilated extracellular matrix. Metastatic cancer is considered to be spreading in each and every nook and corner of the body fluids and tissues, thus causing difficulty in identifying the mischievous cell. Such SPIONic probes, due to the accelerated efforts of National Cancer Institute (NCI), USA have been identified to be the keystones for detection and treatment of such a catastrophic disease.

This chapter is a modest highway rendering complete comprehension of magnetism, peeping towards synthesis of SPIONs as well as understanding their properties over other magnetic substances, probing the resonance of such nanoparticles as contrast agents in MRI, meticulous explanation of heat dissipation mechanisms causing Hyperthermia leading to frying up of the cancerous cells and in the end twittering at the drug-delivery characteristics of SPIONs.

13.2 MAGNETISM

Magnetism encompasses magnetic properties of elements capable of attracting material objects on Earth. William Gilbert (1600) was perhaps the first to realize that, Earth was a giant magnet and the magnets could be made by heating wrought iron. This magnetic seer discovered the magnetic properties of iron and played a key role in the comprehension of Magnetism. Two centuries later Hans Christian Oersted (1820) unraveled the mystery that the magnetism is associated with electricity since an electric wire carrying current leads to deflection in the magnetic compass needle. James Clark Maxwell gave a theoretical explanation on inter-connection between electricity and magnetism, disseminating a range of deceptively simple equations that are the basis of electromagnetic theory today. Thus, Iron is contemplated to be the first element discovered to possess this property of magnetism found in Magnetite, a naturally occurring oxide of iron. Iron is gifted with the characteristic state of ferromagnetism, likewise cobalt, nickel are also showered with the same phenomena. Such materials possess domains with random magnetic moments, which align themselves with the magnetic field and exhibit remanence even if the magnetic field is removed. Likewise, some are attracted to the magnetic field (paramagnetism), others are repelled by magnetic field (diamagnetism), while others have a much more complex relation with applied magnetic field such as anti-ferromagnetism. Pure Oxygen possesses magnetic properties when cooled to a liquid state hence magnetism is temperature-dependent phenomena.

Magnetism is of four different types, which are as follows:

13.2.1 DIAMAGNETISM

Repulsion by a magnetic field and the tendency to oppose an applied magnetic field is a characteristic phenomena exhibited by diamagnetic materials. The modus operandi behind this phenomena is the deficiency of unpaired electrons, thus inherent electron magnetic moments are incapable of producing any bulk effect.

In a classical sense, magnetization, which erupts due to the electrons' orbital motions, can be explained as follows: Electrons which are encircling around the nucleus will experience in addition to the Coulombic attraction, a Lorentz force from the magnetic lines of forces when magnetic field was applied around it. This Lorentz force is responsible to augment the centripetal force on the electrons, which is direction-dependent mechanism pulling the electrons in towards the nucleus, or it causes decrement in the force leading to pushing away from the nucleus. This leads to increment in the orbital magnetic moment that were aligned opposite the field, while there is decrement in the orbital magnetic moment aligned in parallel to the field (Lenz's Law), thus resulting in a small bulk moment with an opposite direction to the magnetic field.

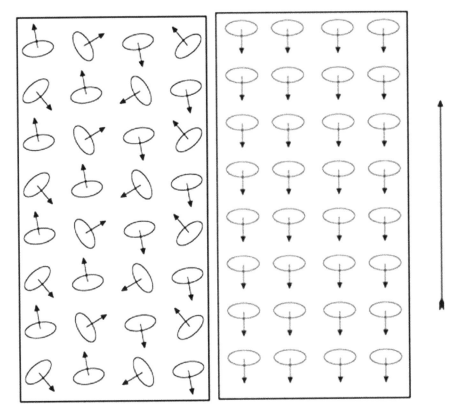

FIGURE 1   Behavior of magnetic moments in (left) absence and (right) presence of magnetic field.

13.2.2   PARAMAGNETISM

Paramagnetic materials have unpaired electrons having one electron in atomic or molecular orbitals. Since paired electrons is the key requirement for Pauli Exclusion Principle to have inherent magnetic moments of the spins, which points in opposite direction leading to cancellation of the magnetic fields, while an unpaired electron is free to align its magnetic moment in any direction.

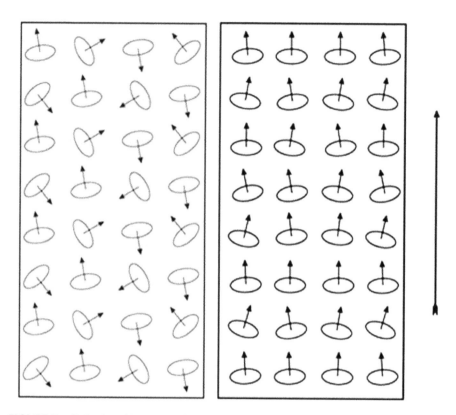

FIGURE 2 Behavior of Paramagnetic substances in (left) the absence and (right) presence of magnetic field.

On application of the external magnetic field in such materials, which possess some magnetic moment, their spins tend to align themselves in the same direction.

13.2.3   FERROMAGNETISM

Ferromagnetic materials, like paramagnetic substances, also possess unpaired electrons, likewise in addition to the electrons' inherent magnetic moment's propensity to be parallel to an applied magnetic, field; they also have an inclination to orient parallel to each other maintaining a lowered energy state. Thus, even when the applied magnetic field is removed, the electrons in the particles tend to maintain a parallel orientation. All ferromagnetic materials possess their own intrinsic temperature called the Curie temperature or Curie point, above which it completely loses its characteristic of being ferromagnetic in nature. This mechanism is due to the dominance of thermal properties over the energy lowering mechanisms. Nickel, gadolinium, iron, cobalt, and their alloys all are known to exhibit detectable ferromagnetic properties.

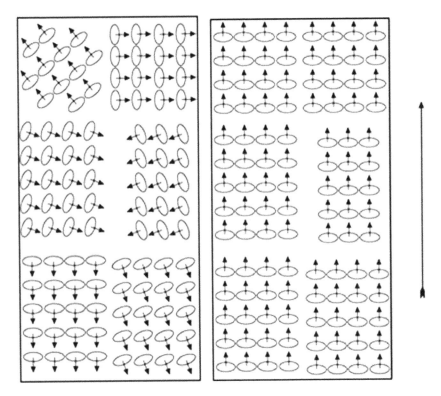

FIGURE 3   Behavior of ferromagnetic substances in (left) the absence and (right) presence of magnetic field.

MAGNETIC DOMAINS IN FERROMAGNETIC MATERIAL

The magnetic moment of atoms align themselves into small regions of uniform alignment called magnetic domains or Weiss domains. When a domain is constituted by too many molecules thus leading to its instability dividing into different domains aligned in opposite directions so that they can stick together in a more stable fashion. On exposure to the magnetic field, there is movement of domain boundaries so that the domains aligned with the magnetic field start growing and becomes dominant in the whole structure. Even if the magnetic field is removed, the domains may not return to an unmagnetized state resulting in remanence leading to the formation of permanent magnet. Strong magnetization causes dominance of the prevailing domains resulting in only one single domain, thus saturating the material magnetically. When such a ferromagnetic material is heated to Curie temperature, the kinetic energy increases causing agitation in such a way that the magnetic domains completely loses its organized behavior inhibiting all its magnetic properties. When such heated material comes back to its original temperature, it spontaneously regains its multi-domain alignment characteristic. (Kittel, 1946).

CURIE TEMPERATURE

Curie temperature or Curie point is the temperature at which a ferromagnetic substance gets converted into paramagnetic one on heating. This effect is a reversible process. Such substances lose their magnetism if heated above the Curie temperature (Curie is named after the scientist Pierre Curie). An analogous temperature Neel temperature is used for anti ferromagnetic materials. Below Curie point, neighboring magnetic spins are aligned parallel within ferromagnetic materials. But as the temperature is increased, the magnetization in each domain decreases. Hence, above the Curie temperature, the substances are paramagnetic in nature due to their completely disordered state. Curie point is a critical point at which magnetic susceptibility is theoretically infinite.

TABLE 1   Curie point of some ferromagnetic elements and alloys of iron.

Substance	Curie Point ^{0}K
Nickel	630
Iron	1043
Cobalt	1390
Gadolinium	293
Fe_2O_3	895

CURIE-WEISS LAW

Above the Curie temperature, the magnetic susceptibility, χ, is given by the Curie-Weiss law:

$$\chi = \frac{C}{T - T_C} \tag{1}$$

Where C is a material-specific Curie constant, T is absolute temperature and Tc is the Curie temperature, both measured in Kelvin. Hence, susceptibility looms around infinity as the temperature advances towards Tc.

13.2.4 APPRAISAL OF MAGNETIC PROPERTIES

Hysteresis loop is a signature marker for studying the magnetic properties of materials. A hysteresis loop shows the relationship between the induced magnetic flux density (B) and the magnetizing force (H), it is often referred to as the B-H loop.

HYSTERESIS LOOP

Hysteresis is seen in ferromagnetic materials in which there is increase in magnetization as the magnetic field increases, but as the magnetic field decreases the magnetization values do not exhibit same behavior, as given follows:

1. Magnetic domains are all aligned at point "1", while even if there is very high magnetizing force, there is infinitesimal rise in magnetic flux, thus point a is considered to be the point of magnetic saturation.
2. As the curve moves from point 1 to point 2, H is then decreased to zero. There is some residual magnetic flux in the material even if the magnetizing force is zero. This point is termed as point of Remanence.
3. The magnetizing force when gets reversed, then the curve turns to point 3, leading to the reduction of magnetic flux to zero, which is called the Point of Coercivity. Coercive force is the force, which is required to remove the residual magnetism from the materials.
4. Since the magnetizing force further increases in the negative direction, the material again saturates in the opposite way (point "4").
5. Point "5" in the curve is an area where magnetization is reduced to zero. Increment in the magnetization will return B to zero. The curve did not enter into the origin of the graph since residual magnetism has to be removed when some force is applied.
6. The loop ceases when the curve takes a different path from a point "6"

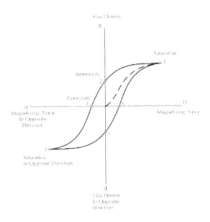

FIGURE 4 Hysteresis loop for ferromagnetic materials.

13.2.45   *SUPERPARAMAGNETISM*

Superparamagnetism is the most celebrated term used for magnetic nanoparticles, which can be exploited in plethora of biomedical applications such as it can act as a contrast agent and as a nanoflotilla for synaphic delivery of drugs. The size of nanoparticles are pertinent for their clinical application (<25 nm) so that they can efficiently act as a magnificent superparamagnetic material (Lee et al., 1996). Above Curie temperature, paramagnetic substances behave in a different manner in which thermal energy overcomes the coupling forces between neighboring atoms thus consequently leading to the random fluctuation in the direction of magnetization resulting in a null magnetic moment. On the contrary, Superparamagnetic substances possess thermal fluctuations leading to the change in the direction of magnetization of entire crystallites. There is compensation of magnetic moments of individual crystallites leading to the cancelling of magnetic moments.

During exposure to external magnetic field, the superparamagnetic substance, which has transformed from a multi-domain structure to a single domain entity, and all the magnetic spins behave like members of a single family, the magnetic moment of entire crystallites gets aligned with the magnetic field. In large nanoparticles as well as bulk ferromagnetic substances, thermodynamic contemplation is favorable for the formation of domain walls. But as the particle size is diced down below a certain value, domain wall vanishes and thus each particle is constituted by single domain. This is a fantastic explanation for superparamagnetic nanoparticles. Further, such SPIONs can also be used for clinical purposes since the external magnetic field is removed, magnetization is completely lost (negligible remanence and coercivity), thus circumventing the problem of platelet aggregation as well as embolization of capillary vessel. The most limiting characteristic of small sized nanoparticles, though they possess superparamagnetic behavior, is low sustenance under very high hemodynamic shear force as well as difficulty to stay in proximity to the target site due to the viscous drag offered by the blood flow. Synaphic delivery of cargoes is proved to be fruitful at sites of lower blood velocities and in particular, when the magnetic field source is closer to the target. (Hofmann-Amtenbrink et al., 2009).

13.3   SYNTHESIS OF SPIONS FOR HYPERTHERMIA

There are many excellent reviews describing various methods of synthesis of magnetic nano particles (MNPs) (Moroz et. al., 2002). Both Physical (For Example, condensation and nano-dispersion of compact material) and Chemical methods (sol-gel method, co-precipitation, thermolysis, decomposition and reduction of metal containing compounds and synthesis at gas-liquid interface) have been found to be suitable for synthesis of MNP. However, since this chapter is devoted to the hyperthermia treatments, here syntheses of only SPIONs that are used for hyperthermia treatment of cancer or bacterial diseases are discussed. Iron oxide NP that are considered for use in hyperthermia, comprise of a core, and a monolayer.

Core is mostly composed of magnetic metals such as ferro, ferri, superparamagnetic materials, simple oxides like magnetite Fe_3O_4, maghemite g-Fe_2O_3, Complex oxides, ferrimagnetic spinels ($Zn_2+xFe_3+(1-x)[Fe_2+(1-x)Fe_3+(1+x)]O_4$), Ferri magnetic composites ($SrFe_{12}O_{19}/\gamma-Fe_2O_3$), and ferromagnetic perovkites ($La1-xSrxMnO_3$).

The organic monolayer, which can be effectively exploited on the surface of iron oxide nanoparticles are silica shells (Bruchez et al., 1998,Santra et al., ,2001), lipids (Bulte et al.,1999, Mulder et al., 2006, Nitin et al., 2006, Jung et al., 2008), polymers (Harris, 2003, Gao, 2004), and amphiphilic ligands (Chan et al., 1998, Song et al., 2005, Lee et al., 2007, Tomlinson, 2007) depending upon the tethering molecules like DNA, proteins, carbohydrates and lipids.

The monolayer provides:

(i)   An interface between the core and the surrounding environment,
(ii) Serve as a barrier between the nanoparticle core and the environment, to protect and stabilize the core because iron oxides are not stable,
(iii) The chemical nature of monolayers dictates the reactivity, solubility and interfacial interactions (Rochelle, 2007), and
(iv) Since iron oxides are insoluble in water, monolayers serve to circumvent this problem.

The synthesis method of SPIONs has great impact on the properties like high saturation magnetization value [emu/g]. Magnetization value depends on the size of the nanoparticles (6–20 nm particle size exhibit highest emu/g (Jun et. al., 2005) Spacing between the nanoparticles (allows each individual MNP to act independently thus enhancing the net magnetism) and the crystalline structure of the iron oxide.

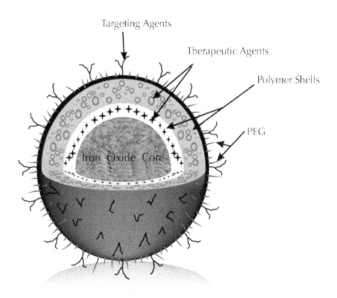

FIGURE 5   Schematic composition of MNP-drug-polymer composite consisting of inner iron oxide core and outer polymer shell.

The first step in synthesizing uniform size iron nano particles is to initiate *Nucleation* using saturated solution. Once the nuclei are formed the concentration of solution drops and the nuclei then grow by diffusion of solutes from the solution onto the nuclear surfaces, until an equilibrium concentration is achieved. In order to achieve monodispersed particles, the two phases of nucleation and growth need to be separated (Boistelle, 1988, Burda, 2005, Pedro et. al., 2006, Laurent, 2008). Size control can be achieved by separating nucleation and growth.

Some of the methods in use are:

1. Chemical precipitation
2. Thermolysis or decomposition or reduction of compounds containing the metal
3. Sol-gel method
4. Synthesis of MNP at gas – liquid interface

Apart from iron NP there is a need for specific methods for preparation of particular type of MNP For Example, (a) hetero-metallic MNP (b) Ferrites (c) NP of rare earth elements, and (d) MNP of anisotropic shapes.

Moreover, stoichiometrically homogeneous MNPs are synthesized by oxidation of NP, chemisorptions and surface modification.

13.4   BIOSYNTHESIS OF INTRACELLULAR MAGNETO-SPATIAL ORGANIZATION OF MAGNETOSOMES

Magnetosomes are signature organelles of magnetotactic bacteria like *Magnetospirillum gryphiswaldense, Magnetospirillum magnetotacticum,* which exploit the functional manifestations of intracellular magnetosomal assembly for the synthesis of magnetic nanoparticles. The discovery of magnetic bacteria was done by Richard P. Blakemore in the year 1975 that orient along the magnetic field lines of Earth's magnetic field. The tendency of aligning all the magnetic crystals inside the bacteria and navigating towards the geomagnetic field with the aid of flagella is the beauty, which is inexplicable. All these magnetic crystals packed in magnetosomal organelle exhibit superparamagnetic nature and have got single domains. This makes them a paradigm for the comprehension of hyperthermal property since their saturation magnetization at 2kOe was found to be 13 G-cm^3/gm. The precise and robust synthesis of nanoparticles with controlled and regulated crystal growth of nano-magnets is an area, which has been explored by many material scientists. They studied the complete modus operandi of their synthesis as well as magnetization properties of nanoparticles. Magnetite synthesis takes place due to the magnetosomes, which act as a nucleation site.

The formation of magnetite crystal inside a magnetosome is a three step phenomena:

• Magnetosome membrane invagination
• Magnetic precursors entry into newly formed vesicle
• Nucleation and growth of the magnetite

Magnetosomes extracted from bacteria can thus be exploited as an efficient magnetic gizmo, which possesses weakly interacting single domains and overwhelming magnetization values as compared to chemically synthesized ones, thus tapping the hyperthermal mystery needs to be unravelled. (Timco et al. 2009).

13.5   HYPERTHERMIA

The classical approach of hyperthermia constitutes of exposing a patient with electromagnetic waves of several hundred MHz frequencies. The tumor can be ablated thermally by an electromagnetic wave emitted by a RF electrode, which is implanted in the pathological area. The most important and significant role of hyperthermia treatment is increase in the perfusion of the tumor tissue thus augmenting the local oxygen concentration. The National Cancer Institute authorizes three different types of hyperthermia treatments.

13.5.1   LOCAL HYPERTHERMIA

Local hyperthermia (or thermal ablation) is used to heat a small area like a tumor. Very high temperatures are used to kill the cancer cells, coagulate the proteins, and destroy the blood vessels. In effect, this cooks the area that is exposed to the heat. Radio waves, microwaves, ultrasound waves, and other forms of energy can be used to heat the area. When ultrasound is used, the technique is called *high intensity focused ultrasound, o*r HIFU.

The heat may be applied using different methods:

*   *External:* High energy waves are aimed at a tumor near the body surface from a machine outside the body.
*   *Internal*: A thin needle or probe is put right into the tumor. The tip of the probe releases energy, which heats the tissue around it.
*   *Regional Hyperthermia*: In regional hyperthermia, various approaches may be used to heat large areas of tissue, such as a body cavity, organ, or limb.
*   *Deep Tissue Approaches:* It may be used to treat cancers within the body, such as cervical or bladder cancer. External applicators are positioned around the body cavity or organ to be treated, and microwave or radiofrequency (RF) energy is focused on the area to raise its temperature.
*   *Regional Perfusion Techniques*: It can be used to treat cancers in the arms and legs, such as melanoma, or cancer in some organs, such as the liver or lung. In this procedure, some of the patient's blood is removed, heated, and then pumped (perfused) back into the limb or organ. Anticancer drugs are commonly given during this treatment.
*   *Continuous Hyperthermic Peritoneal Perfusion (CHPP)*: It is a technique used to treat cancers within the peritoneal cavity including primary peritoneal mesothelioma and stomach cancer. During surgery, heated chemotherapy drugs are circulated through the peritoneal cavity. The peritoneal cavity temperature reaches 106–108°F.
*   *Whole Body Hyperthermia*: It is used to treat metastatic cancer that has spread throughout the body. This can be accomplished by several techniques that raise the body temperature to 107–108°F, including the use of thermal chambers (similar to large incubators) or hot water blankets. The synergistic effect of magnetic hyperthermia and radiation induces the increase in cancer cell killing even at lower temperatures, which is not the case, when hyperthermia is implemented alone. This is called as '*thermal radio-sensitization*'.

This chapter is dealing with the comprehension of modality of cancer treatment using local hyperthermia, which is also called magnetically mediated hyperthermia (SPIONs). The SPIONs are dispersed throughout the target tissue and AC magnetic field is applied of suitable strength and frequency causing the particles to heat by magnetic hysteresis losses or Neel relaxation or other losses as explained later in details.

HOW MNPS CAN BE ADMINISTERED IN VIVO?

Amongst all the different mechanisms of hyperthermal modalities, which include microwave or laser, magnetic hyperthermia has been proved to be the best candidate to selectively target the tumor cells.

The magnetic nanoparticles dispersed in a liquid can be delivered in one of four different ways to the tumor, which are as follows:

1. Arterial injection: The MNPS are delivered in the artery of tumor, which is used as the administration route.
2. Direct injection: The MNPs coated by antibodies allow an easy entry into the interstitial fluid in the body as well as in the blood vessel reaching exactly at the target site.
3. In situ implant formation: Gels entrapping MNPs into a tumor can be usefully exploited as an implant.

Active targeting involves magnetic nanoparticles targeted against tumor using antibody with an external magnetic field.

All the above administrative routes have been exploited for delivery of either stealthy MNPs or MNP-drug conjugates encapsulated inside polymer shells. To efficiently utilize MNPs as drug-delivery system or for hyperthermal therapeutics, the most critical parameter required is the high specific heating power.

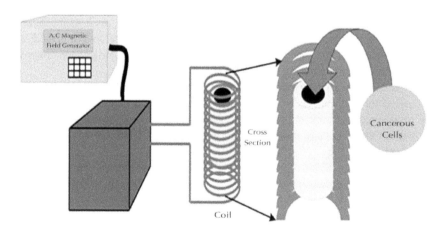

FIGURE 6   A.C. Field Magnetic Generator (Reproduced from reference Laurent S, Forge D, Port.

M et al. (2008) Magnetic iron oxide nanoparticles: synthesis, stabilization, vectorization, physicochemical characterizations, and biological applications. *Chem Rev*, **108**(6), 2064–110).

Mechanism of heat generation in Magnetic Nanoparticles (MNPs).

Heat generation using MNPs has intrigued many fundamental thinkers as to what is the actual mechanism of Hyperthermia? After brainstorming research, this unresolved ambiguity was solved and found that heat arises due to the magnetic losses in an alternating magnetic field. Magnetic losses are considered to be arising further from different magnetization reversal mechanisms, which are as follows:

1. Hysteresis
2. N'eel or Brown relaxation
3. Frictional losses in viscous suspensions

All the above magnetic losses contribute to generation of heat but eddy current induced heating cannot be considered since the size of the MNPs are extremely small.

HYSTERESIS

Hysteresis losses are the most efficient mechanisms due to which heat is generated, which is nothing but the integration of the area of hysteresis loops (It is a measure of energy dissipation per cycle of magnetization reversal). It depends on the field amplitude as well as magnetic field. Above a critical size of nanoparticles, domain walls exist and hysteresis losses of the multidomain particles is purely dependent on the type and configuration of wall pinning centre. With decreased particle size a transition from multi-domain to mono-domain occurs leading to uniform magnetization reversal causing uniform heating. Hysteresis losses are considered to be twice the uniaxial anisotropy, which is due to crystal structure, shape and size of the magnetic nanoparticles. The losses are enhanced further due to the alignment of particle axes align with the external field. Further other complicated modes of magnetization reversal such as curling and fanning also contribute to heat generation

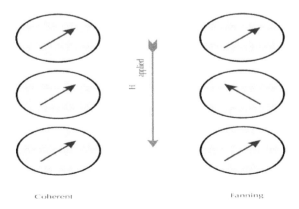

FIGURE 7   Hysteresis losses due to fanning of magnetic moments.

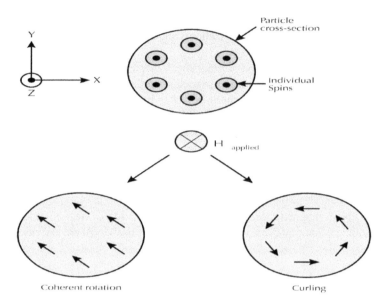

FIGURE 8 Hysteresis losses due to curling of magnetic moments.

BROWNIAN OR NEEL RELAXATION

The delay in the relaxation of the magnetic moment can be due to the rotation of the magnetic moment within the particle (Neel relaxation) or the rotation of the particle as a whole (Brownian relaxation) when AC magnetic field is exposed to MNPs. Heat dissipation occurs due to the magnetic field reversal time shorter than the magnetic relaxation.

$$\tau_N = \tau_0 e^{KVm/\kappa T} \qquad (2)$$

$$\tau_B = \frac{3nV_H}{\kappa T} \qquad (3)$$

$$\tau = \frac{\tau_B \tau_N}{\tau_B + \tau_N} \qquad (4)$$

Where τN is the Néel relaxation time, τB the Brown relaxation time, τ the effective relaxation time if both effects occur at the same time, $\tau0 = 10^{-9}$ s, K the anisotropy constant, V_M the volume of particle, k the Boltzmann constant, T the temperature, η the viscosity, and V_H the hydrodynamic volume of particle. Thermal fluctuations lead to augmented Neel and Brownian relaxations causing magnetization reversal generating large amount heat of during reversal.

Neel relaxation has got an important relationship with anisotropic energy K over thermal energy kT. This leads to the relaxation of the inner magnetic core.

Brownian relaxation is directly related to the viscosity of solution and the hydrodynamic volume of the particle. Thus, both the relaxation mechanisms contribute to Hyperthermia.

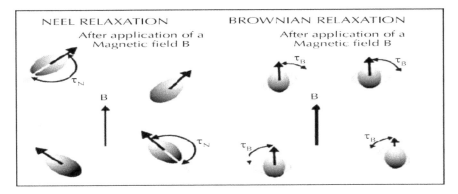

FIGURE 9   Neel Relaxation and Brownian Relaxation.

FRICTIONAL LOSSES IN VISCOUS SUSPENSIONS (VISCOUS LOSSES)

The Brownian relaxation mechanism causes generation of heat due to viscous friction between rotating particles and surrounding medium. Particles which possess remnant magnetization are considered to be active candidate to generate heat, which can be given in the equation given below:

$$T = \mu0 M_R HV \tag{4}$$

Where T is the torque moment, μ0 is the applied magnetic field, M_R is the remnant magnetization, when exposed to a rotating magnetic field *H*. In the steady state the viscous drag in the liquid $12\pi Vf$ is counteracted by the magnetic torque *T* and the loss energy per cycle is simply given by $2\pi T$.

The main parameter which determines the heating of the tissue is the specific absorption rate (SAR), which is defined as the rate at which electromagnetic energy is absorbed by a unit mass of organisms. is expressed in Watts per kilogram. The SAR is directly proportional to the rate of the temperature increases. For the adiabatic situation

$$SAR = 4.1868 \, P\!\big/\!m_e = C_e \, dT\!\big/\!dt$$

Where, P is the electromagnetic wave power absorbed by the sample, m_e is the mass of the sample, and C_e is the specific heat capacity of the sample. (Laurent et. al., 2011).

13.6   MAGNETIC NANOPARTICLES AS CONTRAST AGENTS FOR MRI

Magnetic resonance imaging (MRI) applications have steadily widened over the past decade. Currently, it is the preferred cross-sectional imaging modality in catastrophic diseases such as Cancer.

13.6.1   FROM MRI PHYSICAL PRINCIPLES TO PARAMAGNETIC CONTRAST AGENTS (T1-AGENTS)

The magnificent Nuclear Magnetic Resonance (NMR) giant, Richard Ernst who deciphered the mechanism of proton relaxation, has quenched the need of all medical scientists for the development of MRI, which can distinguish the NMR signal of protons from water in tissues, membrane lipids, proteins, and so on.

The MRI is constituted by the combined effect of static magnetic field B_0 up to 2T and a transverse RF field (5–10 MHz). On exposure to RF, the net magnetization vector is influenced by magnetic field B_0 with an attempt to realign with it in the longitudinal axis as protons have a tendency to return back to equilibrium, which is termed as relaxation.

This relaxation can be further classified into two different independent mechanisms:
1. Longitudinal relaxation, which is the return of longitudinal magnetization in alignment with B_0 and is termed T1-recovery; and
2. Transverse relaxation, which is the vanishing of transverse magnetization and is termed T2-decay.

Hydrogen atoms tend to release a large amount of absorbed energy into the lattice or tissue around it so that they can realign with the magnetic field. This type of T1 recovery is known as spin-lattice relaxation, which is the time required for 63% recovery of the longitudinal magnetization in the tissue. In contrast to T1 recovery, T2 decay is not involved in absorption or evolution or dissipation of energy, instead when RF is exposed to the hydrogen nuclei they all spin in phase with each other but as soon as the exposure halts there is interaction of the magnetic field of all nuclei with each other leading to the exchange of energy during interaction. Nuclei thus tend to lose their phase coherence and spin in a haphazard manner. T2 decay is involved in the exchange of energy between spinning protons, which is known as spin-spin relaxation.

Therefore, T2 is the actual time involved for the transverse magnetization to decrease to 37% of its earlier value.

A tissue can be classified on the basis of their T1 or T2 relaxivity, as given below:

1. Fats possess very high T1 values
2. Cerebrospinal fluid possess T2 decay
3. Fluids possess very long T2 and are frequently related to the pathologies such as cancer lesions, thus T2 weighted images are useful for diagnosing such diseases.

Superparamagnetic substances play an important role in the enhancement of magnetic resonance contrast agents. Thus, such compounds are involved in the augmentation of the relaxation rates of the surrounding proton spins of water and is termed as relaxivity and is known as R1 ~ 1/T1 or R2 ~ 1/T2. Gadolinium chelates are considered to be an efficient T1 contrast agents but it has been found that their proton relaxation rates are lowered. Thus other T1 agents can also be used such as Mn-chelates (hepatobiliary distribution for diagnosing liver lesions), iron and manganese metalloporphyrin (selective retention in tumours), nitroxide radicals, and so on.

13.6.2 SUPERPARAMAGNETIC NANOPARTICLES AS MR CONTRAST AGENTS (T2-AGENTS)

The SPIONs of size 3–10 nm are considered to act as an efficient standard and functional MR imaging. The magnetic moments of SPIONs are found to be higher as compared to paramagnetic substances. Thus, their relaxivities are found to be much higher than the Gd-chelates. They are significant enough to produce T2 relaxation effects leading to signal reduction on T2 weighed images (negative contrast). This phenomenon can be largely decided based on the large magnetic field heterogeneity surrounding the nanoparticle through, which the diffusion of water molecules occurs. Diffusion is concurrently very important since it induces dephasing of the proton magnetic moments leading to T2 shortening. Such contrast agents also possess Magnetic susceptibility. Colloidal T2 agents are called also called ultra-small superparamagnetic iron oxide nanoparticles, which consists of iron oxide cores, whose composition and physicochemical properties vary from magnetite (Fe_2O_3) to maghemite (Fe_3O_4) (Sun, 2008).

13.7 MAGNETIC NANOPARTICLES AS EFFICIENT PIGGYBACKS FOR FERRYING DRUGS AND ANTIBODY FOR SYNAPHIC DRUG-DELIVERY

Cancer therapeutics is a biggest challenge for mankind. Cancer is such a catastrophic disease, which becomes dominant over the whole body and develops mechanisms that can escape out reticuloendothelial clearance. Thus, drug targeting is an avalanche, since they can kill normal cells also, causing inimical side-effects. To refrain from such barriers in cancer therapy, magnetic nanoparticles have been proposed to be efficient three-pronged paraphernalia, which can be exploited as multimodal imaging particles, synaphic delivery of SPIONs using antibodies and on exposure to AC magnetic field drugs can be released in the cancerous milieu. Parameters such as the physicochemical properties of the drug-loaded MNP, field strength and geometry, depth of the target tissue, rate of blood flow, and vascular supply, all play a role in determining the effectiveness of this method of drug delivery. Clinical trials on MNP loaded Epirubicin are directed against solid tumors demonstrating successful congregation in the target site.

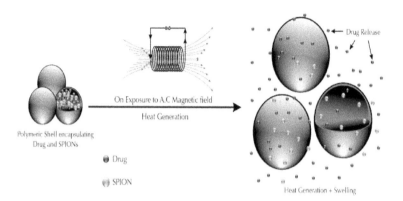

FIGURE 10   Schematic representation of heat dissipation from SPION – Drug – Polymer Conjugate on exposure to A. C. Magnetic field.

Many potential threats such as malignant prostate and breast tumors can be treated using SPIONs as proficient drug delivery vehicle for synaphic delivery of etoposide, doxorubicin, and methotrexate characteristics such as loading capacities and drug release profiles can now be tailored by controlling structural features and chemical bonding within the MNP conjugate. Yang et al. have investigated the synthesis and release characteristics of poly (ethyl-2-cyanoacrylate) (PECA) coated magnetite nanoparticles containing anti-cancer agents cisplatin and gemcitabine. In addition to drug molecules, MNPs have been investigated as carriers of therapeutic proteins and peptides. Herceptin™, also known as trastuzumab, has been conjugated to MNPs as a mAb targeting agent. By incorporating Herceptin™ into magnetite nanoparticle loaded liposomes, Ito et al. demonstrated an antiproliferative effect on breast tumor cells. Apart from drug-and antibody attachment antisense and genes are also attached on the Fe_2O_3 nanoparticles. However, the delivery of genes and their resulting transfection efficiencies are often limited by their short half-life *in vivo*, lack of specificity, and poor diffusion across cell membranes. Magnetofection is a very famous technique, which is exploited for the delivery of antisense oligodioxynucleotides (ODNs) or gene vectors for their use in therapeutic applications. (Sun et al. 2008).

SUMMARY

The predilection of SPIONs due to its less toxicity, synaphicity, and voluminous specificity has led us to pertinently exploit them as an efficient contrast agent as well as a vehicle for drug-delivery. They have been profoundly used as T2 contrast agents since they possess spin-spin relaxation, but possess little less T1 recovery, which leads to spin-lattice relaxation causing dissipation of heat. Thus, SPIONs can be efficiently exploited for hyperthermia. Further, they can also be used as nano-flotillas for delivery of synaphic delivery of drugs with an aid of antibody. Thus clairvoyantly, we can declare SPIONs as targeting agents for induction of hyperthermia in cancerous cells.

Presently, the amount of nanoparticles and hence strength of magnetization is very low at the site of tumor thus antibody targeting is very low for sufficient temperature rise. Hence, there is a need to design stealth nanoparticles, which are able to circulate in the connective tissues such as blood for a long time and the surface grafting of ligands capable enough to specifically internalize in tumor cells. Thus, the size of nanoparticles needs to be optimized for their competency in generation of heat. Likewise, drugs attached on their surface can be released effectively on exposure to AC magnetic field.

KEYWORDS

- **Anisotropic energy**
- **Ayurveda**
- **Diamagnetism**
- **Hyperthermia**
- **Super paramagnetic iron oxide nanoparticles**

REFERENCES

1. Kittel, C. Theory of the structure of ferromagnetic domains in films and small particles. *Phys Rev.*, **70**(11–12), 965–971 (1946).
2. Neel, L. Théorie du trainage magnétique des ferromagnétiques en grains fins avecapplications aux terres cuites. *Ann Géophys.*, **5**, 99–136 (1949).
3. Margarethe, Hofmann-Amtenbrink, Brigitte, von Rechenberg, and Heinrich, Hofmann. *Nanostructured Materials for Biomedical Appl.*, M. C. Tan (Ed.), Transworld Research Network, kerala, pp 119–149 (2009).
4. Moroz, P., Jones, S. K., and Gray, B. N. Magnetically mediated hyperthermia current status and future directions. *Int. J. Hyperthermia*, **18**, 267–284 (2002).
5. Bruchez, M. Jr., Moronne, M.,and Gin, P. *et al.* Semiconductor nanocrystals as fluorescent biological labels. *Science*, **281**(5385), 2013–2016 (1998).
6. Santra, S., Tapec, R., and Theodoropoulou, N. *et al.* Synthesis and Characterization of Silica-Coated Iron Oxide Nanoparticles in Microemulsion: The Effect of Nonionic Surfactants. *Langmuir*, **17**, 2900–2906 (2001).
7. Bulte, J. W., de Cuyper, M., andDespres, D. *et al.* Short- vs. longcirculating magnetoliposomes as bone marrow-seeking MR contrast agents. *J Magn Reson Imaging*, **9**(2), 329–335 (1999).
8. Mulder, W. J., Strijkers, G. J., and van Tilborg, G. A. *et al.* Lipid-based nanoparticles for contrast-enhanced MRI and molecular imaging. *NMR Biomed.*, **19**(1), 142–164 (2006).
9. Nitin, N., LaConte, L. E., and Zurkiya, O. *et al.* Functionalization and peptide-based delivery of magnetic nanoparticles as an intracellular MRI contrast agent. *J Biol Inorg Chem.*, **9**(6), 706–712 (2004).

10. Jung, J., Matsuzaki, T., and Tatematsu, K. *et al*. Bio-nanocapsule conjugated with lipo-somes for in vivo pinpoint delivery of various materials. *J Control Release*, **126**(3), 255–264 (2008).

11. Harris, L., Goff, J., and Carmichael, Y. *et al*. Magnetite Nanoparticle Dispersions Stabi-lized with Triblock Copolymers. *Chem of Mat.*, **15**, 1367–1377 (2003).

12. Gao, X., Cui, Y., and Levenson, R. M. *et al*. In vivo cancer targeting and imaging with semiconductor quantum dots. *Nat Biotechnol.*, **22**(8), 969–976 (2004).

13. Chan, W. C. and Nie, S. Quantum dot bioconjugates for ultrasensitive nonisotopic detec-tion. *Science*, **281**(5385), 2016–2018 (1998).

14. Song, H. T., Choi, J. S., and Huh, Y. M. *et al*. Surface modulation of magnetic nanocrystals in the development of highly efficient magnetic resonance probes for intracellular label-ing. *J Am Chem Soc.*, **127**(28), 9992–9993 (2005).

15. Lee, J. H., Huh, Y. M., and Jun, Y. W. *et al*. Artificially engineered magnetic nanoparticles for ultra-sensitive molecular imaging. *Nat Med.*, **13**(1), 95–99 (2007).

16. Tomlinson, I. D., Gussin, H. A., and Little, D. M. *et al*. Imaging GABA(c) Receptors with Ligand-Conjugated Quantum Dots. *J Biomed Biotechnol.*, **2007**,76514 (2007).

17. Rochelle Arvizo, R., and De, M.. Proteins and Nanoparticles: Covalent and Noncovalent Conjugates. *Nanobiotechnology II. More Concepts and Applications*, C. A. Mirkin and M. C. Niemeyer (Eds.). Wiley-VCH ,Weinheim (2007).

18. Lawaczeck, R. D., Menzel, M., and Pietsch, H. Superparamagnetic iron oxide particles: contrast media for magnetic resonance imaging. *Appl Organomet Chem.*, **18**, 506–551 (2004).

19. Laurent, S., Forge, D., Port, and M. *et al*. Magnetic iron oxide nanoparticles. synthesis, stabilization, vectorization, physicochemical characterizations, and biological applica-tions. *Chem Rev.*, **108**(6), 2064–2110 (2008).

20. Jun, Y. W., Lee, J. H., and Cheon, J. Nanoparticle Contrast Agents for Molecular Magnetic Resonance Imaging. Nanobiotechnology II. More Concepts and Applications. C. A. Mir-kin. and M. C. Niemeyer (Eds.). *Wiley-VCH* ,Weinheim (2007).

21. Jun, Y. W., Huh, Y. M., and Choi, J. S. *et al*. Nanoscale size effect of magnetic nanocrystals and their utilization for cancer diagnosis via magnetic resonance imaging. *J Am Chem Soc.*, **127**(16), 5732–5733 (2005).

22. Boistelle, R. and Astier, J. P. Crystallization mechanisms in solution, *J Crys Growth*, **90**, 14-30 (1988).

23. Burda, C., Chen, X., and Narayanan, R. *et al*. Chemistry and properties of nanocrystals of different shapes. *Chem Rev.*, **105**(4), 1025–1102 (2005).

24. Pedro, T., P, M. M., and Sabino, V. V. *et al*. Synthesis, Properties and Biomedical Appli-cations of Magnetic Nanoparticles. *Handbook of Magnetic Materials*, K. H. J. Buschow (Ed.), Elsevier B V, Netherlands (2006).

25. Laurent, S., Forge, D.and Port, M. *et al*. Magnetic iron oxide nanoparticles. synthesis, sta-bilization, vectorization, physicochemical characterizations, and biological applications. *Chem Rev.*, **108**(6), 2064–2110 (2008).

26. Timko, M., zarova, A. D., Kovac, J., Kopcansky, P., Gojzewski, H., and Szlaferek, A. Magnetic Properties of Bacterial Nanoparticles, *Acta Physica Polonica A.*, **115**(1), (2009).

27. Laurent, S., Dutz, S., Häfeli, U. O., and Mahmoudi, M. Magnetic fluid hyperthermia. Focus on superparamagnetic iron oxide nanoparticles, *Advances in Colloid and Interface Science*, **166**, 8–23 (2011).

28. Sun, C., Lee, J. S. H., and Zhang, M. Magnetic nanoparticles in MR imaging and drug delivery. *Advanced Drug Delivery Reviews*, **60**, 1252–1265 (2008).

CHAPTER 14

SUSTAINED AND RESPONSIVE RELEASE OF CORROSION INHIBITING SPECIES FROM MULTILAYERED ASSEMBLY OF ZINC PHOSPHOMOLYBDATE CONTAINERS WITH APPLICATION IN ANTICORROSIVE PAINT COATING FORMULATION

S. E. KAREKAR, B. A. BHANVASE, and S. H. SONAWANE

CONTENTS

14.1   INTRODUCTION

Corrosion occurs due various phenomena's for example, reaction of metal surface with oxygen and water, contact of metal surface with acid, base, salt, or due to any electrochemical reaction. There is no such industry, which does not have corrosion, mainly corrosion takes place in chemical and petrochemical industries. Any metal surface can be protected from corrosion by using three various mechanisms such as cathodic protection, anodic protection (passivation technique), and barrier mechanism (Revie, 2008). Almost in every industry barrier technique is more popular due to ease of application and main thing is that it forms a barrier of anticorrosive coatings in between metal surface and corrosive environment, which also results in low penetration of corrosive chemicals on metal surface (Schweitzer, 2006). A corrosion inhibiting compound mainly used as a barrier coating, which forms a barrier film on the surface of metal by attaching on the surface of metal, which results in preventing the corrosive reaction from occurring and can be easily added into paint coating formulation. Borisova et al. (2011) put forth that ACP that is active corrosion protection is a system in which anticorrosive liquids can be introduced into the coatings. Murphy et al. (2010) predicted that the direct application of anticorrosive liquid in anticorrosive coatings is not preferable, which have an effect on the performance of formed coatings. Bare corrosion inhibitor pigment can be encapsulated by the formation of multilayered structure of different anticorrosive species for the responsive release with respect to outside environment and which offers good corrosion resistance performance than alone bare corrosion inhibitor pigment (Wu et al., 2008, Tedim et al., 2010, Zheludkevich et al., 2007, Hu et al., 2005, Evaggelos et al., 2011, Zheludkevich et al., 2005, Suryanarayana et al., 2008, Nesterova et al., 2011). Polyelectrolyte encapsulated nanocontainer can be used in a variety of applications such as biomedical (Benito et al., 2003), controlled release drugs (Yang et al., 2008), reaction rate accelerator catalyst (George et al., 2011), textile industries, and so on. The polyelectrolyte layer is a pH dependent formulation, which forms a matrix layer on encapsulated material thus can be applicable in number of fields and to release species in responsive manner. Tedim et al. (2010) invented the synthesis of nanocontainer and encapsulation of core corrosion inhibiting pigment (Zheludkevich et al., 2007, Evaggelos et al., 2011, Zheludkevich et al., 2005, Suryanarayana et al., 2008, Nesterova et al., 2011, Kumar et al., 2006, Aramaki,2002, Paliwoda-Porebska et al., 2005, Shchukin et al., 2006, Kartsonakis and Kordas, 2009, Shchukin et al., 2008). The multilayered structure of ZMP nanocontainer is useful because of its responsive and sustained release of active anticorrosion species. Liquid corrosion inhibitors generally react with the coating when it is directly used, therefore it can be used in the form of layered structure of micro/nanocontainer. Encapsulated corrosion inhibiting species can release in responsive manner but, which dependent upon pH, temperature, mechanical stimulation, and their layers combinations.

Nanocontainer can be used for corrosion inhibition when the following conditions are precisely maintained:

(1) Nanocontainer should be dispersed in paint formulation uniformly and it should remain inactive when not in application,

(2) Nanocontainer should be compatible with paint formulation for uniform formation of coating.

(3) Nanocontainer which incorporated in paint coating formulation should not disturb other properties of coatings such as adherence as so on,

(4) Self healing is an important function of nanocontainer so for that nanocontainer show fast release kinetics and responsive release, and

(5) Nanocontainer should be added in required quantity to maintain pigment volume concentration (PVC ratio).

Shchukin et. al., (2006) put-forth that layer by layer (LbL) synthesis of nanocontainer by encapsulation with polyelectrolyte. The PAA is a matrix like structure, which encapsulates the corrosion inhibiting pigment in multilayered synthesis of nanocontainer, which shows responsive and controlled release based on outside environment. A multifunctional shell assembly of nanocontainer can be prepared using deposition of oppositely charged species (polyelectorlytes and nanoparticles) on the surface of the core material. Shchukin et. al., (2006) have used inorganic nanotubes called as halloysite nanoclay of inside diameter 25 nm for encapsulation of corrosion inhibitor that is benzotriazole. Above nanoclays are pH based basically made up of kaolin on upper surface of which negative and inside, which positive charge exist at pH 8. This halloysite nanoclay has an ability to absorb species inside the lumen and shows release according to external environmental pH conditions. Shchukin et. al. (2008) used halloysite nanoclay with inside lumen loaded with corrosion inhibitor such as benzotriazole. These halloysite tubes were encapsulated with polyelectrolyte by using Sol–gel method, which shows responsive and sustained release. Shchukin et al. have reported the preparation of core and shell structure of halloysite nanoclay using emulsion method (Shchukin et. al., 2008, Shchukin and Möhwald, 2007). Responsive release of halloysite nanocontainer depends upon external pH conditions (Shchukin et .al., 2008). Corrosion is a destructive attack of metal by external environmental conditions and which occurs due to formation of carbonic acid (H_3O). When any metal surface comes in contact with oxygen and water, corrosion occurs due to formation of carbonic acid, therefore zinc phosphomolybdate forms a passive layer in contact with water and oxygen, this results in inhibition of the flow of electrons from anode to cathode (Chico et. al., 2008, Kalendova and Vesely, 2009) as well as metal surface get protected from rusting. Some chromate and lead bearing pigments are environmentally restricted due to their toxicity to nature but Zn_2PMoO_7 is efficient pigment because of optimum cost and environmental friendly behavior. From all such a characteristics Zn_2PMoO_7 is used as an inorganic core in the synthesis of layered structure of ZMP nanocontainer and to avoid rusting at edge of metal and polyaniline and PAA are important because these are liquid corrosion inhibitors which attaches to metal surface and inhibit the corrosion even if other film is removed from the surface. A Zn_2PMoO_7 act as an anodic-type inhibitor for metal surface, which was encapsulated in between multilayered structure of ZMP nanocontainer also it is excellent nano-filler.

The Zn_2PMoO_7 nanocontainers were prepared by deposition of polyaniline/benzotriazole/polyacrylic acid layers on Zn_2PMoO_7 nanoparticles and incorporated in alkyd resin and that act as excellent corrosion inhibitor for metal surface. Benzotriazole is encapsulated in between the two polyelectrolytes (polyaniline and PAA) layers.

Polyaniline (PANI) used as a polyelectrolyte, has inhibiting effect due to the creation of a compact iron/dopant complex assembly at the metal-coating interface, which acts as a passive protective layer having a redox potential to go through a continuous charge transfer reaction at the metal-coating boundary, in which PANI is reduced from emeraldine salt form (ES) to an emeraldine base (EB) (Jose et. al., 2005). The above synthesized nanocontainer has an excellent corrosion inhibiting efficiency. Estimation of release of benzotriazole at various pH was investigated using UV spectroscopic analysis. Corrosion inhibiting ability of Zn_2PMoO_7 nanocontainer on MS has been confirmed by various analysis as well as responsive release of benzotriazole at different pH.

The current book chapter consists of synthesis of zinc phosphomolybdate (ZMP) by ultrasound assisted sonochemical synthesis and ZMP nanocontainer by the formation of multilayered structure of oppositely charged species on the surface of ZMP nano pigment by ultrasonic emulsion synthesis. To ease the application of deposition of oppositely charged layers ZMP nanoparticles were functionalized with myristic acid (MA). Benzotriazole is a good corrosion inhibitor for copper containing alloys so it is adsorbed in between two polyelectrolyte layers (polyaniline and polyacrylic acid). The average particle size of ZMP nanoparticles, which are synthesized by sonochemical method, was found to be 68.4275 nm. Lesser particle size obtained is due to efficient mixing and maximum amplitude of ultrasound horn. The ZMP nanocontainer and ZMP nanoparticles characterized by XRD, PSD, FTIR, zeta potential, and TEM, which are evident to show successful formation of multilayered structure of ZMP nanoparticles with ZMP nanoparticles at the core. The UV-vis spectroscopic analysis was used to estimate the release rate, release flux, and diffusivity of benzotriazole in water with respect to time and at varying pH environment. The obtained results showed that the benzotriazole release is found to be decreased with an increase in pH value. Also, the estimated diffusivity is found to be higher at lower pH that is 2.78×10^{-14} cm^2/min at pH 2. From above results it can be predicted that the multilayered structure of ZMP nanoparticle can be used as good anticorrosive coating formulation in paint. The ability of ZMP nanocontainer to sustain against the corrosive environment was tested by dispersing ZMP nanocontainers in alkyd resin. From these results, it is observed that the corrosion resistance ability of ZMP nanocontainer is dependent upon the amount incorporation of it in alkyd resin. It is found that with an increase in the loading of ZMP nanocontainer (0–5 wt% of total paint volume composition) in alkyd resin, the corrosion rate was found decreased significantly. Also it is observed that the corrosion rate was higher in 5% NaOH solution compared to 5% HCl and NaCl solution, which indicates that ZMP nanocontainer has more corrosion resistant ability in HCl solution. Tafel plot results shows nanocontainer has improvement in the anticorrosive behavior.

14.2   EXPERIMENTAL

14.2.1   MATERIALS

For the synthesis of ZMP nanocontainer, analytical grade sodium molybdate (Na_2MoO_4), zinc sulphate ($ZnSO_4$), potassium dihydrogen phosphate (KH_2PO_4), 16 N nitric acid, 50 % caustic lye ,ammonium persulphate (APS, $(NH_4)_2S_2O_8$), as an initiator, and sodium dodecyl sulfate (SDS, $NaC_{12}H_{25}SO_4$) as a surfactant were procured from S. D. Fine Chem. and used as received without further purification. HCl, NaOH, NaCl, benzotriazole, and ethanol these are all of analytical grade chemicals procured from Sigma Aldrich and were used as received. Polyacrylic acid (PAA, $M_w = 50000$ g mol^{-1}) procured from Sigma Aldrich and were used as received. The monomer aniline (analytical grade, M/s Fluka) was distilled two times before actual use. Millipore water was used as a medium throughout the all experimentation.

14.2.2   PREPARATION OF ZINC PHOSPHOMOLYBDATE NANOCONTAINERS

The Zn_2PMoO_7 as an anticorrosive pigment in nanoform was synthesised using sonochemical irradiation method. At first aqueous solutions of sodium molybdate dihydrate, zinc sulphate heptahydrate, and potassium dihydrogen phosphate were prepared separately by adding 153.5 g sodium molybdate dihydrate in 50 ml distilled water, 100 g zinc sulphate heptahydrate in 50 ml distilled water, and 8.0 g potassium dihydrogen phosphate in 20 ml distilled water. The solutions of zinc sulfate heptahydrate and potassium dihydrogen phosphate were added drop wise simultaneously to the sodium molybdate dihydrate solution under ultrasonic irradiation within time span of 10 min. After complete addition of zinc sulfate heptahydrate and potassium dihydrogen phosphate solution, 16 N nitric acid is added as an oxidizing agent under sonication. After addition of nitric acid the aqueous clear reaction mixture was turned to yellowish initially. During the continuation of the reaction the pH of solution was adjusted to 7 by the addition of 50% caustic lye solution under constant sonication, which results into dense white precipitate. The total reaction time was about 1 hr under sonication at room temperature ($24 \pm 2°°C$). To separate out the synthesized product after completion of reaction, the reaction mixture was kept in water bath at 150°C for 10 min, which results into formation of ZMP (whitish in nature) nanoparticles. Synthesis of Zn_2PMoO_7 nanocontainers have been carried out in a subsequent steps, which are summarized below:

(1) Doping of PANI Layer on Nano Zn_2PMoO_7 by Sonochemical Emulsion Polymerization Method: The ZMP nanocontainer was synthesized using Zn_2PMoO_7 as a core material due to its multifunctional and environmental legislation acceptable nature and excellent anticorrosive characteristics. Hydrophobic property incorporated in ZMP nanoparticles (5 g in 100 ml water) for that these were initially functionalized with 0.5 g myristic acid solution in 10 ml methanol at 60°C under sonication at 60 min. After modification of ZMP with myristic acid it will create negative charges onto the surface of ZMP nanoparticles as $C_{13}H_{27}COO^-$ functional group adsorbed on ZMP

nanoparticle surface. The ZMP nanoparticles having negative charge on the surface and which are functionalized with the help of myristic acid were encapsulated by polyaniline (PANI, positively charged) layer by sonochemical microemulsion polymerization method reported by Bhanvase and Sonawane (2010). According to method proposed by Bhanvase and Sonawane (2010), at first surfactant solution of sodium dodecyl sulfate was prepared by adding 3 g of SDS and 0.2 g of Zn_2PMoO_7 (on basis of monomer aniline) in 50 ml water, which was directly transferred to sonochemical reactor (Hielscher Ultrasonics GmbH, 22 KHz frequency 240W power). The APS, that is ammonium persulfate used as an initiator for the polymerization reaction, its solution was prepared separately with the addition of 3.5 g APS in 20 ml of deionized water and which was mixed to the reactor. Aniline in amount of 5 g was added in semibatch mode within time span of 30 min, after completion of addition the reaction was kept continued further for 1 hr (total reaction tine 1.5 hr) in the presence of ultrasound at 4°C temperature. The synthesized product was separated by using centrifugal sedimentation method (Remi Instruments Supply 220/230 V, 50Hz, 1 AC) at 8000 rpm for 10 min and further washed with ultrapure water to take out un-reacted material and impurities. The separated product was dried with the help of oven for and at 60°C.

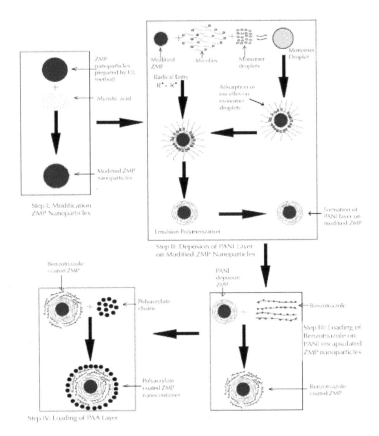

FIGURE 1   Schematic representation of the formation of ZMP nanocontainer.

(2) Loading of Benzotriazole (Corrosion Inhibitor) Layer on PANI Encapsulated Zn_2PMoO_7: Benzotriazole layer loaded in between two polyelectrolyte layers that is PANI and PAA layer. This is carried out by adding 2 g PANI loaded Zn_2PMoO_7 (synthesized in step 1) in 0.1 N NaCl solution prepared within 100 ml water. Positively charged benzotriazole (third layer) layer deposited on the surface of PANI loaded Zn_2PMoO_7 nanoparticles by adding 2 mg ml^{-1} of benzotriazole in acidic media at pH 3 in sonochemical assisted culture for 20 min. Synthesized product was separated by using centrifugal sedimentation method at 8000 rpm continued for 10 min, dried at 60°C for a time period of 3 hr and which was used again for loading of PAA layer.

(3) Deposition of PAA Polyelectrolyte Layer on Benzotriazole Loaded PANI Encapsulated Zn_2PMoO_7 Nanoparticles: The PAA polyelectrolyte layer was adsorbed on benzotriazole loaded PANI encapsulated ZMP nanoparticles to attain sustained and responsive release of corrosion inhibitor at varying pH environment and to make ZMP nanocontainer suitable to prepare alkyd resin based paint formulation. For the adsorption of PAA layer, 2 mg ml^{-1} of concentration solution of PAA was prepared in 0.1 N NaCl solution by using of sonochemical irradiation for time duration of 20 min. The prepared product was separated by centrifugal sedimentation at 8000 rpm and for time span of 10 min, washed with deionized water and dried in oven at 60°C for 48 hr. Formed nanocontainer has multilayered structure, which contain myristic acid functionalized ZMP nanoparticles as a core and benzotriazole corrosion inhibitor was entrapped in between two polyelectrolyte layers that is PANI and PAA layer. Sonochemical microemulsion method polymerization was used to make sure the complete coverage of ZMP nanoparticles surface with PANI. The formation mechanism of ZMP nanocontainer is reported in Figure 1.

14.2.3   PREPARATION OF ZINC PHOSPHOMOLYBDATE NANOCONTAINER/ALKYD MULTIFUNCTIONAL COATINGS

Pigment muller was used to incorporate ZMP nanocontainer particles in alkyd resin. Zn_2PMoO_7 nanocontainers/alkyd multifunctional coatings has been prepared by dispersing ZMP nanocontainer with an concentration of 0–5.0 wt % of total paint volume concentration (10 g) in alkyd resin with help of pigment muller (Sheen Instruments at 400 RPM). To make uniform coating and to relieve the procedure of application above prepared coating was carefully mixed with acetone solution and applied on MS panel by using bar coater panels having dimensions $50 \times 40 \times 1$ mm.

14.2.4   CHARACTERIZATION OF ZINC PHOSPHOMOLYBDATE NANOCONTAINER AND ZINC PHOSPHOMOLYBDATE NANOCONTAINER/ALKYD COATINGS

The X-ray diffraction analysis (Rigaku Mini-Flox, USA) was performed in oder to identify the types of phases exist, crystallite size in Zn_2PMoO_7 nanoparticles and Zn_2PMoO_7 nanocontainer. The morphology of Zn_2PMoO_7 nanocontainer was performed by using transmission electron microscopy (TEM) (Technai G20 working at 200 kV). Responsive release of (corrosion inhibitor) benzotriazole was measured at various pH values of 2, 4, 7, and 10 using UV-vis spectrophotometer (SHIMADZU

160A model). To observe the kinds of bonding present in molecules infrared spectroscopic analysis of samples was carried out using SHIMADZU 8400S analyzer in the spectral range of 500–4000 cm^{-1}. In order to find out the surface charge present at the surface that is zeta potential and to observe particle size distribution present in molecules, Malvern Zetasizer Instrument (Malvern Instruments, Malvern, UK) was used. To test corrosion inhibiting ability of ZMP nanocontainer MS plate (density 7.86 gm/cm^3) was used. Coating film of ZMP nanocontainer was maintained in the thickness of 50 μm on the surface of MS plate and further these plates were used to carry out dip test in acid. Above plates were dipped in acid, base and salt (5 wt% for each) solution for nearly 750 hrs. Ability to resist corrosion of ZMP nanocontainer with respect to loss in weight of MS was estimated by calculating the corrosion rate (V_c) in cm/year for each one of the samples (Bhanvase and Sonawane, 2010) by using an fundamental expression:

$$V_C = \frac{\Delta g}{Atd} \tag{1}$$

Where Δg is the loss in weight of each species in gram calculated by difference in weight of each species before dipping and after dipping, A is the exposed area of the sample in cm^2 in corrosive media, t is the time of exposure in years under dip test, and d is the density of the metallic species in g/cm^3 (mild steel that is 7.86 gm/cm^3). Loss in weight of each species was measured by washing the samples with ultrapure water after 750 hrs and removing loosen material from species and then drying was carried in oven at 60° C (± 1) out to obtain equilibrium weight by removing moisture present at surface of species because of washing. Electrochemical corrosion potential that is Tafel plot analysis (log $|I|$ vs. E) of 0%, 2%, and 4% (wt% of total paint composition) ZMP nanocontainer coated plate of M.S and was carried out in 5% NaCl solution as an electrolyte at room temperature (25°C) and this characterization was performed on computerized electrochemical analyzer (supplied by Autolab Instruments, Netherlands). Three different MS plates of Zn$_2$PMoO$_7$ nanocontainer having composition 0, 2.0, and 4.0 wt% of coated with alkyd resin containing were used as working electrode, as Pt and Ag/AgCl were used as counter and reference electrodes respectively. The plate area used for characterization was about 1 cm^2 for sample testing. The electrochemical window was − 1.0 V to + 1 V with 2 mV/s scanning rate.

14.3 RESULTS AND DISCUSSIONS

14.3.1 FORMATION MECHANISM, ZETA POTENTIAL AND PARTICLE SIZE DISTRIBUTION OF ZINC PHOSPHOMOLYBDATE NANOCONTAINER

Figure 1 explains the formation mechanism of ZMP nanocontainer. The ZMP nanoparticles were prepared by sonochemical irradiation method explained in Figure 1. The ZMP nanoparticles were functionalized by using myristic acid (C$_{13}$H$_{27}$COO$^-$) to achieve hydrophobicity of ZMP nanoparticles for efficient encapsulation of ZMP nanoparticles in PANI layer. The MA has ability to improve the hydrophobicity and create negative potential on the surface by the adsorption of negatively charged $C_{13}H_{27}COO^-$ functional group. After modification step, functionalized ZMP nanoparticles were encapulated by PANI layer and which was achieved by sonochemical micro-emulsion polymerization process. The PANI layer on ZMP nanoparticles was successfully adsorbed because of hydrophobic nature of ZMP nanoparticles due to myristic acid modification step and the presence of negative charge on ZMP nanoparticles. The adsorption of next layer that is benzotriazole (corrosion inhibitor) was successfully carried out on PANI/MA/ZMP particles. Finally, above Benzotriazole/ PANI/MA/ZMP structure was encapsulated by using deposition of the negatively charged PAA layer (forth layer) and which was carried out after the formation of the layer of benzotriazole by sonochemical microemulsion polymerization method. As sonochemical irradiation method is used for the synthesis of ZMP nanocontainer, which results into distinct reduction in particle size and formation of ZMP nanoparticles without agglomeration.

Nanocontainer is a multilayered structure, due to deposition of number of layers ionic strength of liquid changes, which affects stability of nanocontainer. Zeta potential gives the value of the charge present at surface after each layer deposition. The value of Zeta potential after each layer deposition of synthesized ZMP nanocontainer reported in Figure 2(a). The zeta potential value of bare ZMP nanoparticles is – 21.2 mV. After modification of ZMP nanoparticles with myristic acid, the negative value of zeta potential increases and becomes –25.7 mV, because myristic acid, which is having $C_{13}H_{27}COO^-$ functional group get adsorbed on bare ZMP nanoparticles. After deposition of PANI layer negative value of zeta potential decreases slightly and becomes – 24.3 mV. It is seems that the negative value of Zeta potential decreases after loading PANI layer. Zeta potential shows negative value due to presence of sodium dodecyl sulfate (SDS), which was used during emulsion polymerization, though PANI layer introduces positive charges on the surface. During emulsion polymerization, aniline comes in contact with the micelle and occupies the space near to the head group of the adsorbed SDS rather than the hydrophobic tail region (Barkade et al., 2011). Due to this reduction in the value of zeta potential takes place after loading into PANI layer. The phenyl moiety is located within the hydrophobic region and the polar group remains between the SDS head groups put forth by Kim et al. (2001). After deposition of PANI layer loading of benzotriazole (corrosion inhibitor) takes place, which results in decrease in the negative value of zeta potential that is –19.9 mV.

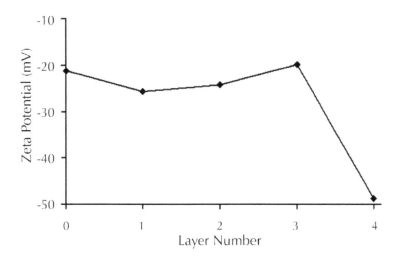

FIGURE 2   (a) Zeta Potential of ZMP nanocontainer in water. Layer number 0: initial Zn$_2$PMoO$_7$, 1: Myristic acid treated ZMP, 2: Zn$_2$PMoO$_7$/PANI, 3: Zn$_2$PMoO$_7$/PANI/ Benzotriozole, 4: Zn$_2$PMoO$_7$/PANI/ Benzotriozole/PAA.

Benzotriazole loaded ZMP nanoparticles encapsulated with PAA layer, after en-capsulation with PAA layer negative value of zeta potential value suddenly increases up to –48.9 mV. This sudden increase in value of zeta potential is due to adsorption of negatively charged PAA chains. Charged species are adsorbed on bare ZMP nanopar-ticles are confirmed by zeta potential and which also shows intra-particle interaction in the ZMP nanocontainer. From Figure 2(b), it is observed that after addition of each layer, the average particle size goes on increasing gradually. From the results of par-ticle size distribution analysis, it is confirmed that there is a formation of multilay-ered structure of ZMP nanocontainer. The particle size of bare ZMP nanoparticle was found to be around 68.43 nm and it is found to be decreased marginally to 65.32 nm after deposition of myristic acid layer. The adsorption of PANI layer by the emulsion polymerization onto ZMP nanoparticles results into an increase in the particle size to an average value of 544.95 nm. The thickness of PANI layer is found to be 98.05 nm. After deposition of benzotriazole that is corrosion inhibitor, the particle size increases to 863.69 nm with layer thickness 159.37 nm, from above it is observed that there is more adsorption of benzotriazole molecules on PANI loaded ZMP nanoparticles. Finally deposition of PAA on benzotriazole layer takes place by emulsion polymeriza-tion method, which results in increase in average particle size up to 1492.23 nm with an average layer thickness of 314.26 nm. Increase in particle size after deposition of each layer signifies that successful formation of ZMP nanocontainer.

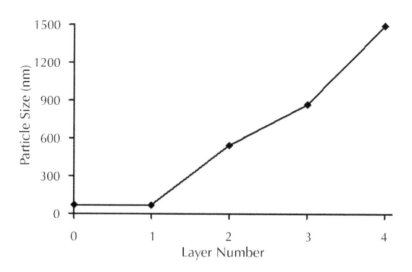

FIGURE 2   (b) Growth in particle size of ZMP nanocontainer. Layer number 0, initial Zn_2PMoO_7, 1, Zn_2PMoO_7/PANI, 2, Zn_2PMoO_7/PANI/Benzotriozole, 3, Zn_2PMoO_7/PANI/Benzotriozole/PAA.

14.3.2   MORPHOLOGY ANALYSIS OF ZINC PHOSPHOMOLYBDATE NANOCONTAINER

A transmission electron microscopy image of zinc phosphomolybdate nanocontainer containing zinc phosphomolybdate nanoparticles at core along with polyelectrolyte material surrounding the zinc phosphomolybdate nanoparticles is depicted in Figure 3. The TEM images indicate the distinct formation of zinc phosphomolybdate nanocontainer with distorted spherical morphology. Further, the presence of bright intensity layered structure around the dark zinc phosphomolybdate nanoparticles in TEM image has confirmed the layer formation of PANI, benzotriazole and PAA. The presence of benzotriazole and polyelectrolytes around zinc phosphomolybdate nanoparticles results in an increase in the average particles size of zinc phosphomolybdate nanocontainer, which is around 1250 nm. These results are consistent with the average particle size of zinc phosphomolybdate nanocontainer obtained from the particle size analyzer. Change in the scattering of light around each particles show that multilayer assembly (PANI/benzotriozole/PAA layers) is established on the zinc phosphomolybdate nanoparticles.

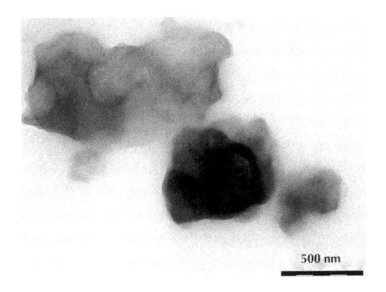

FIGURE 3   The TEM image of ZMP nanocontainers.

14.3.3   FTIR ANALYSIS OF ZINC PHOSPHOMOLYBDATE NANOCONTAINER

Figure 4 signifies the FTIR spectrum of pure Zn_2PMoO_7 (pattern A), Modified Zn_2PMoO_7 (pattern B), Zn_2PMoO_7 loaded with PANI (pattern C), Zn_2PMoO_7 loaded with PANI and Benzotriazole (pattern D) and Zn_2PMoO_7 loaded with PANI-Benzotriazole-Polyacrylic acid (pattern E). Figure 4 (Pattern A) shows FT-IR spectra of ZMP nanoparticles. Very strong Mo–O stretching vibration in $[MoO_4]^{2-}$ was detected at 825–936 cm^{-1} and weak Mo–O bending vibration was found at 437, 494 cm^{-1} (Isac and Ittyachen, 1992, Phuruangrat et al., 2010). Further, modification of ZMP nanoparticles was done by ultrasound assisted method by using myristic acid (MA). Figure 4 (B) shows an FTIR spectrum of myristic acid modified ZMP nanoparticles. Due to addition of myristic acid to synthesized ZMP nanoparticles characteristics peaks are observed on the bare ZMP nanoparticles at 2924, 2853, and 1541 cm^{-1} shows stretching vibration of the C–H, which came from the $–CH_3$ and $–CH_2$ in the myristic acid respectively. Bending of –OH bond attributed by the characteristic peak at 1541 cm^{-1}. Figure 4 (C) shows the characteristic peak due to addition of PANI layer to the modified ZMP nanoparticles, which are at 1230 cm^{-1} is due to (C–N) stretching mode of the amine group and the peak at 1433 cm^{-1} reflects C=C stretching mode of the quinoid rings and C=C stretching of benzenoid rings respectively (Sun et al., 1990). Secondary =N–H bending showed by the characteristics peak at 1526 cm^{-1}. Above mentioned peaks reflects the formation of polyaniline layer on the modified ZMP nanoparticles. Figure 4 (D) reflects the FTIR spectrum of benzotriazole loaded PANI loaded modified myristic acid ZMP nanoparticles. Benzotriazole layer of PANI coated ZMP nanoparticles shows the characteristic peaks at 1520, 1260, and 750 cm⁻¹, which reflects the effective formation benzotriazole layer. The bands which are close to 750 cm⁻¹ are typical of the benzene ring vibration and the band near to 1520 cm⁻¹ is characteristic of the aromatic and the triazole rings stretching vibration (Mennucci et al., 2009). Finally benzotriazole layer loaded myristic acid modified ZMP nanoparticles coated with PAA layer the FTIR spectrum of which is depicted in Figure 4 (E). The characteristic peak at 1732 cm⁻¹ reflects the adsorption of PAA layer, which is attributed to carbonyl C=O stretching in PAA (Lu et al., 2005).

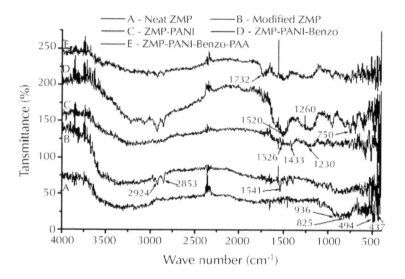

FIGURE 4   The FTIR spectra of (A), Neat ZMP (B), Modified ZMP (C), ZMP loaded with PANI, (D), ZMP loaded with PANI and benzotriazole, and (E) ZMP loaded with PANI, benzotriazole and PAA (Polyacrylic acid).

14.3.4   XRD ANALYSIS OF ZINC PHOSPHOMOLYBDATE NANOCONTAINER

Figure 5 depicts the XRD patterns of ZMP nanoparticle and ZMP nanocontainers. Figure 5 (pattern A) signifies the XRD pattern of Zn_2PMoO_7 nanoparticles. It is observed that the synthesized ZMP nanoparticles are crystalline in nature (Marques et al., 2006) and observed phase of ZMP nanoparticles is scheelite phase (Ryu et al., 2005). In above XRD pattern the diffraction peaks at 2θ value of 25.2, 27.9, 29.1, 31.7, 34.2, 40.3, 51.8, and 52.8° are corresponds to the planes (112), (004), (114), (211), (200), (220), (312), and (224) (Raj et al., 2002). The peaks correspond to 2θ values at 31.7, 34.2, 36.1, 56.1, 62.4, and 68.2° shows the characteristics peaks of phosphate addition. The crystallite size of ZMP nanoparticle was estimated as equal to 74.15 nm at

$2\theta = 25.2°$ by Debye Scherrer's formula. The particle size of ZMP nanoparticles is consistent with PSD data reported above. Afterwards due to uncalcined state of zinc phosphomolybdate some impurities peaks of Na_2SO_4 and KNO_3 even after number of hot water washing cycles is observed. Further in XRD pattern of ZMP nanocontainer (Figure 5(B)), it is found that particles are completely covered with adsorbed layers of polyelectrolyte and corrosion inhibitors. The presence of XRD peaks at 25.2, 36.1, 52.8° clearly shows the presence of zinc phosphomolybdate at the core of the nano-container.

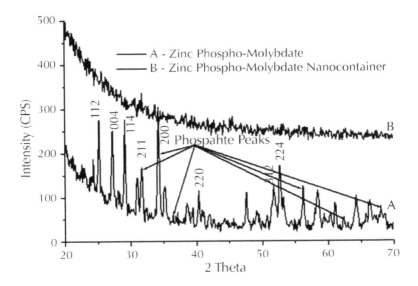

FIGURE 5 The XRD pattern of (A) Zn_2PMoO_7 nanoparticles and (B) Zn_2PMoO_7 nanocontainers.

14.3.5   TGA AND DTA OF ZINC PHOSPHOMOLYBDATE NANOCONTAINER

The TGA plot of zinc phosphomolybdate nanocontainers shown in Figure 6 (a), which shows weight loss in ZMP nanoparticle at four different steps. Initially, the weight loss of 23 wt% is observed in the range of 40–270°C (Section I) due to desorption of physically adsorbed as well as hydrated water in the ZMP nanocontainer. Polyacrylic acid is adsorbed on the surface of benzotriazole loaded ZMP nanoparticles, due to burning of PAA layer second weight loss (21 wt. %) is observed in the range of 271–440°C (Section II). The third loss in weight of benzotriazole that is about (14 wt. %) between 441 and 525°C (Section III) is due to oxidative degradation of the corrosion inhibitor (benzotriazole) that is enclosed into the two layers of polyelectrolytes in the ZMP nanocontainers. As weight loss is observed in the section 14.3 it does confirms that more adsorption of corrosion inhibitor in the nanocontainer. From above figure it is observed that the adsorbed PANI layer is burned off after 525°C (Section IV), and weight loss observed about 22 wt%. The overall weight loss is observed from 271–725°C is due to oxidative degradation of hydrocarbon moieties present in the polymer. From above thermo gravimetric analysis it is observed that ZMP nanoparticles are more stable at high temperature.

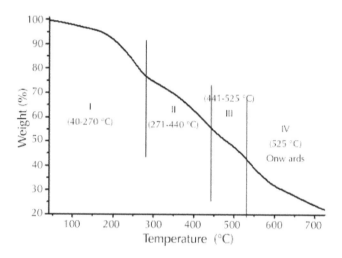

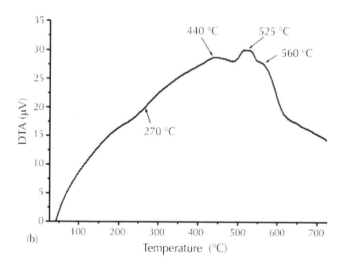

FIGURE 6   (a) TGA and (b) DTA analysis of Zn_2PMoO_7 nanocontainer.

Figure 6 (b) depicts the DTA plot of ZMP nanocontainers. The ZMP nanocontainer DTA plot shows endothermic peak at 270°C, which is due to the removal of physically adsorbed water (In the range from 200–300°C the water molecules interacting with ZMP nanocontainer surface such as hydroxyls groups releases from the ZMP nanocontainer structure) (Takeuchi et. al., 2005, Yoshino et. al., 2005). The exothermic peaks at 400–500°C are due to multistage decomposition of PAA, PANI and Benzotriazole (corrosion inhibitor), which is also confirmed by TGA analysis. A intense peak in the range of 500–550°C corresponds to oxidative degradation of PAA, PANI and benzotriazole (Kartsonakis et al., 2008). The endothermic peak in the range of 550–600°C is due to loss in weight of ZMP nanocontainer after deposition of benzotriazole layer due to excess removal of moisture. The endothermic peak at 650–700 observed, which signifies loss in weight due to removal of moisture from PAA layer.

14.3.6 RELEASE STUDY OF CORROSION INHIBITOR FROM ZINC PHOSPHOMOLYBDATE NANOCONTAINER

Figure 7 depicts the release rate and release flux of benzotriazole. It is well known that the benzotriazole is a good inhibitor for ferrous metal under acidic environment (Matheswaran and Ramasamy, 2010, Selvi et. al., 2003, Popova, and Christov, 2006) also in neutral environment (Cao et. al., 2002, Ramesh and Rajeswari, 2004). Cao et. al. (2002) put-forth that benzotriazole forms the compact passive layer by the adsorption of benzotriazole in its molecular or protonated form. The purpose of this study is to observe the effect of pH of the aqueous medium (2, 4, 7, and 9) on the release rate of the benzotriazole trough ZMP nanocontainer. Final layer of nanocontainer that is PAA is pH dependent, due to this it is found that release rate as well as release flux of benzotriazole is goes on increasing gradually with respect to time initially but as time passes it goes on decreasing as the concentration of corrosion inhibitor that is benzotriazole increases in the surrounding medium. As the concentration of benzotriazole goes on increasing in the surrounding pH solution, which results into decrease in the diffusion rate of benzotriazole due to saturation stage with respect to time. It is clearly observed that release rate and release flux of benzotriazole get decreased with respect to time and increasing pH value. The release rate gets decreases from 0.0065 to 0.085 mgL^{-1}/g of ZMP nanocontainer.min for increase in the pH from 2 to 9 at the end of 10 min. As well as release flux is found to be decreased from 0.00375 to 0.0007 mgL^{-1}/g of ZMP Nanocontainer cm^2.min for increase in the pH from 2 to 9 at the end of 10 min. From above figure it is observed that more release is observed at more acidic pH that is at pH 2 due to ability of PAA to expand its matrix structure at more acidic environment. Due to above ability of benzotriazole it forms passive layer on the surface of ferrous metal leads to more corrosion inhibition in acidic medium.

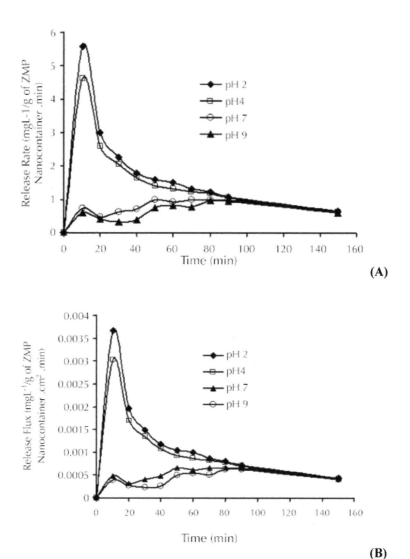

FIGURE 7   (A) Release rate and (B) Release flux of benzotriazole from ZMP nanocontainers at different pH values.

The diffusive ability of the benzotriazole in pH environment was estimated by the equation as follows:

$$J_A = -D_{AB} \frac{dC_A}{dr} \qquad (2)$$

Where J_A = *Release flux of benzotriazole,* D_{AB} = *Diffusivity of benzotriazole,* C_A = *Concentration of benzotriazole and r = radial position in spherical nanocontainer.*

From Figure 8 it is observed that the diffusivity of benzotriazole gets decreased from 2.78×10^{-14} to 1.5×10^{-14} cm²/min with respect to increase in the pH value from 2 to 9. Due to ability of PAA to expand its matrix structure in case of highly acidic pH that is at pH 2, benzotriazole shows more sustained release at acidic pH as compare to basic pH..

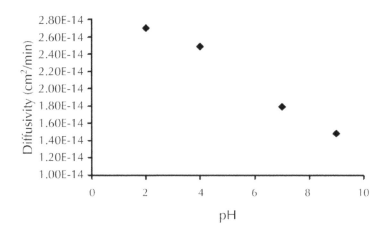

FIGURE 8   Diffusivity of benzotriazole from nanocontainers versus pH.

14.3.7   CORROSION RATE ANALYSIS OF ZINC MOLYBDATE NANOCONTAINER/ALKYD COATINGS

Figure 9 indicates the corrosion rate per year at different wt% loading of ZMP nanocontainers in alkyd resin coatings. At first ZMP nanocontainers were incorporated in the alkyd resin with the help of pigment Muller and formulation was coated on MS plates for study corrosion rate analysis. Experimentation was carried out by dipping coated MS in HCl, NaCl, and NaOH (5 wt% each) solutions for a period of 750 hr. The loss in weight of each species was measured by gravimetric analysis for the analysis of corrosion rate through dip test. Corrosion rate for 0 % loading of ZMP nanocontainer is 0.0765 cm/yr, 0.0849 cm/yr and 0.1207 cm/yr in case of 5% HCl, NaOH, and NaCl solution respectively. From the figure, it can be concluded that the corrosion rate was maximum for NaCl solution and minimum for HCl solution. The corrosion rate was changing predominantly at varying percent loading of nanocontainer in all cases. Increase in percentage loading of ZMP nanocontainer from 2 to 5 wt%, decrease in the corrosion rate takes place in all cases of HCl, NaCl, and NaOH solution, which results into minimum corrosion rate per year observed. From above it can be predicted that corrosion resisting ability of ZMP nanocontainer is maximum in HCl solution compared to NaCl and NaOH solution, which attributes to more passivation in acidic medium resulted from more release of benzotriazole from the ZMP nanocontainer.

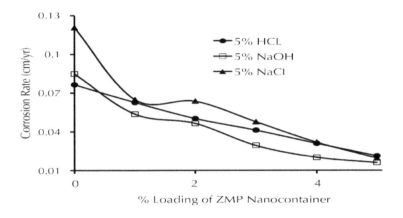

FIGURE 9   Effect of loading of nanocontainer on the corrosion rate after loading into the alkyd resin.

14.3.8   ELECTROCHEMICAL CHARACTERIZATION OF ZINC PHOSPHOMOLYBDATE NANOCONTAINER/ALKYD COATINGS

Figure 10 signifies the electrochemical analysis (Tafel plot) of MS panels coated with neat alkyd resin, and coated with 2 and 4 wt % loading of ZMP nanocontainer incorporated in alkyd resin. Above analysis was carried out in 5 wt % aqueous NaCl solution at room temperature. The Tafel plot is plotted as log (current density) as a function of applied potential. In Tafel plot analysis current density is measured in corrosion process for simultaneous redox reactions occurs at the surface of cathode and anode of MS plate. I_{corr} that is corrosion current density and E_{corr} that is corrosion potential, values were found from the Tafel plot analysis. It is observed that corrosion current density was decreased from 0.027 for pure alkyd resin to 0.012 A/cm^2 for 2 wt% ZMP nanocontainers incorporated in pure alkyd resin coatings and it gets decreased upto value of 0.01 A/cm^2 when percentage loading of ZMP nanocontainers is increased to 4 wt %. E_{corr} values goes on increasing with increase in percentage loading of ZMP nanocontainer and which is found to be shifted to positive side from 7.42 V to 1.034 V with the addition of 4 wt % of ZMP nanocontainers in alkyd resin.

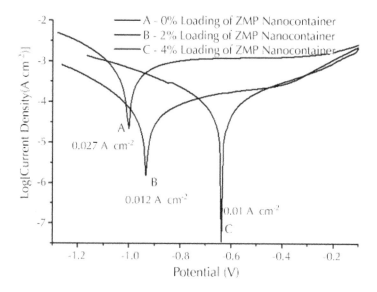

FIGURE 10   Tafel plots of MS panels coated with alkyd/ZMP nanocontainer coatings at different loading of ZMP nanocontainer in alkyd resin carried out in 5 % NaCl solution as an electrolyte at room temperature (25°C).

From above results it is observed that release rate, corrosion rate, and electrochemical (Tafel plot) analysis shows that the percentage loading of 4 wt% ZMP nanocontainers in alkyd resin shows a considerable improvement in the anticorrosive properties compare to pure alkyd resin.

14.4   CONCLUSION

The present work confirmed the layer by layer assembly of ZMP nanocontainers and the release mechanism of corrosion inhibitor. FTIR and TEM study confirms the successful formation of ZMP nanocontainer as a layer by layer system with the aid of ultrasonic irradiation. Zeta potential and particle size analysis also shows the formation of layered structure and shows appropriate change in the surface charge, which could be responsible for the release mechanism initiated by the change in pH. Release study and corrosion results from Tafel plot and corrosion rate analysis showed significant improvement in the anticorrosion properties of coatings due to the optimum loading of the ZMP nanocontainers.

KEYWORDS

- **Corrosion inhibitor**
- **Nanocontainers**
- **Release flux**
- **Release rate**
- **Zn_2PMoO_7 nanoparticles**

ACKNOWLEDGMENT

S H Sonawane acknowledge the support of Department of Science and Technology (DST), Government of India, for providing the Funds through Project Grant Reference No: SR/S3/CE/0060/2010.

REFERENCES

1. Aramaki, K. Self-healing mechanism of an organosiloxane polymer film containing sodium silicate and cerium (III) nitrate for corrosion of scratched zinc surface in 0.5 M NaCl. *Corrosion Science*, **44**(7), 1621–1632 (2002).
2. Barkade, S. S., Naik, J. B., and Sonawane, S. H. Ultrasound assisted miniemulsion synthesis of polyaniline/Ag nanocomposite and its application for ethanol vapor sensing. *Colloids and Surfaces A: Physicochemical and Engineering Aspects*, **378**(1–3), 94–98 (2011).
3. Benito, S., Graff, A., Stoenescu, R., Broz, P., Saw, C., Heider, H., Marsch, S., Hunziker, P., and Meier, W. Polymer nanocontainers for biomedical applications. *European Cells and Materials*, **6**(1), 21–22 (2003).
4. Bhanvase, B. A. and Sonawane, S. H. New approach for simultaneous enhancement of anticorrosive and mechanical properties of coatings: application of water repellent nano $CaCO_3$–PANI emulsion nanocomposite in alkyd resin. *Chemical Engineering Journal*, **156**(1), 177–183 (2010).
5. Borisova, D., Mohwald, H., and Shchukin, D. G. Mesoporous silica nanoparticles for active corrosion protection. *ACS Nano*, **5**(3), 1939–1946 (2011).
6. Cao, P. G., Yao, J. L., Zheng, J. W., Gu, R. A., and Tian, Z. Q. Comparative study of inhibition effects of benzotriazole for metals in neutral solutions as observed with surface enhanced Raman spectroscopy. *Langmuir*, **18**(1), 100–104 (2002).
7. Chico, B., Simancas, J., Vega, J., Granizo, N., Diaz, I., De la Fuente, D., and Morcillo, M. Anticorrosive behavior of alkyd paints formulated with ion-exchange pigments. *Progress in Organic Coatings*, **61**(2–4), 283–290 (2008).
8. Evaggelos, M., Ioannis, K., George, P., and George, K. Release studies of corrosion inhibitors from cerium titanium oxide nanocontainers. *Journal of Nanoparticle Research*, **13**(2), 541–554 (2011).

9. George, C., Dorfs, D., Bertoni, G., Falqui, A., Genovese, A., Pellegrino, T., Roig, A., Quarta, A., Comparelli, R., Curri, M. L., Cingolani, R., and Manna, L. A cast-mold approach to iron oxide and Pt/iron oxide nanocontainers and nanoparticles with a reactive concave surface. *Journal of American Chemical Society*, **133**(7), 2205–2217 (2011).

10. Hu, Y., Chen, Y., Chen, Q., Zhang, L., Jiang, X., and Yang, C. Synthesis and stimuli-responsive properties of chitosan/poly (acrylic acid) hollow nanospheres. *Polymer*, **46**(26), 12703–12710 (2005).

11. Isac, J. and Ittyachen, M. A. Growth and characterization of rare-earth mixed single crystals of samarium barium molybdate. *Bulletin of Materials Science*, **15**(4), 349–353 (1992).

12. Jose, E., Pereira, S., Susana, I., Cordoba, T., and Roberto, M. T. Polyaniline acrylic coatings for corrosion inhibition the role played by counter-ions. *Corrosion Science*, **47**(3), 811–822 (2005).

13. Kalendova, A. and Vesely, D. Study of the anticorrosive efficiency of zincite and periclase-based core-shell pigments in organic coatings. *Progress in Organic Coatings*, **4**(1), 5–19 (2009).

14. Kartsonakis, I. A. and Kordas, G. Synthesis and characterization of cerium molybdate nanocontainers and their inhibitor complexes. *Journal of the American Ceramic Society*, **93**(1), 65–73 (2009).

15. Kartsonakis, I., Daniilidis, I., and Kordas, G. Encapsulation of the corrosion inhibitor 8-hydroxyquinoline into ceria nanocontainers. *Journal of Sol-Gel Science and Technology*, **48**(1–2), 24–31 (2008).

16. Kim, B. J., Oh, S. G., and Im, S. S. Investigation on the solubilization locus of aniline–HCl salt in SDS micelles with 1H-NMR spectroscopy. *Langmuir*, **17**, 565–266 (2001).

17. Kumar, A., Stephenson, L. D., and Murray, J. N. Self-healing coatings for steel. *Progress in Organic Coatings*, **55**(3), 244–253 (2006).

18. Lu, X., Yu, Y., Chen, L., Mao, H., Wang, L., Zhang, W., and Wei, Y. Poly (acrylic acid)-guided synthesis of helical polyaniline microwires. *Polymer*, **46**(14), 5329–5333 (2005).

19. Marques, A. P. A., Melo, D. M. A., Paskocimas, C. A., Pizani, P. S., Joya, M. R., Leite, E. R., and Longo, E. Photoluminescent BaMoO$_4$ nanopowders prepared by complex polymerization method (CPM). *Journal of Solid State Chemistry*, **179**(3), 658–678 (2006).

20. Matheswaran, P. and Ramasamy, A. K. Influence of benzotriazole on corrosion inhibition of mild steel in citric acid medium. *E-Journal of Chemistry*, **7**(3), 1090–1094 (2010).

21. Mennucci, M. M., Banczek, E. P., Rodrigues, P. R. P., and Costa, I. Evaluation of benzotriazole as corrosion inhibitor for carbon steel in simulated pore solution. *Cement and Concrete Composites*, **31**(6), 418–424 (2009).

22. Murphy, E. B. and Wudl, F. The world of smart healable materials. *Progress in Polymer Science*, **35**(1–2), 223–251 (2010).

23. Nesterova, T., Dam-Johansen, K., and Kiil, S. Synthesis of durable microcapsules for self-healing anticorrosive coatings a comparison of selected methods. *Progress in Organic Coatings*, **70**(4), 342–352 (2011).

24. Paliwoda-Porebska, G., Stratmann, M., Rohwerder, M., Potje-Kamloth, K., Lu, Y., and Pich, A. Z. On the development of polypyrrole coatings with self-healing properties for iron corrosion protection. *Corrosion Science*, **47**(12), 3216–3233 (2005).

25. Phuruangrat, A., Thongtem, T., and Thongtem, S. Synthesis of nanocrystalline metal molybdates using cyclic microwave radiation. *Materials Science-Poland*, **28**(2), 557–563 (2010).

26. Popova, A. and Christov, M. Evaluation of impedance measurements on mild steel corrosion in acid media in the presence of heterocyclic compounds. *Corrosion Science*, **48**(10), 3208–3221 (2006).

27. Raj, A. M. E. S., Mallika, C., Swaminathan, K., Sreedharan, O. M., and Nagaraja, K. S. Zinc (II) oxide-zinc (II) molybdate composite humidity sensor. *Sensors and Actuators B: Chemical*, **81**(2–3), 229–236 (2002).

28. Ramesh, S. and Rajeswari, S. Corrosion inhibition of mild steel in neutral aqueous solution by new triazole derivatives. *Electrochimica Acta*, **49**(5), 811–820 (2004).

29. Revie, R. W. *Corrosion and corrosion control*, 4th ed., John Wiley & Sons, New Jersey (2008).

30. Ryu, J. H., Yoon, J. W., Lim, C. S., Oh, W. C., and Shim, K. B. Microwave-assisted synthesis of camoo₄ nano-powders by a citrate complex method and its photoluminescence property. *Journal of Alloys and Compound*, **390**(1–2), 245–249 (2005).

31. Schweitzer, P. A. *Paint and coatings applications and corrosion Resistance*, CRC Press (2006).

32. Selvi, S. T., Raman, V., and Rajendran, N. Corrosion inhibition of mild steel by benzotriazole derivatives in acidic medium. *Journal of Applied Electrochemistry*, **33**(12), 1175–1182 (2003).

33. Shchukin, D. G. and Möhwald, H. Surface-engineered nanocontainers for entrapment of corrosion inhibitors. *Advanced Functional Materials*, **17**(9), 1451–1458 (2007).

34. Shchukin, D. G., Lamaka, S. V., Yasakau, K. A., Zheludkevich, M. L., Ferreira, M. G. S., and Mohwald, H. Active anticorrosion coatings with halloysite nanocontainers. *The Journal of Physical Chemistry C*, **112**(4), 958–964 (2008).

35. Shchukin, D. G., Zheludkevich, M., Yasakau, K., Lamaka, S., Ferreira, M. G. S., and Mowald, H. Layer-by-layer assembled nanocontainers for self healing corrosion protection. *Advanced Materials*, **18**(13), 1672–1678 (2006).

36. Sun, Y., Macdiarmid, A. G., and Epstein, A. J. Polyaniline: synthesis and characterization of pernigraniline base. *Journal of the Chemical Society, Chemical Communications*, **7**, 529–531 (1990).

37. Suryanarayana, C., Rao, K. C., and Kumar, D. Preparation and characterization of microcapsules containing linseed oil and its use in self-healing coatings. *Progress in Organic Coatings*, **63**(1), 72–78 (2008).

38. Takeuchi, M., Martra, G., Coluccia, S., and Anpo, M. Investigations of the structure of H_2O clusters adsorbed on TiO_2 surfaces by near-infrared absorption spectroscopy. *The Journal of Physical Chemistry B*, **109**(15), 7387–7391 (2005).

39. Tedim, J., Poznyak, S. K., Kuznetsova, A., Raps, D., Hack, T., Zheludkevich, M. L., and Ferreira, M. G. S. Enhancement of active corrosion protection via combination of inhibitor-loaded nanocontainers. *Applied Materials and Interfaces*, **2**(5), 1528–1535 (2010).

40. Wu, D. Y., Meure, S., and Solomon, D. Self-healing polymeric materials: a review of recent developments. *Progress in Polymer Science*, **33**(5), 479–522 (2008).

41. Yang, J., Lee, J., Kang, J., Lee, K., Suh, J., Yoon, H., Huh, Y., and Haam, S. Hollow silica nanocontainers as drug delivery vehicles. *Langmuir*, **24**(7), 3417–3421 (2008).

42. Yoshino, K., Fukushima, T., and Yoneta, M. Structural, optical and electrical characterization on ZnO film grown by a spray pyrolysis method. *Journal of Material Science Materials in Electronics*, **16**(7), 403–408 (2005).
43. Zheludkevich, M. L., Serra, R., Montemor, M. F., and Ferreira, M. G. S. Oxide nanoparticle reservoirs for storage and prolonged release of corrosion inhibitors. *Electrochemistry Communications*, **7**(8), 836–840 (2005).
44. Zheludkevich, M. L., Shchukin, D. G., Yasakau, K. A., Möhwald, H., and Mario, G. S. Anticorrosion coatings with self-healing effect based on nanocontainers impregnated with corrosion inhibitor. *Chemistry of Materials*, **19**(3), 402–411 (2007).

CHAPTER 15

EMERGING MULTIFUNCTIONAL NANOMATERIALS FOR SOLAR ENERGY EXTRACTION

MAYANK BHUSHAN, NIRMAL GHOSH O. S.,
and A. KASI VISWANATH

CONTENTS

15.1   INTRODUCTION

Energy is one of the prime needs to our society. Due to the fast development in industry and technology, there is a huge demand for energy and meeting these energy demands in future, is a challenging issue for the humanity. At present, we are mainly depending up on the fossil fuels and conventional energy resources for the energy generation. The recent statistical studies and analysis give an alarming conclusion that in the near future there will be a huge gap between energy demand and supply, and with the depleting conventional energy resources it is not possible to fill this gap. The fossil fuels will liberate the greenhouse gases during the energy generation, which is one of the major reasons for global warming and the related environmental pollution followed by the ecological imbalance. These problems lead to the fall in agricultural production and climate changes, which are having a direct impact on the society and sustainability of ecosystem.

In this scenario of rising energy demands of the increasing population and transportation, there is a need for the development of clean and sustainable energy (Pfeffer et al. 2002). The greenhouse gas emissions from hydrocarbon fuels are a threat to the environment and are a major cause of global climate change and they should be avoided for the survival of humanity on the earth. Renewable energy resources including Solar, Wind, Tide, and Geothermal resources can provide the sufficient energy to meet the energy demands in future. Solar energy is a clean and sustainable source of energy and utilization of solar energy and conversion into electrical energy has now become one of the major research areas due to the critical importance of clean energy sources. In this regard switching over to the energy generation from renewable energy resources is a good alternative. Earth is getting 127500 TW energy/day, so if we are able to extract the solar energy efficiently, we can meet our energy demand in future without affecting the environment. Solar energy conversion is possible in different ways and photovoltaic and solar thermal energy conversions are the major methods that are used for this purpose. In this regard the solar photovoltaic conversion is a good solution for the clean energy generation and currently the silicon solar cells are contributing the major share of solar photovoltaic energy conversion due to the easy processing technology and standard conversion efficiencies around 15% (Guldin et. al., 2011). But the cost of pure silicon for the solar cell fabrication is high that restricts the switch over from conventional energy generation and power production to solar photovoltaic energy conversion. To meet these challenges a new range of hybrid nanomaterials are emerging for the development of solar energy extraction systems.

15.2   QUANTUM DOT SENSITIZED SOLAR CELLS

Semiconductor nanostructures have several applications in light emitting diodes, semiconductor lasers, photovoltaics and photochemical energy conversion (Viswanath , 2000, 2001, 2006). Quantum dot (QD) sensitized solar cells are an emerging field in solar cell research that uses quantum dots as the photovoltaic material, as opposed to better-known bulk materials such as silicon, copper indium gallium selenide (CIGS) or CdTe (Lee et. al., 2009, Ruhle et. al., 2010, Scholes et. al., 2011). The QDs have bandgaps that are tunable across a wide range of energy levels by changing the QD size (Nagpal et al., 2011). This is in contrast to bulk materials, where the bandgap is fixed by the choice of material composition. This property makes quantum dots attractive for multi-junction solar cells, where a variety of different energy levels are used to extract more power from the solar spectrum.

The QDs are semiconductor nanostructures having zero dimension due to three dimensional electronic confinement. The QDs have discrete electronic energy states, high quantum yield and they undergo impact ionization followed by multi exciton generation. Due to this multiexciton generation the quantum dots are used to sensitize the solar cells. In a QD very low energy is required for the energy transfer from highest occupied molecular orbital (HOMO) to lowest unoccupied molecular orbital (LUMO) transition.

15.2.1   MULTIPLE EXCITON GENERATION

The QDs having a triplet state having energy $Et = E1$, and a singlet state $Es = 2 * E1$, When the photon energy is greater than the absorption threshold the singlet fission occurs and due to this process the multi exciton generation occurs. Multi Exciton Generation occurs at, $Ephoton = 2ET$.

PHOTOCURRENT MULTIPLICATION BY IMPACT IONIZATION

When the photon energy of the incident light is equal to or greater than the band gap of the quantum dot, two excitons are produced by one photon and this phenomena is called as impact ionization. This is due to the singlet fission to form two triplet states followed by the generation of two excitons. The electron ejected in to the conduction band creates a hole in the valence band. This causes the production of an electron-hole pair. But if the excited electron is having high energy then it will undergo singlet fission to form triplet electonic state that causes the production of another electron in the conduction band and accordingly a hole is created in the valence band. This will give rise to the photocurrent multiplication. For the highest efficiency the band gap of the material should be 0.05 eV. To tune this band gap (Satoh et al., 2008) introduces a number of intermediate energy levels by doping or band gap tuning of the QDs. Figure 1 shows the schematic of photocurrent multiplication by impact ionization.

Highest Efficiency = 84.9% at Eg = 0.05 eV

Highest Photon Energy in AM 1.5 Solar Spectrum is 4.4 eV

$M_{max} = E_{max}/Eg$,

M_{max} – Maximum number of excitons, E_{max} –Highest photon energy, Eg – Band gap of QD. For PbS and PbTe QDs quantum yield is 300%.

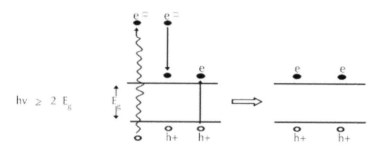

FIGURE 1   Photo impact ionization.

Because of photo impact ionization 1 photon yields 2 (or more) e– – h+ pairs, that results in a drastic increase in quantum yield and conversion efficiencies. Using impact ionization and multiple exciton generation (MEG) we can overcome the schottky quissner limit and achieve the maximum theoretical efficiency of 84.9 %. The photoanode can be sensitized by using QDs as given below. Due to the unique optical properties of QDs they can absorb whole range of solar spectrum by varying the size of the QD (Talgorn et. al., 2011). Thus by introducing intermediate bandgap between the conduction band and valence band we can harvest the maximum solar radiation power and it helps to fabricate highly energy efficient solar cells. Figure 2 shows I–V characteristics of QD sensitized TiO$_2$ photo anode.

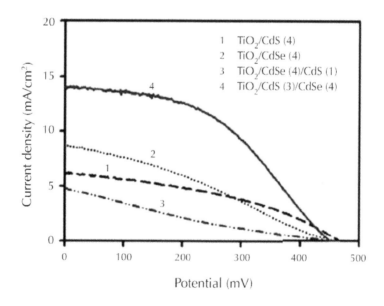

FIGURE 2   *(Continued)*

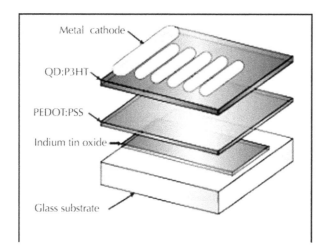

FIGURE 2   The I–V characteristics of QD sensitized TiO2 photo anode (Jabbour and Doderer, 2010).

Due to the photo impact ionization the exciton density increases and it leads to improve the photocurrent followed by the enhanced quantum yield of the anode material. Figure 3 shows the schematic of increase in density of states during MEG process.

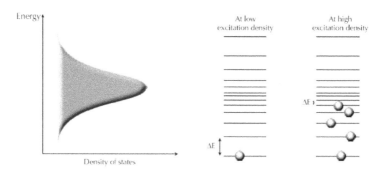

FIGURE 3   Density of states increases during MEG process (Talgorn et al., 2011).

15.2.2   HOT ELECTRON CAPTURE MECHANISM TO IMPROVE EFFICIENCY

The main cause for the loss of solar cell efficiency is the heat generation due to the excess photon energy and during the exciton generation the remaining energy after exciton formation is lost as heat energy. This is known as hot electron energy loss. This energy loss can be overcome by capturing the heat of hot electrons in the form of phonon vibrations to thermally relax the hot electron and extract the heat generated after exciton formation. This method is known as hot electron relaxation (Mora-Sero et. al., 2009). So by using this technique we will be able to extract the maximum energy of solar spectrum using QD sensitized solar cells. Figure 4 explains the mechanism of hot electron relaxation process and Figure 5 shows the contour diagram of exciton induced femtosecond photon energy transfer.

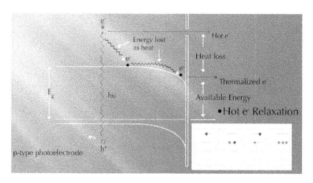

FIGURE 4   Hot electron relaxation mechanism.

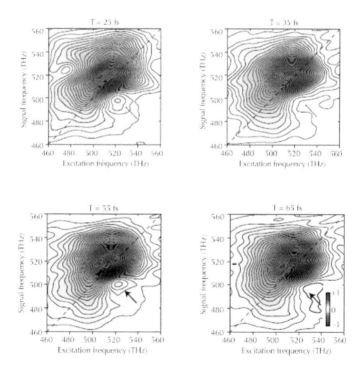

FIGURE 5   Contour diagram of exciton induced femtosecond phonon energy transfer (Ruhle et. al., 2010).

By using phonon mediated energy transfer we will be able to transfer the excess energy efficiently to produce more excitons in the intermediate energy levels thereby increasing the solar cell efficiency up to the maximum of 84.9% .

15.3   DYE SENSITIZED SOLAR CELL (DSSC)

Dye-sensitized Solar Cells (DSSCs) have received widespread attention owing to their low cost, easy fabrication, and relatively high solar-to-electricity conversion efficiency for future large-scale clean electrical energy production. Typically, DSSCs consist of Ru-based complex dye sensitized TiO2 photoanode, Pt-coated ITO as the counter electrode, and liquid electrolyte with I/I3– redox couple. DSSCs have already exhibited efficiency > 11% and offer an appealing alternative to conventional solar cells.

The key components of DSSC as are follows:

(i) Nano crystalline photoanode, which provides large surface area, high porosity, pore size distribution, light scattering, and higher scale of electron percolation for example (Anatase TiO2, ZnO, Nb2O5, and SnO2.

(ii) Sensitizer which is usually a dye distributed on the semiconducting anode surface (Gratzel M., 2003, Zhang et al., 2010). It consists of anchoring groups like carboxylates or phosphonates. Usually polypyridyl, porphyrin, and phthalocyanin complexes are used as sensitizers.

(iii) Electrolyte having ionic conductivity is used as an ion transporter and it should have redox potential. The electrolyte acts as a separator between sensitized photoanode and counter electrode. It provides the interfacial contact for dye, photo anode and counter electrode.

15.3.1   GENERAL OPERATING PRINCIPLES OF DYE SENSITIZED SOLAR CELLS

A new type of solar cell based on dye-sensitized nanocrystalline titanium dioxide has been developed (Hardin et al., 2010). Remarkably high quantum efficiencies have been reported for this type of solar cell (also called DSSC), with overall conversion efficiencies up to 11%. This fact, in combination with the expected relatively easy and low cost manufacturing makes this new technology an interesting alternative for existing solar cell technologies. Realisation of stable efficiencies in the order of 10 % in production, however, requires a lot of effort on the research and development side. For this reason, the first application of this type of solar cell will probably be one in which only a low power output is required, since this is easier to achieve. Less stringent efficiency requirements leave room for increased flexibility in the manufacturing process of these cells, that is the requirements that are put on the materials used are less severe. These results in lower manufacturing costs and more flexibility in materials choice, opening the way for an alternate to high cost silicon solar cells.

Conventional solar cells convert light into electricity by exploiting the photovoltaic effect that use semiconductor junctions. They are thus closely related to transistors and integrated circuits. The semiconductor performs two processes simultaneously absorption of light, and the separation of the electric charges ("electrons" and "holes "), which are formed as a consequence of that absorption. However, to avoid the premature recombination of electrons and holes, the semiconductors employed must be highly pure and defect-free. The fabrication of this type of cell presents numerous difficulties, preventing the use of such devices for electricity production on an industrial scale.

15.3.2   PRINCIPLE OF DSSC

The DSSC are photo electrochemical cells which are used for conversion of solar energy in to electrical energy at low cost. The energy payback time of DSSC is very less and due to the flexibility and transparency it is having a potential market value in portable and consumer electronics. Figure 6 shows schematic of DSSC.

The DSSC work on a different principle, whereby the processes of light absorption and charge separation are differentiated. Due to their simple design, the cells offer hope for a significant reduction in the cost of solar electricity. Light absorption is performed by a monolayer of dye (S) adsorbed chemically at the semiconductor surface. After having been excited (S*) by a photon of light, the dye usually a transition metal complex whose molecular properties are specifically engineered for the task is able to transfer an electron to the semiconductor (TiO_2) (the process of "injection"). The electric field inside the bulk material allows extraction of the electron.

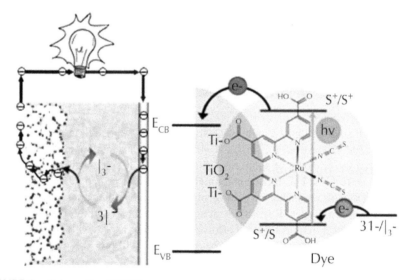

FIGURE 6   Schematic of DSSC.

Positive charge is transferred from the dye (S+) to a redox mediator ("interception") present in the solution with which the cell is filled, and then to the counter electrode. *Via* this last electron transfer, in which the mediator is returned to its reduced state, the circuit is closed. The theoretical maximum voltage that such a device could deliver corresponds to the difference between the redox potential of the mediator and the Fermi level of the semiconductor.

15.3.3 PHOTOANODE OF DSSC

The efficiency of the DSSC depends up on various parameters including current density, operating voltage, fill factor, dye loading, stability of dye molecules and efficiencies of electrodes and the electrolytes (Sivaranjani et. al., 2012). Since, the photoanode provides the surface interface for exciton generation and electron hole separation it is having a major role in current efficiency of the DSSC. Titania is the best material for photoanode due to its suitable band gap low cost, non-toxic and abundant in nature. The dimensional and structural morphologies of the photoanode have a crucial impact in the DSSC efficiency (Hardin et. al., 2010). The photoanode contains a number of interfaces. These include the interface between the semiconducting anode material and the transparent conducting metal oxide coating, dye and anode material interface. Using the optimized structural morphologies we can tune the band gap and efficiency of DSSC. The TiO$_2$ nanocrystalline has been reported (Chen et. al., 2001) as the promising material for photoanode of DSSC.

15.3.4 ANTHOCYANIN BASED NATURAL DYES

The ability of sensitizers in the natural dye is linked to anthocyanin's properties. Anthocyanin molecule in the form of carbonyl and hydroxyl which occurs naturally in fruit, leaf and flowers is responsible to show different colours in red–blue spectrum. Recent studies confirmed that the natural dyes can be good alternatives to ruthenium based dyes. Cyanin complexes have reported by (Hardin et. al., 2009, 2010) as alternate sensitizers to the high cost ruthenium based N719 dyes. In a recent work, (Zhou et. al., 2011) have reported the extraction of natural dye using acid hydrolysis method. Figure 7 shows schematic of natural dye extraction.

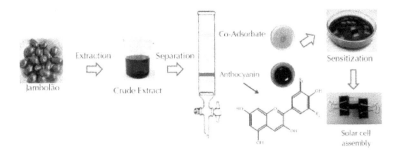

FIGURE 7   Schematic of natural dye extraction.

In recent past, many groups (Zhou et al., 2011, Dumbrava et al., 2008, Chen and Mao 2007, Jasieniak et al., 2004, and Meng and Kaxiras 2010) have reported the extraction of natural dyes like Anthocyanin from different fruits such as Jambool and Cherry showing a positive result in the absorption of solar spectrum. Acid hydrolysis method is used for the extraction of natural dye from the resources. Due to the photo bleaching, anthocyanin based dyes are photochemically unstable. Anthocyanin based dyes have good solar absorption capability because of the functional groups present in them. Figure 8 shows the schematics of structure of N3, N719, and Black dye.

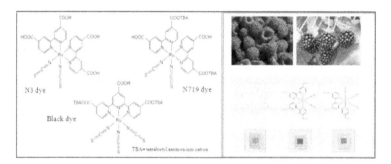

FIGURE 8   Structure of N3, N719, and Black dye (Ryan M., 2009).

15.3.5 OPERATION

Sunlight enters the cell striking the dye on the surface of the TiO_2. Photons striking the dye with enough energy to be absorbed create an excited state of the dye, from which an electron can be "injected" directly into the conduction band of the TiO_2. From there it moves by diffusion (as a result of an electron concentration gradient) to the clear anode on top. Meanwhile, the dye molecule has lost an electron and the molecule will decompose if another electron is not provided. The dye strips one from iodide in electrolyte below the TiO_2, oxidizing it into tri iodide. This reaction occurs quite quickly compared to the time that it takes for the injected electron to recombine with the oxidized dye molecule, preventing this recombination reaction that would effectively short-circuit the solar cell. The tri iodide then recovers its missing electron by mechanically diffusing to the bottom of the cell, where the counter electrode re-introduces the electrons after flowing through the external circuit.

To enhance electron collecting efficiency, improve interfacial contact area, structural stability, a high specific surface area for dye adsorption, the methodologies such as electrochemical fabrication of vertically aligned, ordered 1D structure top-end-open nanotube array architectures of nanocrystalline semiconducting metal oxides (for example TiO_2, ZnO) are used. In order to minimize recombination of photo induced electrons with I3− in the electrolyte, "double-layered photoanodes" could be synthesized via novel solvothermal method. Subsequently thin film solar cells with controlled thickness can be fabricated by doctor-blade, dip-coating techniques (Fu et. al., 2012, Meng et. al., 2008, Lee et. al., 2009, Choi et. al., 2012). By using transition metal binary alloy or metal sulfide consisting of Mo, W, Fe, Co, Ni, Ti, Bi, and Cu with or without incorporating into "core-shell" nanostructured conducting poly(3,4-ethyl-enedioxy thiophene) (PEDOT), 1D (MWCNT) and 2D Graphene nanosheets hybrid materials the scients are replaced the "Pt-FTO-free" counter electrodes.

15.4 POLYMER SOLAR CELL

In polymer solar cells bulk hetero-junctions between an organic polymer and inorganic molecule acts as an electron acceptor. Currently, fullerene embedded into conjugated polymer nanohybrids are used for the fabrication of the polymer solar cells. Polymer solar cells have the attractive advantages like lightweight, disposable nature, inexpensive to fabricate, flexible, designable on the molecular level, and have little potential for negative environmental impact. Present best efficiency of polymer solar cells lies near 5% but, the cost is roughly one-third of that of traditional silicon solar cell technology. The band gap of this type of cells is nearer to 2 eV. After the excitation in photoactive polymer, the electron is transferred to the C60 due to the high electron affinity of carbon atoms in the fullerene structure (Eder D., 2010, Dennler et al., 2009). Thus the photo induced quasiparticle (polaron P+) is formed on the polymer chain and fullerene ion radical C60. Figure 9 shows the schematic of the structure of the nano hybrid designs for polymer solar cell.

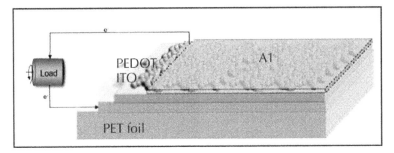

FIGURE 9   Structure of the nano hybrid designs for polymer solar cell.

15.4.1   ORGANIC-INORGANIC BULK HETERO-JUNCTION (BHJ) PHOTOVOLTAIC CELL

The efficiency of organic photovoltaic (PV) cells depends upon the use of bulk hetero-junctions, in which semiconductors with equivalent energy levels, interpenetrate at about 10 nm length scale.

Most of the reported bulk hetero-junction PV cells have been made by casting solutions containing the two semiconductors for example,

  (i)   A conjugated polymer,
  (ii) CdSe nanocrystals,
  (iii) Titania nanocrystals,
  (iv) ZnO nanocrystals, and
  (v)   To make blends. These blends are easy to fabricate, and it is desirable to use this simple process to manufacture PV cells.

There are problems with the disordered nanostructures that are typically created. In some cases, the two semiconductors phase separate on too large a length scale, as a result, some of the excitons (electron–hole pairs) generated after light absorption are unable to diffuse to an interface to be dissociated by electron transfer before they decay (Liang et. al., 2010). In other cases, there are dead ends in one of the phases, which prevent charge carriers from reaching electrodes. In most cases, the rate of charge transport is not fast enough to enable the charge carriers to reach the electrodes before the recombination of excitons, unless the films are made so thin that they cannot absorb entire incident light.

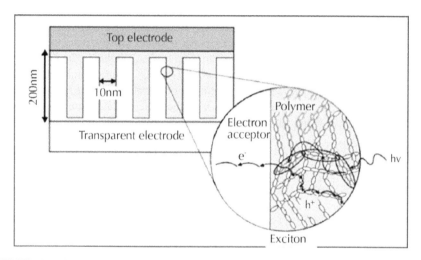

FIGURE 10   Organic inorganic bulk hetero-junctions.

To fabricate ordered bulk hetero-structure PV cells, such as the one sown in Figure 10, is more difficult than disordered blends, but the former has several advantages over the later one. First, both phases can be dimensionally controlled to ensure that every spot in a film is within an exciton diffusion length of an interface between the two semiconductors. Second, there are no dead ends in the structure. After excitons are dissociated by electron transfer, the electrons and holes have straight pathways to the electrodes. This geometry ensures that the carriers escape the device as quickly as possible, which minimizes recombination. Third, alignment of conjugated polymer chains can be possible in an ordered structure, which in turn increases the mobility of their charge carriers. An additional advantage of ordered structures is its feasibility for modeling and easy understanding, which is the requirement of current stage research on organic photovoltaic cell.

15.4.2   FABRICATION OF ORDERED BULK HETERO-JUNCTIONS

Current advancement in nanotechnology makes it possible to fabricate films at the 10 nm length scale. Due to the various properties, the film is required to have to be used to fabricate PV cell, it restricts the choice of material to be used. The phases should b straight, aligned perpendicular to the electrodes, and ideally connected only to the appropriate electrodes. Furthermore, to absorb most of the solar spectrum, the energy levels of semiconductors must be selected appropriately, excitons are dissociated by electron transfer, and a significant voltage is generated. One good approach is to use a block copolymer, which forms an array of cylinders oriented perpendicular to the substrate by self-assembly process. Nonconjugated block copolymers, such as poly(methyl-meth-acrylate)-polystyrene can be used to make well ordered films with the desired nanostructures. But unfortunately, keeping the desired structure and switching over to conjugated blocks is difficult because low polydispersities can not be achieved by most of the polymerization routes used to make conjugated polymers, also many conjugated blocks cannot be attached to each other, and stiff conjugated blocks do not form the same structures as flexible blocks (Yaha and Rusop, 2012, Diguna et. al., 2007, Han et. al., 2009). This problem can be solved by adapting an alternate approach of making a nanoporous film or fabricating an array of nanowires with an inorganic semiconductor, such as titania (TiO_2), zinc oxide, or cadmium sulfide, and then fill in the pores or, the spaces between the wires with a conjugated polymer (Choi et. al., 2012, Sista et. al., 2011). Figure 11 shows schematic of a model multilayer polymer solar cell and Figure 12 shows I–V characteristics of a tandem solar cell (Kim et. al., 2007).

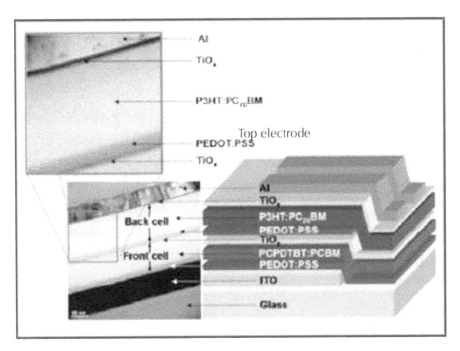

FIGURE 11   Model of a multilayer polymer solar cell (Sista et. al., 2011).

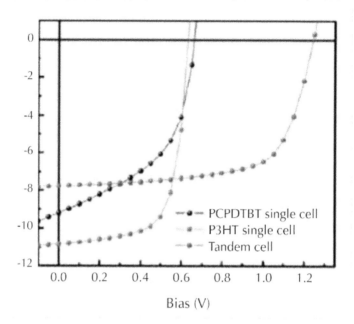

FIGURE 12   The I V Characteristics of a tandem solar cell (Kim et al., 2007).

15.5   BIO SENSITIZED SOLAR CELLS

15.5.1   ANTHOCYANIN BASED NATURAL DYE

Anthocyanins are one of the most abundant natural pigments available. Anthocyanins are the vacuolar pigments found in almost every part of higher plants. The word "anthocyanin" is derived from two Greek words, *anthos* (flower) and *kyanos* (blue). There are six types of anthocyanins all exhibiting different stability profile based on their structure, supporting compounds, process and chemicals used for extraction. Anthocyanin pigments change their color with the change in the pH. So based on the pH, anthocyanins can be red, blue, or purple. Because of this property of changing anthocyanin color, this pigment is also used as a pH indicator. Anthocyanin pigments change from blue in bases and red from acids. Anthocyanin pigments are odorless and flavorless and fond in every part of the higher plants like stems, roots, leaves, fruits, and flowers. Anthocyanins are glycosylated polyhydroxyl derivatives of 2-phenylbenzopyrylium salts or better known as flavylium salts. They are made up of three six-membered rings, that is, an aromatic ring bonded to heterocyclic aromatic ring that contains oxygen and carbon, which is also bonded to another aromatic ring. It is said that anthocyanins can exist in various chemical forms, that is, quinonoidal base, flavylium cation, carbinol or pseudobase, and chalcone, which depend on the pH of the solution. These pigments undergo reversible structural transformations with a change in pH manifested by strikingly different absorbance spectra. Berries are the rich source of anthocyanin color and natural anthocyanin pigments.

15.5.2   BACTERIORHODOPSIN (BR)

The bR grows in purple membrane of salt marsh bacterium, Halobacterium salinarium, consist of 7-trans membrane α-helices. It contains chromophore called all-trans retinal and this will undergo 13-*cis*–trans isomerization causes a series of protein intermediate states to be formed having different absorption spectra (Thavasi et al., 2008). It is having an isomerization cycle capacity of ten lac cycles without degradation.

Quantum coherence plays a crucial role in this process, by enabling the energy of the absorbed photons to spread out and sample the physical space occupied by number of light absorbing molecules in order to find the right place for electron transfer charge separation. The use of QDs as photo-absorbers is well studied and several advantages are apparent, most prominent being the ability to tune the absorption spectrum by altering the size of the QD. However, current photovoltaic devices, which uses QDs suffer from low efficiency (approx. 10%), because QDs re-emit much of the absorbed light through fluorescence, thereby losing potential photo-current. In the Bacteriorhodopsin Sensitized Solar Cell (BSSC), the QD absorption layer is replaced by a mixture of QDs and bR molecules, and the interaction between the QD and bR is used to increase efficiency. This results from the coupling between the QD and bR, which quenches re-emissionof light from the QD, and thereby increase the amount of trapped incident energy available for conversion to photo-current. The mechanism for this coupling is Forster Resonant Energy Transfer (FRET) (Jin et al., 2008, Griep et al., 2010), which provides an efficient avenue for energy transfer between molecules separated by distances of up to 10 nm. Bacteriorhodopsin is known as an efficient trap for light energy, effectively locking the energy in place by a conformational distortion of the retinal molecule due to cis-trans isomerization. Thus the QD acts as the antenna and the bR as the storage device. bR alone can play the role of the dye in an initial version of the BSCC.

The maximal theoretical efficiency for a bR cell could reach 25%, which is comparable to silicon. Moreover, this work also reported that the average specific power for bR is 2,103 W/kg compared to 32 W/kg in Si. Addition of QDs into the bR photovoltaic system extends the absorption spectrum. The QDs can be tailored to re-emit the absorbed light at the wavelength that is most efficiently absorbed by bR, thus increasing the quantity of solar energy captured. When QDs are attached to bR via a biotin-streptavidin linkage, and deliver a 35 % photoelectric increase in the bR/QD photovoltage over that of just bR. There are two possible mechanisms available for subsequent conversion of optical excitations to photocurrent direct electron transport through bR (Yum et. al., 2011) leading to enhanced photocurrent and Auger mediated de-excitation in the Au layer (Alfinito and Reggiani, 2009). An efficient production of ballistic electrons in the Au, which can cross the Au layer and then collected by the underlying semiconductor layer can be done by covalent bonding or chemisorptions (Meulenberg et. al., 2009). Figure 13 shows the absorption spectra of bR.

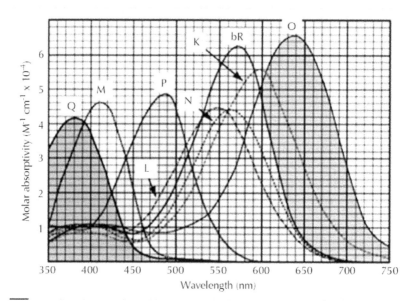

FIGURE 13   The bR absorption wave lengths (Birge et al., 1989).

15.5.3   VIRUS TEMPLATE ASSEMBLY OF PORPHYRINS IN TO LIGHT HARVESTING NANO ANTENNAE

In the natural light-harvesting complexes of cyanobacteria and plants, energy transfer is essential for increasing the exciton flux into the reaction centre, where the electron transfer is triggered to drive photosynthesis. Sophisticated self-organization of the natural photosystems serves as a model for artificial photosynthetic systems that require efficient energy and electron transfers. Accordingly, the synthesis and supramolecular self-assembly of a variety of pigments have been widely explored with the aim of constructing photochemical and photoelectronic devices, including photovoltaics, nonlinear optical materials, and photo switched conductors. Dendritic systems have been widely studied for multipigment arrays; however, it is very challenging and time-consuming to produce such complex molecules a reasonable yield. Recent studies have shown that biological materials can serve as templates guiding nanoscale organization of pigments via chemical linkage or electrostatic interactions. Viruses are particularly attractive scaffolds because of their highly ordered coat protein structures (Su et. al., 2011). M13 viruses have a filamentous structure composed of ~2700 copies of R-helical coat proteins, named pVIII, which are symmetrically arrayed on the viral DNA to compose the ~880 nm long coat that is ~6.5 nm in diameter. The chemical modification of the M13 coat proteins is straightforward because of the exposed N-terminus and Lys residue on the viral surface, which can be conjugation sites. The total number of the primary amines available for conjugation is ~5400 per virus.

Locations of primary amines are indicated a, b, c, and d for the N-termini and o for the Lys. The calculated average distances between the N-termini and the Lys are oa ☐ 1.0 nm, ob ☐ 1.6 nm, oc ☐ 2.4 nm, and od ☐ 2.4 nm. Such close distances between the primary amines on the viral surface allow the energy transfer to occur between neighbouring pigments attached to the virus. It has well known optical properties, water solubility and pendant carboxylic groups. The zinc porphyrins were conjugated to pVIII *via* a carbodiimide coupling reaction. Two samples, ZP-M13-1 and ZP-M13-2, were prepared using different ratios of ZnDPEG and M13 viruses. Approximately, 1564 and 2900 porphyrins were conjugated per virus for ZP-M13-1 and ZP-M13-2, respectively, as measured from inductively coupled plasma-atomic emission spectrometry. ZP-M13 exhibited a very straight structure. Such a morphological change in ZP-M13 is thought to be the result of the interactions of the hydrophobic, p-conjugated macrocycles of the porphyrin with the viral coat protein complexation of virus and Single Waal Nano Tube (SWNT). Various starting SWNTs were either prepared according to the previously reported methods or purchased from Nano Integris. A calculated amount of SWNT-binding virus solution was mixed with a calculated volume of SWNTs dispersed by 2 wt% sodium cholate in water. The mixed solution was dialysed against water (10m M NaCl, pH ¼ 5.3) for two days, with frequent solution changes. After two days of dialysis, the pH of the dialysing solution was increased to 10. A dialysis membrane, with a molecular weight cut-off of 12,000–14,000 SpectraLabs.com), was used for all dialysis procedures. Biomineralization of TiO_2 on the surface of virus/ SWNT complexes: once negative charges were induced on the surface of virus/SWNT complexes, TiO_2 biomineralization was completed using an alkoxide precursor.

The dye absorbs photons and generates electron–hole pairs, and instant charge separation then occurs at the dye/TiO_2 interface. Semiconducting SWNTs improve electron collection at FTO electrodes, whereas incorporation of metallic SWNTs results in recombination and back reaction. In a typical experiment, 50 ml of titanium n-butoxide (Sigma Aldrich) was dissolved in 30 ml ethanol, and the solution was stirred at 220.8°C. A 10 ml volume of aqueous solution of each different virus/SWNT complex, pre-cooled at 48°C, was poured into the ethanol solution under vigorously stirring (700 r.p.m.). The final solution typically comprised 25% water and 75% ethanol. The SWNT/TiO_2 weight ratio was 1/100 for the virus-to-SWNT 1:5 sample, and the template/TiO_2 ratio was fixed when the virus-to-SWNT ratio was changed to 1:2.5 or 1:10. After 1 hr of stirring, the precipitates were centrifuged at 3,000 r.p.m., washed with ethanol twice and water twice, then dried in a vacuum oven at room temperature overnight. The yield of biomineralized TiO2 was higher than 90%. The templated nanowire morphology was observed using transmission electron icroscopy (TEM) under 200 kV (JEOL 200CX TEM and JEOL 2010F TEM) (Nam et. al., 2010).

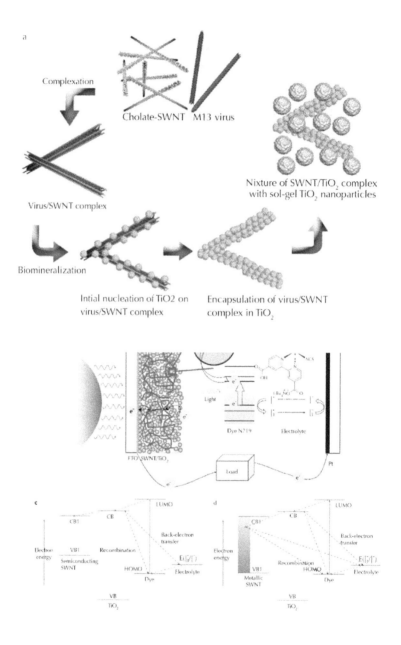

FIGURE 14   Schematic diagram of virus-enabled SWNT/TiO$_2$ DSSCs (Dang et. al., 2011).

(a) Process of virus/SWNT complexation and biomineralization of TiO$_2$ on the surface of the virus/SWNT complex, (b) Scheme of DSSCs incorporating the SWNT/

TiO_2 complex, and (c and d) Energy diagrams of DSSCs incorporating semiconducting SWNTs(c) and metallic SWNTs(d).

SUMMARY AND OUTLOOK

In summary, we have discussed various nanohybrid materials formed between conducting polymers, and inorganic oxides with carbon nanotube demonstrate the excellent solar energy conversion capacity. The solar energy conversion biosystems containing organic–inorganic functional nanohybrid materials are promising candidates for low cost ecofriendly biosolar cells. This results in a dramatic increase in the surface area and increase in electronic conductivity, and electrolyte accessibility of the nanoporous structure, which undoubtedly, improve the performance of the nanohybrid photoanode and cathode as a solar energy extraction material. The results on nanocomposites also suggest that nanoporous composites of MWCNTs and conducting polymers hold great promise for energy conversion devices such as biosolar cells and QD sensitized solar cells. The deposition of conducting polymer is a feasible way to enhance the performance of the bulk hetero junction and layered polymer solar cells. The DSSCs are a promising alternative to the silicon solar cells and in future the emerging nanohybrid materials will influence and change the course of solar energy extraction systems in a dramatic way.

KEYWORDS

- **Bacteriorhodopsin**
- **Dye sensitized solar cells**
- **Polymer solar cells**
- **Quantum dot**
- **Sensitized solar cells**

REFERENCES

1. Alfinito, E. and Reggiani, L. Charge transport in bacteriorhodopsin monolayers: The contribution of conformational change to current-voltage characteristics. Europhysics Letters, 85, 68002 (1 6) (2009).
2. Birge, R. R., Zhang, C. F., and Lawrence, A. F. Optical Random Access Memory Based on Bacteriorhodopsin. F. Hong (Ed.), Molecular Electronics, Plenum Press, New York, 369 379 (1989).
3. Chen, S. G., Chappel, S., Diamant, Y., and Zaban, A. Preparation of Nb2O5 Coated TiO2 Nanoporous Electrodes and Their Application in Dye-Sensitized Solar Cells. Chemistry of Materials, 13(12), 4629 4634 (2001).
4. Chen, X. and Mao Samuel, S. Titanium Dioxide Nanomaterials: Synthesis, Properties, Modifications, and Applications. Chemical Review, 107(7), 2891 2959 (2007).

5. Choi, H., Santra Pralay, K., and Kamat Prashant, V. Synchronized Energy and Electron Transfer Processes in Covalently Linked CdSe-Squaraine Dye-TiO2 Light Harvesting Assembly. American Chemical Society Nano, 6(6), 5718 5726 (2012).
6. Choi, Yoon-Young, Kang Seong, Jun, Kim, Han-Ki, Choi, Won Mook, and Na Seok, In Multilayer graphene films as transparent electrodes for organic photovoltaic devices. Solar Energy Materials and Solar Cells, 96, 281–285 (2012).
7. Dang, Xiangnan, Yi, Hyunjung, Ham, Moon-Ho, Qi, Jifa, Yun, Dong Soo, Ladewski, Rebecca, Strano, Michael S., Hammond, Paula T., and Belcher, Angela M. Virus-templated self-assembled single-walled carbon nanotubes for highly efficient electron collection in photovoltaic devices. Nature Nanotechnology, 6, 377–384 (2011).
8. Dennler, Gilles, Scharber, Markus C., and Brabec, Christoph J. Polymer-Fullerene Bulk-Heterojunction Solar Cells. Advanced Materials, 21, 1323–1338 (2009).
9. Diguna, Lina. J., Shen, Qing, Kobayashi, Junya, and Toyoda, Taro. High efficiency of CdSe quantum-dot-sensitized TiO2 inverse opal solar cells. Applied Physics Letters, 91, 023116 (2007).
10. Dumbrava, A., Georgescu, A., Damache, G., Badea, C., Enache, I., Oprea, C., and Girtu, M. A. Dye-sensitized solar cells based on nanocrystalline TiO2 and natural pigments. Journal Of Optoelectronics and advanced materials, 10(11), 2996–3002 (2008).
11. Eder, Dominik. Carbon Nanotube-Inorganic Hybrids. Chemical Review, 110, 1348–1385 (2010).
12. Fu, Nian Qing, Xiao, Xurui, Zhou, Xiaowon, Zhang, Jing Bo, and Lin, Yuan. Electrodeposition of Platinum on Plastic Substrates as Counter Electrodes for Flexible Dye-Sensitized Solar Cells. Journal of Physical Chemistry C., 116, 2850–2857 (2012).
13. Gratzel, Michael. Dye-sensitized solar cells. Journal of Photochemistry and Photobiology C. Photochemistry Reviews, 4, 145–153 (2003).
14. Griep, Mark. H., Walczak, Karl. A., Winder, Eric. M., Lueking, Donald. R., and Friedrich, Craig. R. Quantum dot enhancement of bacteriorhodopsin-based electrodes. Biosensors and Bioelectronics, 25, 1493–1497 (2010).
15. Guldin, S., Huttner, S., Tiwana, P., Orilall, M. C., Ulgut, B., Stefik, M., Docampo, P., Kolle, M., Divitini, G., Ducati, C., Redfern, S. A., Snaith, H. J., Wiesner, U., Eder, D., and Steiner, U. Improved conductivity in dye-sensitised solar cells through block-copolymer confined TiO2 crystallisation. Energy and Environmental Science, 4(1), 225–233 (2011).
16. Han, Jishu, Zhang, Hao, Tang, Yue, Liu, Yi, Yao Xi, and Yang, Bai. Role of Redox Reaction and Electrostatics in Transition-Metal Impurity-Promoted Photoluminescence Evolution of Water-Soluble ZnSe Nanocrystals. J. Phys. Chem. C., 113(18), 7503–7510 (2009).
17. Hardin, Brian E., Hoke1, Eric T., Armstrong, Paul B., Yum, Jun-Ho, Comte, Pascal, Torres, Toma´s, Fre´chet, Jean M. J., Nazeeruddin, Md Khaja, Gratzel, Michael, and McGehee, Michael D. Increased light harvesting in dye-sensitized solar cells with energy relay dyes. Nature Photonics, 3, 406–411 (2009).
18. Hardin, B. E., Yum, J. H., Hoke, E. T., Jun, Y. C., Pechy, P., Torres, T., Brongersma, M. L., Nazeeruddin, M. K., Gratzel, M., and McGehee, M. D. High Excitation Transfer Efficiency from Energy Relay Dyes in Dye-Sensitized Solar Cells. Nano Lett., 10, 3077–3083 (2010).
19. Jabbour, Ghassan E. and Doderer, David. quantum dot solar cells The best of both worlds. Nature Photonics, 4, 604–605 (2010).
20. Jasieniak, Jack, Johnston, Martin, and Waclawik, Eric R. Characterization of a Porphyrin-Containing Dye-Sensitized Solar Cell. Journal of Physical Chemistry B., 108(34), 12962–12971 (2004).

21. Jin, Yongdong, Honig, Tal, Ron, Izhar, Friedman, Noga, Sheves, Mordechai, and Cahen, David. Bacteriorhodopsin as an electronic conduction medium for biomolecular electronics. Chemical Society Reviews, 37, 2422–2432 (2008).

22. Kim, Jin Young, Lee, Kwanghee, Coates, Nelson E., Moses, Daniel, Nguyen, Thuc-Quyen, Dante, Mark, and Heeger, Alan J. Efficient Tandem Polymer Solar Cells Fabricated by All-Solution Processing. Science, 317(5835), 222–225 (2007).

23. Lee, Byung Hong, Song, Mi Yeon, Jang, Sung-Yeon, Jo, Seong Mu, Kwak, Seong, Yeop, and Kim, Dong Young. Charge Transport Characteristics of High Efficiency Dye-Sensitized Solar Cells Based on Electrospun TiO2 Nanorod Photoelectrodes. Journal of Physical Chemistry C., 113(51), 21453–21457 (2009).

24. Lee, Yuh-Lang and Lo, Yi-Siou. Highly Efficient Quantum-Dot-Sensitized Solar Cell Based on Co-Sensitization of CdS/CdSe. Adv. Funct. Mater., 19, 604–609 (2009).

25. Liang, Yongye, Xu, Zheng, Xia, Jiangbin, Tsai, Szu-Ting, Wu, Yue, Li, Gang, Ray, Claire, and Yu, Luping. For the Bright Future—Bulk Heterojunction Polymer Solar Cells with Power Conversion Efficiency of 7.4%. Advanced Energy Materials, 22, 135–138 (2010).

26. Meng, Sheng, Ren, Jun, and Kaxiras, Efthimios. Natural Dyes Adsorbed on TiO Nanowire for Photovoltaic Applications: Enhanced Light Absorption and Ultrafast Electron Injection. Nano Letters, 8(10), 3266–3272 (2008).

27. Meng, Sheng and Kaxiras, Efthimios. Electron and Hole Dynamics in Dye-Sensitized Solar Cells: Influencing Factors and Systematic Trends. Nanoletters, 10, 1238–1247 (2010).

28. Meulenberg, Robert W., Lee, Jonathan, R. I., Wolcott, Abraham, Zhang, Jin Z., Terminello, Louis. J., and Buuren, Tony Van. Determination of the Exciton Binding Energy in CdSe Quantum Dots. ACS Nano, 3(2), 325–330 (2009).

29. Mora-Sero, Ivan, Fabregat-Santiago, Sixto, Gime Nez,Francisco, Shen, Roberto, Gomez,Qing, Toyodha, Taro, and Bisquert, Juan. Recombination in Quantum Dot Sensitized Solar Cells. Accounts Of Chemical Research, 42(11), 1848–1857 (2009).

30. Nagpal, Prashant and Klimov, Victor. Role of mid-gap states in charge transport and photoconductivity in semiconductor nanocrystal films. http://www.nature.com/ncomms/journal/v2/n9/full/ncomms1492.html (accessed on 13 Feb. 2013) (2011).

31. Nam, Yoon Sung, Shin, Taeho, Park, Heechul, Magyar, Andrew P., Choi, Katherine, Fantner, Georg, Nelson, Keith A., and Belcher, Angela M. Virus-Templated Assembly of Porphyrins into Light-Harvesting Nanoantennae. JACS Comm., 132(5), 1462–1463 (2010).

32. Pfeffer, R. A. And Macon, W. A. Emerging Energy Requirements for Future, C4ISR. http://www.dtic.mil/cgi-bin/GetTRDoc?AD=ADA467640 (accessed on 13 Feb. 2013) (2002).

33. Ruhle, Sven, Shalom, Menny, and Zaban, Arie. Quantum-Dot-Sensitized Solar Cells. Chem Phys Chem., 11, 2290–2304 (2010).

34. Ryan, M. PGM HIGHLIGHTS: Progress in Ruthenium Complexes for Dye Sensitised Solar Cells. Platinum Metals Rev., 53(4), 216–218 (2009).

35. Satoh, Norifusa, Nakashima, Toshio, Kamikura, Kenta, and Yamamoto, Kimihisa. Quantum size effect in TiO2 nanoparticles prepared by finely controlled metal assembly on dendrimer templates. Nature Nanotechnology, 3, 106–111 (2008).

36. Scholes Gregory D., Fleming Graham R., Alexandra, Olaya-Castroand, Grondelle Rienk van. Lessons from nature about solar light harvesting. Nature Chemistry, 3, 763–774 (2011).

37. Sista, Srinivas, Hong, Ziruo, Chen, Li-Min, and Yang, Yang. Tandem polymer photovoltaic cells - current status, challenges and future outlook. Energy and Environmental Science, 4, 1606–1620 (2011).

38. Sivaranjani, Kumarsrinivasan, Agarkar, Shrithi, Ogale, Satischandra B., and Gopinath, Chinnakonda S. Toward a Quantitative Correlation between Microstructure and DSSC Efficiency A Case Study of TiO2ÀxNx Nanoparticles in a Disordered Mesoporous Framework. Journal of Physical Chemistry C., 116(3), 2581–2587 (2012).

39. Su, Wu, Bonnard, Vanessa, and Burley, Glenn A.. DNA-Templated Photonic Arrays and Assemblies: Design Principles and Future Opportunities. J. European Chem., 17, 7982–7991 (2011).

40. Talgorn, Elise, Gao, Yunan, Aerts, Michiel, Kunneman, Lucas, T., Schins, Juleon M., Savenije, T. J., Huis, Marijn A. van, Zant, Herre S. J. van der, Houtepen, Arjan J., and Siebbeles, Laurens D. A. Unity quantum yield of photogenerated charges and band-like transport in quantum-dot solids. Nature Nanotechnology, 6, 733–739 (2011).

41. Thavasi, Velmurugan, Lazarova, Tzvetana, Filipek, Slawomir, Kolinski, Michal, Querol, Enric, Kumar, Abhishek, Ramakrishna, Seeram, Padros, Esteve, and Renugopalakrishnan, V. Study on the Feasibility of Bacteriorhodopsin as Bio-Photosensitizer in Excitonic Solar Cell A First Report. Journal of Nanoscience and Nanotechnology, 8, 1–9 (2008).

42. Viswanath, A. K. Photoluminescence in GaN, InGaN and their applications in photonics. In Handbook of Advanced Electronic and Photonic Materials and Devices, H. S. Nalwa (Ed.), Academic Press, USA., 1, 109–150 (2000).

43. Viswanath, A. K. Surface and interfacial recombination in semiconductors. In Handbook of Surfaces and Interfaces of Materials, H. S. Nalwa (Ed.), Academic Press, USA, 1, 218–271 (2001).

44. Viswanath, A. K. Photoluminescence and Ultrafast Phenomena in III-V Nitride Quantum Structures. In Handbook of Semiconductor Nanostructures and Nanodevices, A. A. Balandin and K. L. Wang (Eds.), American Scientific Publishers, Los Angeles, California, USA, 4, 46–110 (2006).

45. Yahya, Nurul Zayana and Rusop, Mohamad. Investigation on the Optical and Surface Morphology of Conjugated Polymer MEH-PPV:ZnO Nanocomposite Thin Films. Journal of Nanomaterials, 2012(793679), 1–4 (2012).

46. Yum, Jun-Ho, Hardin, Brian E., Hoke, Eric T., Baranoff, Etienne, Zakeeruddin, Shaik M., Nazeeruddin, Mohammad K., Torres, Tomas, McGehee, Michael D., and Gratzel, Michael. Incorporating Multiple Energy Relay Dyes in Liquid Dye-Sensitized Solar Cells. Journal of Chemical Physics and Physical Chemistry, 12(3), 657–661 (2011).

47. Zhang, Yan-Zhen, Jao, Xia, Wang, Li-Xin, Xu, Hui, Hou, Qian, Zhou, Wei-Lie, and Chen, Jian-Feng. Novel ZnO-Based Film with Double Light-Scattering Layers as Photoelectrodes for Enhanced Efficiency in Dye-Sensitized Solar Cells. Chemistry of Materials, 22(3), 928–934 (2010).

48. Zhou, H., Wu, L., Gao, Y., and Ma, T. Dye-sensitized solar cells using 20 natural dyes as sensitizers. J. Photochemistry and Photobiology A: Chemisry, 219, 188–194 (2011).

INDEX